Exercises in Statistical Reasoning

Students cultivate learning techniques in school that emphasize procedural problem solving and rote memorization. This leads to efficient problem solving for familiar problems. However, conducting novel research is an exercise in creative problem solving that is at odds with a procedural approach; it requires thinking deeply about the topic and crafting solutions to unique problems. It is not easy to move from a topic-based, carefully curated curriculum to the daunting world of independent research, where solutions are unknown and may not even exist. In developing this book, we considered our experience as graduate students who faced this transition.

Exercises in Statistical Reasoning is a collection of exercises designed to strengthen creative problem-solving skills. The exercises are designed to encourage readers to understand the key points of a problem while seeking knowledge, rather than separating out these two activities. To complete the exercises, readers may need to reference the literature, which is how research-based knowledge is often acquired.

Features of the Exercises

- The exercises are self-contained, though several build upon concepts from previous problems.
- Each exercise opens with a brief introduction that emphasizes the relevance of the content. Then, the problem statement is presented as a series of intermediate questions.
- For each exercise, we suggest one possible solution, though many may exist.
- Following each solution, we discuss the historical background of the content and points of interest.
- For many exercises, a brief demonstration is provided illustrating relevant concepts.

There is an abundance of high-quality textbooks that cover a vast range of statistical topics. However, there is also a lack of texts that focus on the development of problem-solving techniques that are required for conducting novel statistical research. We believe that this book helps fill the gap. Any reader familiar with graduate-level classical and Bayesian statistics may use this book. The goal is to provide a resource that such students can use to ease their transition to conducting novel research.

Michael R. Schwob, Yunshan Duan, Beatrice Cantoni, and Bernardo Flores-López are Ph.D. candidates in the Department of Statistics and Data Sciences at The University of Texas at Austin. Michael will join the Department of Statistics at the Virginia Polytechnic Institute and State University as a faculty member in Fall 2025.

Stephen G. Walker is a professor in the Department of Mathematics and the Department of Statistics and Data Sciences at The University of Texas at Austin. He is the holder of the Paul D. and Betty Robertson Meek and American Petrofina Foundation Centennial Professorship in Business in the McCombs School of Business.

Exercises in Statistical Reasoning

Michael R. Schwob, Yunshan Duan,
Beatrice Cantoni, Bernardo Flores-López,
and Stephen G. Walker

CRC Press
Taylor & Francis Group
Boca Raton London New York

CRC Press is an imprint of the
Taylor & Francis Group, an **informa** business
A CHAPMAN & HALL BOOK

First edition published 2025
by CRC Press
2385 Executive Center Drive, Suite 320, Boca Raton, FL 33431, U.S.A.

and by CRC Press
4 Park Square, Milton Park, Abingdon, Oxon, OX14 4RN

CRC Press is an imprint of Taylor & Francis Group, LLC

ISBN: 978-1-032-79710-6 (hbk)
ISBN: 978-1-032-78486-1 (pbk)
ISBN: 978-1-003-49347-1 (ebk)

DOI: 10.1201/9781003493471

Typeset in URW Palladio font
by KnowledgeWorks Global Ltd.

Publisher's note: This book has been prepared from camera-ready copy provided by the authors.

Dedication

Michael R. Schwob: "To my parents, whose love and support have fueled my pursuit of knowledge and happiness."

Yunshan Duan: "致我的家人，感谢你们始终如一的支持与鼓励。"

Beatrice Cantoni: "A Enrico e Marialuisa, anime curiose."

Bernardo Flores-López: "Para Migue; que nunca nadie te recuerde ausente."

Stephen G. Walker: "Ne chiege kod nyithinda mabeyo."

Contents

Preface

Students cultivate learning techniques in school that emphasize procedural problem solving and rote memorization. This leads to efficient problem solving for familiar problems. However, procedural problem solving is troublesome when one is confronted with an unfamiliar problem. Thus, when graduate students transition to conducting novel research, many are unequipped with the tools for autodidactic study. Research is an exercise in creative problem solving, which is at odds with a procedural approach; it requires thinking deeply about the topic and crafting solutions to unique problems. It is not easy to move from a topic-based, carefully curated curriculum to the daunting world of independent research, where solutions are unknown and may not even exist. In developing this book, we considered our experience as graduate students facing this transition.

This book is a collection of statistical exercises that train creative and holistic problem solving. These exercises appeared on statistics preliminary exams that demarcated the transition from coursework to research activity. They were designed to encourage exploratory thought and discourage rote mechanics and memorization. Exercises start with a seemingly trivial concept, which then develops into a significant topic in statistics. The path to some solutions may seem circuitous, while other paths may be elegantly simple. Each exercise is a practice of research methodology and encourages the reader to explore the true meaning and intuition behind many well-known statistical concepts.

Expected Audience

The expected readers are graduate or advanced undergraduate students that are interested in developing a research-oriented mindset. However, anyone with the appropriate prerequisites may benefit from this practice in creative problem solving. An understanding of graduate-level classical and Bayesian statistics is assumed. More specifically, we assume an intermediate understanding of the following topics: the exponential family, hypothesis testing, likelihood estimation, linear models, mathematical statistics, probability theory, Markov processes, and Markov chain Monte Carlo algorithms.

Structure of This Book

We grouped the exercises into eight chapters, which are named after particular statistical topics. Though the ordering of the chapters may suggest a linear reading order, each chapter is self-contained, so the reader may explore the topics in whichever order they wish. The chapters concerning probability, hypothesis testing, and asymptotics are divided into two sections: the classical perspective and the Bayesian perspective. We structured these chapters this way to explicitly compare the two frameworks.

The exercises are self-contained, though several build upon concepts from previous problems. Exercises that are particularly difficult have an asterisk next to their names. Many exercises bring several topics together to make progress on a single research problem. Hence, being widely read is important; understanding ideas beyond any specific research agenda will always be useful.

Each exercise opens with a brief introduction that emphasizes the relevance of the content. Then, the problem statement is presented as a series of intermediate questions; solutions to these questions typically depend on the previous steps. For each exercise, we suggest one possible solution, though many may exist. These solutions can be found at the end of each section. However, the reader needs to understand concepts and ideas in their own way to ensure that they will feel comfortable adapting to different problems that they have not yet seen. This is, in fact, the nature of research. Following each solution, we discuss the historical background of the content and points of interest. For many exercises, we provide a brief demonstration that illustrates relevant concepts. Finally, there are "Miscellaneous" points for several exercises that provide details on mathematical or numerical techniques used to complete the exercises.

The exercises in this book aim to check for an understanding of foundational statistical topics. Our approach does not heavily emphasize a theorem-proof structure nor does it focus on detailed regularity conditions upon which some technical result strictly holds. We do not argue over whether a function needs to be twice or thrice differentiable or debate the strict conditions under which a series expansion can be suitably stopped. Rather, this book explores how statistical concepts work, revisiting even the most basic ideas that may have been overlooked or omitted in previous coursework.

How to Use This Book

We recommend that readers mindfully work through this book, drafting solutions and deliberately noting any successes and obstacles. We encourage readers to attempt each exercise on their own before reviewing the proposed solution. The aim of this book is to teach problem-solving skills – not content. We do not present all the material needed to answer the questions. To complete an exercise, readers may need to reference the literature, which is how research-based knowledge is often acquired. In fact, the exercises are designed to encourage readers to understand the key points of a problem while seeking knowledge, rather than separating out these two activities. Finally, we highly encourage readers to look through the further reading for exercises of particular interest.

Why We Wrote This Book

We collaborated on this book because we shared a singular vision – that actively solving problems unlocks a far deeper understanding of statistical concepts than simply memorizing theorems, proofs, and definitions. The exercises in this book originally appeared as a series of preliminary exam questions for PhD students. The unique nature of the questions prompted us (the student authors) to compile a summary of solutions, along with points of interest containing further readings and research questions. The exam setter (Prof. Stephen Walker) was then brought in on the project to ensure that the solutions and questions were checked for full correctness. Hence, the original solutions were the students', the original questions were the professors', and all the rest a joint effort. To complete the book, we added brief backgrounds to each exercise and added figures to illustrate particular concepts of interest. Aside from correcting errors, the questions and answers remain as they were originally constructed.

There is an abundance of high-quality textbooks that cover a vast range of statistical topics. However, there is also a lack of texts that focus on the development of problem-solving techniques that are required for conducting novel statistical research. We believe that this book helps fill the gap.

A Perspective from Prof. Walker

Graduate students often learn advanced material and engage in research by participating in a structured master's or PhD program. The term "program" suggests a hint of homogeneity – that all students can learn in the same way. This assumption is likely not the case. In most graduate programs, the first year (possibly more) involves a regular schedule of courses, homework, and exams. If homework and exams shortly follow after a concept or idea is introduced, it is quite reasonable to ask, "where is the time to understand going to come from?" An alternative program design might check knowledge after students have had sufficient time to understand the material. From a practical experience, I attended an undergraduate program in which eight exams (three hours each) were administered over the final four days of the program, after 3 years. How was it possible to memorize all the material for these exams? The only way to deal with things was to understand the material. Some concepts took days to master, while others took weeks or even months. Given three years, there was more than enough time to deeply understand the content, and the most successful approach to this was to tackle relevant problems and seek help when getting stuck after sufficient time figuring it out. The idea of reading course materials may not be enough to fully understand or be able to move on. Hence, the book is designed to assist with the understanding through problem solving. This would involve self study and creating the time to do problems, not unlike my undergraduate and graduate programs.

A Perspective from the Students

The origins of this book can be traced back to Spring 2022, when we began to prepare for our preliminary exam. We initially struggled with the questions, finding them difficult to prepare for because they demanded a deeper level of understanding than the "reading, memorizing, and replicating" cycle that is prevalent in higher education. Success on the exam hinged on our ability to adapt to unfamiliar questions and apply our knowledge to new scenarios. Thus, we needed to move beyond memorization and cultivate the problem-solving skills necessary to tackle statistical problems that were new to us.

To prepare for the exam, we deeply engaged with questions from past exams every week. We intentionally approached the questions with a slow pace to foster a deeper understanding of the content and explore related concepts. After several weeks, all of us recognized the value of the questions and our more profound understanding of the material. After a couple of months, our problem-solving skills felt sharper than ever before.

Education often prioritizes breadth of knowledge over depth, leaving students with a superficial understanding of many topics. As a result, the learned content is quickly forgotten. The traditional learning-through-memorization paradigm indeed has a major flaw: its efficacy significantly decreases with time. An approach that aims to build problem-solving skills and a deeper understanding of content ages much better. The critical thinking and problem-solving skills that we strengthened have become essential tools in our broader academic and professional pursuits.

Acknowledgments

We are grateful to our professors and peers for their impacts on our academic journeys. In particular, we would like to thank Nhat Ho, Mevin Hooten, Peter Müller, Mary Parker, Abhra Sarkar, Purna Sarkar, James Scott, Cory Zigler, Tianqi Chen, Amber Day, Rosalía Hernández, Emily Hsiao, Khai Nguyen, and Jack Xiao. We also thank Angelica Faye Alcantara Mayor for designing the cover image and Denise Mayang Wong for designing the overall cover of our book.

1

Probability

Statisticians use an extensive amount of probability theory, both for constructing models and for finding properties of statistical procedures. For example, probability motivates the use of the sample mean for estimating a population mean; establishing properties of such estimators is done so by treating the sample as random variables. Many frequentist properties are based on the large sample behavior of an estimator. Thus, the convergence of random variables becomes important to statistical inference. In nonparametric problems, it is the convergence of random functions that is the focus of attention.

In Bayesian analysis, probability underpins the prior and posterior distributions, which are fundamental for Bayesian inference. Recently, many Bayesian models have been estimated using sampling-based techniques, such as Markov chain Monte Carlo samplers. As a consequence, much work on the convergence properties of Markov chains appears in the Bayesian literature. New sampling ideas include the use of diffusion processes, such as Langevin diffusion, which may be defined to have a specific stationary distribution. Other samplers include Hamiltonian Monte Carlo or particle filters, which are suitable for hidden Markov models.

Additionally, Bayesian nonparametrics is highly dependent on a strong knowledge of probability. Probability measures are constructed on spaces of functions, such as for density and distribution functions. For example, the Dirichlet process relies on a suitable definition of finite dimensional distributions but can also be understood directly via the use of stochastic processes behaving as a distribution function with probability one. Then, probability serves an essential role for deriving the correct posterior distribution and establishing conditions for consistency.

The questions in this chapter focus on the use of probability for statistics and stochastic processes. Special attention is given to the study of stochastic processes, which is often where statisticians meet probability for which they may not be so familiar.

Question 1.1.1 investigates the central limit theorem and its convergence in probability, which are useful for looking at sample means. Question 1.1.2 considers infinitely divisible random variables, which can be represented as the sum of n independent and identically distributed random variables for all n; such variables include the gamma, Gaussian, and Poisson. Finally, Question 1.1.3 is on the maximal inequality for a martingale sequence, which could be seen as an extension of the well-known Markov inequality.

The questions on stochastic processes mostly concern Markov processes. For example, Question 1.2.1 is on the simple random walk, whereby a walk on the integers goes up one or down one with equal probability. A primary interest is how long the process will take to return to the starting point while also reaching a given height. Question 1.2.2 includes a simple birth process, where any individual alive at a given generation gives birth to an independent set of a random number of offspring according to a probability mass function. The sum of these offspring determines the size of the next generation, and so on. The question of interest is the probability that the population goes extinct.

Two questions are concerned with Markov chains in discrete time, which are represented as transition matrices. Question 1.2.3 explains fundamental properties for a chain

DOI: 10.1201/9781003493471-1

with just two states; the eigenvalues of the transition matrix determine how the chain behaves and also establishes the rate of convergence to the stationary probability, assuming it exists. Question 1.2.4 explores a larger state space with focus placed on properties of the eigenvalues and the corresponding properties of the chain.

Finally, there are two questions on the Poisson process. Question 1.2.5 concerns the construction of the Poisson process via changes in the process during an arbitrarily small amount of time. Describing changes of random mechanisms over a small interval of time is often reasonable in practice, as this is where the information would be available. The subsequent plan is to find how the changes then look over a larger interval of time. Question 1.2.6 concerns the construction of the Poisson process via independent exponential random variables, which determine how long the process remains at any given height.

Q1.1 Questions – Probability Theory

Q1.1.1 – Types of convergence of a sample mean

Introduction. This question examines probabilistic properties of a sample mean. The first result is that the sample mean converges in distribution to a normal random variable when it is rescaled to have a fixed variance; this is regardless of the number of samples. The proof of this can be found using Laplace transforms. The next step is to show that the sample mean converges in probability to the mean of the population, which implies that a deterministic sequence of subsamples converges almost surely. The exercise does not include the almost sure convergence of $\overline{X}_n$ to the mean (and the conditions under which this happens) because it is quite difficult to prove.

For a random sequence $\{X_n\}$, there are associated sequences: the sequence of probabilities on events $\{P_n(A_n)\}$ and the sequence of distribution functions $\{F_{X_n}(x)\}$. These different types of sequences lead to the different types of convergence for $\{X_n\}$.

Question. Suppose $X_1, \ldots, X_n$ are independent and identically distributed random variables with mean zero and variance 1. Define

$$\overline{X}_n = \frac{1}{n} \sum_{i=1}^{n} X_i.$$

(i) Show that $\mathrm{E}(\overline{X}_n) = 0$ and $\mathrm{Var}(\overline{X}_n) = 1/n$.

(ii) By considering the Laplace transform $\phi_n(t) = \mathrm{E}\{\exp(t\sqrt{n}\overline{X}_n)\}$ with $t > 0$, show why $Z_n = \sqrt{n}\overline{X}_n$ is approximately a standard normal random variable for large n.

(iii) Prove the Markov inequality: if Y is a positive continuous random variable, then $\mathrm{P}(Y > a) < \mathrm{E}(Y)/a$ for all $a > 0$.

(iv) For all sequences $\varepsilon_n \to 0$ for which $n\varepsilon_n^2 \to \infty$, show that $\mathrm{P}(|\overline{X}_n| > \varepsilon_n) \to 0$.

(v) Show there exists a deterministic sequence $\{n_i\}_{i \geq 1}$ for which $|\overline{X}_{n_i}| \to 0$ almost surely as $i \to \infty$.

Q1.1.2 – Infinite divisibility of random variables

Introduction. Introduced by Bruno de Finetti in 1929, the concept of infinitely divisible distributions plays an important role in modern probability theory and financial modeling, where it has found applications in limit theorems and Lévy and additive processes. Many known distributions are infinitely divisible, such as the gamma, Gaussian, and Poisson. Thus, the study of infinitely divisible distributions and their decompositions are of broad interest. This question concerns the use of moment generating functions to determine whether distributions are infinitely divisible. In particular, the exercise investigates the gamma and Poisson distributions.

Question. A random variable X is infinitely divisible if, for every integer $n = 1, 2, 3, \dots$, it is possible to write

$$X = \sum_{i=1}^{n} X_i,$$

where the $\{X_i\}$ are independent and identically distributed. The distribution of the $\{X_i\}$ may depend on n. For example, if X is standard normal, then each X_i is normal with mean 0 and variance $1/n$.

(i) Suppose $X \sim \mathrm{Ga}(a, 1)$. Find the moment generating function (Laplace transform) for X; i.e., $\phi_X(\theta) = \mathrm{E}\left(e^{-\theta X}\right)$ for $\theta > 0$.

(ii) If X is infinitely divisible, explain why $\phi_X(\theta) = \{\phi_{X_i}(\theta)\}^n$ for each $i = 1, \dots, n$.

(iii) Determine whether the gamma variable is infinitely divisible, and if so, find the distribution for each X_i.

(iv) Show that $Z = \mathrm{Pois}(\lambda)$ is infinitely divisible.

(v) If X is $\mathrm{Ga}(a, 1)$, Z is $\mathrm{Pois}(\lambda X)$, and Y is $\mathrm{Ga}(a + Z, 1 + \lambda)$, show that Y is marginally infinitely divisible.

Q1.1.3* – A maximal inequality for a martingale sequence

Introduction. This question outlines a proof for the Doob martingale inequality, which provides an upper-bound for the probability that the maximum value of a martingale sequence exceeds any given value over a particular interval of (discrete) time. This inequality may be viewed as an extension of the Markov inequality.

Question. Suppose $\{M_n : n \in N\}$ is a non-negative martingale such that $\mathrm{E}\left(M_{n+1} \mid M_{1:n}\right) = M_n$. Define the events

$$E_n = \{M_1 < \epsilon, \dots, M_{n-1} < \epsilon, M_n > \epsilon\}$$

for $n = 1, \dots, K$.

(i) Show that the set of events $\{E_n : n = 1, \dots, K\}$ is mutually disjoint.

(ii) Show that

$$\{\max\{M_1,\ldots,M_K\} > \epsilon\} \equiv \bigcup_{n=1}^{K} E_n.$$

(iii) Using the martingale property, show that

$$\int_{E_n} M_K\, f(M_1,\ldots,M_K)\, dM_1 \ldots dM_K = \int_{E_n} M_n\, f(M_1,\ldots,M_K)\, dM_1 \ldots dM_K$$

for any $n \leq K$, where $f(M_1,\ldots,M_K)$ represents the joint density function for $\{M_1,\ldots,M_K\}$.

(iv) Show that $\int_{E_n} M_n\, f(M_1,\ldots,M_K)\, dM_1 \ldots dM_K > \epsilon\, \mathrm{P}(E_n)$.

(v) Show that $\mathrm{P}(\max\{M_1,\ldots,M_K\} > \epsilon) \leq \mathrm{E}(M_K)/\epsilon$.

S1.1 Solutions & Further Reading – Probability Theory

S1.1.1 – Types of convergence of a sample mean

(i) The mean and variance are

$$\mathrm{E}(\overline{X}_n) = \mathrm{E}\left(\frac{1}{n}\sum_{i=1}^{n} X_i\right) = \mathrm{E}(X_1) = 0,$$

$$\mathrm{Var}(\overline{X}_n) = \mathrm{Var}\left(\frac{1}{n}\sum_{i=1}^{n} X_i\right) = \frac{1}{n}\mathrm{Var}(X_1) = \frac{1}{n},$$

which follow from the independence of the $\{X_i\}$. If the $\{X_i\}$ are not independent, then covariances would need to be included.

(ii) This can be solved through a Taylor expansion of e^x (ignoring higher order terms of $1/n$) and then taking an expectation. This is a standard practice for establishing the central limit theorem:

$$\begin{aligned}
\phi_n(t) &= \mathrm{E}\left\{\exp(t\sqrt{n}\overline{X}_n)\right\} = \mathrm{E}\left\{\exp\left(t\sqrt{n}\frac{1}{n}\sum_{i=1}^{n} X_i\right)\right\} \\
&= \mathrm{E}\left\{\exp\left(t\frac{1}{\sqrt{n}}\sum_{i=1}^{n} X_i\right)\right\} \\
&= \left[\mathrm{E}\left\{\exp\left(t\frac{1}{\sqrt{n}}X_i\right)\right\}\right]^n \\
&= \left\{\mathrm{E}\left(1 + tX_i/\sqrt{n} + \tfrac{1}{2}t^2X_i^2/n + o(1/n)\right)\right\}^n \\
&= \left(1 + 0 + \tfrac{1}{2}t^2/n + o(1/n)\right)^n \to \exp\left(\tfrac{1}{2}t^2\right),
\end{aligned}$$

where the fourth equality is obtained because the $\{X_i\}$ are independent and identically distributed. The final line uses the exponential approximation $(1 + x/n)^n \to e^x$. Note that $o(1/n)$ represents the terms for which $n\,o(1/n) \to 0$, so $(1 + a/n + 1/n^2)^n \to e^a$. By Lévy's continuity theorem, pointwise convergence of the Laplace transform of Z_n implies convergence in distribution to a standard normal random variable as $n \to \infty$.

(iii) The key starting point is $a\,1(Y > a) \leq Y1(Y > a) \leq Y$. Taking expectations with respect to Y, it is seen that $a\,\mathrm{P}(Y > a) \leq \mathrm{E}(Y)$, where $\mathrm{E}\{1(Y > a)\} = \mathrm{P}(Y > a)$.

(iv) To obtain a useful Markov inequality, consider

$$\mathrm{P}(|\overline{X}_n| > \epsilon_n) = \mathrm{P}(|Z_n| > \sqrt{n}\epsilon_n) = \mathrm{P}(Z_n^2 > n\epsilon_n^2) \leq \mathrm{E}(Z_n^2)/(n\epsilon_n^2),$$

which goes to 0 as $n\epsilon_n^2 \to \infty$.

(v) Because $|\overline{X}_n|$ converges in probability to 0, there exists a n_i large enough such that $P(|\overline{X}_{n_i}| > 1/i) \leq 1/i^2$ for any $i \geq 1$. Hence, for any $\epsilon > 0$,

$$\sum_{i>1/\epsilon} P(|\overline{X}_{n_i}| > \epsilon) < \infty.$$

Thus, by the Borel–Cantelli lemma, $|\overline{X}_{n_i}| < \epsilon$ for all large i. Because ϵ is arbitrary, it must be that $|\overline{X}_{n_i}|$ converges to 0 almost surely.

Further Reading

Historical Background. The law of large numbers states that the sample mean converges to the population mean, if it exists and $E(|X_1|) < \infty$. The concept of the law of large numbers was first put forth by Italian mathematician Gerolamo Cardano in the sixteenth century. A rigorous analysis was then provided by Swiss mathematician Jacob Bernoulli over the span of 20 years, eventually being published in 1713. Though Bernoulli named it the "Golden Theorem," French mathematician Siméon D. Poisson later referred to the theorem as the law of large numbers in his native French: "la loi des grands nombres."

Points of Interest.

1. The sample mean has a shrinking variance (i.e., $1/n$), which drives the convergence to the mean in probability. A rescaling to $\sqrt{n}\overline{X}_n$ maintains a level variance, resulting in a weaker form of convergence to a finite random variable. This is known as the central limit theorem, which can best be understood using Laplace transforms due to how easily it handles the independence of the variables.

2. Sample means are closely related to random walks. Scaling limits of random walks (the random walk being represented by $\sum_{i=1}^{n} X_i$ at time n) is a common theme in modern probability theory (Pollard, 1984). For random walks, the key lies in the distribution of "jumps." Consider the simple random walk on the non–negative integers $\{S_n\}_{n\geq 0}$ with $S_0 = 0$. The walk moves up or down by 1 with equal probability and independently of all other moves. Thus,
$$S_n = X_1 + \cdots + X_n,$$
where the $\{X_i\}$ are mutually independent and $P(X_i = -1) = P(X_i = 1) = \frac{1}{2}$. To extend this concept to a continuous time process, consider the time between jumps to shrink to 0 and the size of the jumps to be rescaled. To achieve a variance of $t = nh$ at time t, define
$$S_n(t) = \sqrt{h}\,(X_1 + \cdots + X_n),$$
so the jumps have been rescaled to $\sqrt{h}$. Think of $S_n(t)$ as being piecewise constant on intervals of length h, resulting in a continuous path as $h \to 0$; this is in fact Brownian motion. Fig. 1.1 shows two random paths; one is Brownian motion and the other is a simple random walk.

3. The convergence of the finite dimensional distributions follows from the central limit theorem, and convergence of these processes in an appropriate function space follows from tightness (Billingsley, 1968). In general, a discrete process converging to a continuous process requires that the scaling sequence converges fast enough to 0 to compress the discontinuities. If the jump distribution is, for instance, any stable law, the random walk would jump too wildly, and the rescaled version would instead converge weakly to a discontinuous process.

4. There is typically interest in the almost sure convergence of the sample mean $\overline{X}_n$. Consider the event that $\overline{X}_n$ converges to 0. To say $\overline{X}_n$ converges to 0 almost surely means that the event happens with probability one, which may be written as $P(|\overline{X}_n| > \epsilon \text{ infinitely often}) = 0$ for all $\epsilon > 0$.

5. Following on from the previous point, if A_n is the event $\{\overline{X}_n > \epsilon\}$, then

$$\limsup_{n\to\infty} A_n = \{A_n \text{ i.o.}\} = \bigcap_n \bigcup_{m>n} A_m,$$

where i.o. denotes "infinitely often." The implication being that the elements in $\{A_n \text{ i.o.}\}$ are the events that occur in $\bigcup_{m>n} A_m$ for all n. If $P(\{A_n \text{ i.o}\}) = 0$, then $\{A_n \text{ i.o.}\} = \emptyset$, meaning that $\bigcup_{m>n} A_m$ is empty for some n. Alternatively,

$$\liminf_{n\to\infty} A_n = \bigcup_n \bigcap_{m>n} A_m,$$

which is a subset of $\limsup_{n\to\infty} A_n$. The elements in this event are in all the $\{A_n\}$ except for a finite number of them.

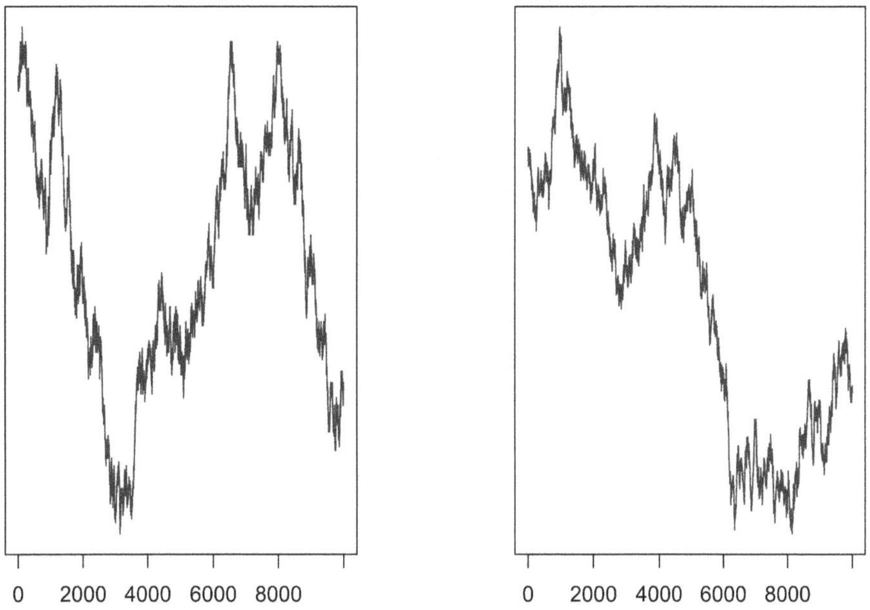

FIGURE 1.1
One of these images represents Brownian motion and the other a simple random walk. The overall pattern of these processes are indistinguishable from each other. The horizontal axis represents time, and the vertical axis is the height of the process which is unlabeled because it is arbitrary.

Miscellaneous. This question made use of the Markov inequality, despite it being a rather blunt tool. For example, suppose Y is a standard exponential random variable. Then, $\mathrm{P}(Y > \epsilon) \approx 1 - \epsilon$ for small ϵ. However, the Markov inequality yields the upper bound $1/\epsilon$, which is useless as it is greater than 1. The Markov inequality does provide useful results (i.e., tighter upper bounds) in other situations. For example, if $\mathrm{E}(Y_n) < e^{-n\epsilon}$, then

$$\sum_{n=1}^{\infty} \mathrm{P}(Y_n > e^{-n\epsilon/2}) < \infty.$$

Consequently, the Borel–Cantelli lemma states that for all large n, it is that $Y_n < e^{-n\epsilon/2}$ with probability 1 (Feller, 1966). The combined use of the Markov inequality and the Borel–Cantelli lemma is a standard technique in asymptotic studies of statistical procedures, notably in Bayesian consistency proofs. An interesting use of such techniques is to demonstrate that a likelihood ratio converges almost surely to 0 at an exponential rate.

S1.1.2 – Infinite divisibility of random variables

(i) This question asks for a Laplace transform, which in this case is a moment-generating function with a negative parameter:

$$\begin{aligned} \phi_X(\theta) &= \mathrm{E}\left(e^{-\theta X}\right) = \int_0^\infty \frac{1}{\Gamma(a)} x^{a-1} e^{-x} e^{-\theta x} dx \\ &= \int_0^\infty \underbrace{\frac{1}{\Gamma(a)} x^{a-1} e^{-x(1+\theta)}}_{\text{kernel of } \mathrm{Ga}(a,1+\theta)} dx \\ &= (1+\theta)^{-a}, \quad \theta > -1. \end{aligned}$$

(ii) Let X be infinitely divisible, and write $X = \sum_{i=1}^n X_i$. Because the $\{X_i\}$ are independent and identically distributed,

$$\begin{aligned} \phi_X(\theta) &= \mathrm{E}(e^{-\theta X}) = \mathrm{E}[\exp\{-\theta(X_1 + X_2 + \cdots + X_n)\}] \\ &= \mathrm{E}(e^{-\theta X_1}) \cdots \mathrm{E}(e^{-\theta X_n}) \quad \text{(independence)} \\ &= \left\{\mathrm{E}(e^{-\theta X_1})\right\}^n \quad \text{(identically distributed)} \\ &= \{\phi_{X_1}(\theta)\}^n. \end{aligned}$$

(iii) From part (i), $\phi_X(\theta) = (1+\theta)^{-a}$, so

$$\phi_X(\theta) = (1+\theta)^{-an/n} = \left\{(1+\theta)^{-a/n}\right\}^n = \{\phi_{X_i}(\theta)\}^n,$$

where $X_i \sim \mathrm{Ga}(a/n, 1)$ for $i = 1, \ldots, n$. Thus, the gamma random variable is infinitely divisible.

(iv) The Laplace transform of Z is given by

$$\begin{aligned}\phi_Z(\theta) &= \mathrm{E}(e^{-\theta Z}) = \textstyle\sum_{z=0}^{\infty} \lambda^z e^{-\lambda} e^{-\theta z}/z! \\ &= e^{-\lambda} \textstyle\sum_{z=0}^{\infty} (\lambda e^{-\theta})^z/z! = e^{-\lambda} e^{\lambda e^{-\theta}} \quad \text{(exponential series)} \\ &= \exp\{\lambda(e^{-\theta}-1)\}.\end{aligned}$$

Therefore, a Poisson random variable with mean λ has the Laplace transform $\phi_Z(\theta) = \exp\{\lambda(e^{-\theta}-1)\}$. Now, raise the Laplace transform to the power $1 = n/n$, to give

$$\phi_Z(\theta) = \left[\exp\{\lambda(e^{-\theta}-1)\}\right]^{n/n} = \left[\exp\left\{\tfrac{\lambda}{n}(e^{-\theta}-1)\right\}\right]^n = \{\phi_{Z_i}(\theta)\}^n,$$

where $Z_i \sim \text{Pois}(\lambda/n)$, for $i = 1, \ldots, n$. Therefore, $Z \sim \text{Pois}(\lambda)$ is infinitely divisible.

(v) First, search for a relationship between X, Z, and Y using previously obtained results. The distributions $p_X(x)$ and $p_{X|Z}(x \mid z)$ are known, the latter of which is obtained using Bayes theorem:

$$p(x \mid z) \propto p(z \mid x)\, p(x) \propto \frac{e^{-\lambda x}(\lambda x)^z}{z!} \cdot x^{a-1} e^{-x} \propto x^{a+z-1} e^{-x(1+\lambda)},$$

which means $p(x \mid z) \equiv \text{Ga}(a+z, 1+\lambda)$; this is the exact same distribution as Y. That is, $p_{Y|Z}(\cdot \mid z) = p_{X|Z}(\cdot \mid z)$. Because it was already shown that X is infinitely divisible, Y is also marginally infinitely divisible.

An alternative approach would be to write the solution out by construction. For example, to find $\{Y_i\}_{i=1:n}$ for which $Y = \sum_{i=1}^{n} Y_i$, consider $X_i \sim \text{Ga}(a/n, 1)$, $Z_i \sim \text{Pois}(\lambda X_i)$, and $Y_i \sim \text{Ga}(a/n + Z_i, 1+\lambda)$.

Further Reading

Historical Background. Infinite divisibility was a concept introduced in a series of Italian papers in 1929. Soviet mathematician Aleksandr Khintchine later provided a formal definition: an infinitely divisible random variable can be represented as the sum of n independent and identically distributed random variables for all non-negative integers n. Such variables form the basis of many stochastic processes, such as Brownian motion and the Poisson process.

Several well-known distributions are infinitely divisible, including the Cauchy, chi-squared, gamma, Gaussian, inverse gamma, Lévy, log-normal, negative binomial, and t distributions. For all of them, infinite divisibility can be shown using the moment-generating function method. Two examples of known distributions that are not infinitely divisible are the uniform and binomial distributions. In fact, the only distribution with bounded support that is infinitely divisible is the one-point distribution (Steutel and van Harn, 2004).

Points of Interest.

1. Distributions that are infinitely divisible benefit from a broad generalization of the central limit theorem. That is, as $n \to \infty$, a sum $S_n = X_{n1} + \cdots + X_{nn}$ of independent uniformly asymptotically negligible (UAN) random variables within a triangular array approaches an infinitely divisible distribution. An array of random variables

$\{X_{nk}\}_{n,k\leq n}$ is UAN if it satisfies

$$\lim_{n\to\infty}\max_{1\leq k\leq n} \mathrm{P}(|X_{nk}| > \epsilon) = 0$$

for every $\epsilon > 0$. If the UAN condition is satisfied with an appropriate scaling of identically distributed random variables, weak convergence to a stable distribution is guaranteed. In addition, if the variance is finite, weak convergence to the normal distribution is guaranteed via the central limit theorem (Feller, 1966).

2. Infinite divisibility is also a key component of Lévy processes. Lévy processes, including Brownian motion and Poisson processes, are continuous-time analogues of a random walk and are of great importance in probability theory. A Lévy process can exist solely with jumps. For example, the independent increment homogeneous gamma process has jumps over an interval $(s,t]$ of size $\mathrm{Ga}(t-s,1)$; this is true for all such intervals. The process exists for the simple reason that $\mathrm{Ga}(a,1)+\mathrm{Ga}(b,1) =_d \mathrm{Ga}(a+b,1)$. Thus, the distribution of the process at any time t is unambiguous and is a $\mathrm{Ga}(t,1)$ random variable, assuming that the process starts at 0 at time 0. Further details are provided in Sato (1999).

3. Lévy processes appear fundamentally in Bayesian nonparametric models. The idea is that the Bayesian prior is a probability on random functions that behaves with probability one as either a distribution function, survival function, cumulative hazard function, etc. The original class of such priors was based on independent increment processes without Gaussian components. For example, a classic Bayesian nonparametric prior, the Dirichlet process, can be viewed as a normalized gamma process, which is an independent increment process with gamma increments. This process can be characterized by Laplace transforms: if Z_t is the process for $t \geq 0$, then

$$-\log\{\mathrm{E}(e^{-\theta Z_t})\} = t\int_0^\infty (1-e^{-\theta s})dN(s),$$

where N is known as the Lévy measure and, in the case of the gamma process, is given by $dN(s) = s^{-1}e^{-s}\,ds$. The infinite divisibility of the gamma distribution provides the properties and the existence of the process.

S1.1.3* – A maximal inequality for a martingale sequence

(i) For any n and n' such that $n > n'$, E_n will have $M_{n'} < \epsilon$, whereas $E_{n'}$ has $M_{n'} > \epsilon$. Thus, the intersection of E_n and $E_{n'}$ is empty, and the set of events is mutually disjoint.

(ii) The usual strategy to show that two sets A and B are equivalent is to show that $A \subseteq B$ and $B \subseteq A$. For example, $A \subseteq B$ is proved by showing that if $a \in A$, then $a \in B$. Let $A = (\max\{M_1,\ldots,M_K\} > \epsilon)$ and $B = \cup_{n=1}^K E_n$. For an element $\{M_1,\ldots,M_K\} \in A$, define n to be the smallest integer for which $M_n > \epsilon$. Such an $n \in \{1,\ldots,K\}$ must exist and defines the event E_n, which is a member of B. To show $B \subseteq A$, consider that any event E_n indicates that the maximum of $\{M_1,\ldots,M_K\}$ is larger than ϵ.

(iii) The left side of the given equality can be written as

$$\int_{E_n} \int M_K f(M_K \mid M_1, \ldots, M_n) f(M_1, \ldots, M_n) \, dM_K \, dM_1 \ldots dM_n,$$

where the inner integral is M_n, using the martingale property:

$$\mathrm{E}(M_K \mid M_1, \ldots, M_n) = M_n.$$

The right side of the given equality can be written as

$$\int_{E_n} M_n f(M_1, \ldots, M_n) \, dM_1 \ldots dM_n$$

by integrating out $M_{n+1}, \ldots, M_K$.

(iv) Because the integral is over E_n, it follows that $M_n \geq \epsilon$. Having used this inequality, the left over term (i.e., having integrated out $M_{n+1}, \ldots, M_K$) is

$$\int_{E_n} f(M_1, \ldots, M_n) \, dM_1 \ldots dM_n = \mathrm{P}(E_n).$$

(v) From parts (i) and (ii), it is that

$$\mathrm{P}(\max\{M_1, \ldots, M_K\} > \epsilon) = \textstyle\sum_{n=1}^{K} \mathrm{P}(E_n)$$

$$< \epsilon^{-1} \textstyle\sum_{n=1}^{K} \int_{E_n} M_K f(M_1, \ldots M_K) \, dM_1 \ldots dM_K$$

$$= \epsilon^{-1} \int M_K f(M_1, \ldots M_K) \, dM_1 \ldots dM_K = \epsilon^{-1} \mathrm{E}(M_K).$$

Further Reading

Historical Background. Martingales are an important class of process, both in discrete and continuous time. There exists a continuous-time version of this question that would be interesting to pursue. Many common processes, such as centered Poisson processes or Brownian motion, are martingales. See Doob (1940a) for a review of martingale properties and Williams (1991) and Rogers and Williams (2000) for fundamental books on the topic.

Doob discovered several fundamental results concerning martingales, including a convergence theorem: if $\{M_n\}$ is a non-negative martingale as described in the question and $\sup_n \mathrm{E}(M_n) < \infty$, then M_n converges with probability one to a random variable M_∞ with a finite mean. Doob demonstrated this result on the consistency of a Bayesian model (Doob, 1940b), and the results in Doob's paper were recently developed by Fong et al. (2024) to provide an interpretation for Bayesian analysis.

Points of Interest.

1. Denote a standard Brownian motion $\{B_t\}$, which is a continuous-time martingale. Applying Doob's maximal inequality gives a maximal inequality for Brownian motion as

$$\mathrm{P}\left(\sup_{0 \leq t \leq T} B_t \geq \lambda \right) \leq \exp(-\tfrac{1}{2}\lambda^2 / T)$$

for any fixed $T > 0$. It is left as an exercise to get this result.

2. An interesting application of martingales involves a proof of the strong law of large numbers under minimal assumptions. The sequence $\{M_n\}$, $n = 1, 2, \ldots$, is a reverse martingale with respect to the filtration $\{\mathcal{G}_n\}_{n=1}^{\infty}$, where $\mathcal{G}_n = \sigma(M_n, M_{n+1}, \ldots)$, if $\mathrm{E}\left(M_m \mid \mathcal{G}_n\right) = M_n$ for all $m \leq n$. The reverse martingale theorem states the following: (i) there exits M_∞ such that $M_n \rightarrow M_\infty$ with probability 1; (ii) $M_n \rightarrow M_\infty$ in mean; and (iii) $M_\infty = \mathrm{E}(M_1 \mid \mathcal{G}_\infty)$. This theorem can be used to prove the law of large numbers for an independent and identically distributed sample $\{X_n\}$ with common mean μ because

$$M_n = \frac{1}{n} \sum_{i=1}^{n} X_i$$

can be written as a reverse martingale with respect to

$$\mathcal{G}_n = \sigma\{M_n, M_{n+1}, \ldots\} = \sigma\{M_n, X_{n+1}, X_{n+2}, \ldots\}.$$

Thus, $\mathrm{E}(X_1 \mid \mathcal{G}_n) = \mathrm{E}(X_1 \mid M_n) = M_n$ because the $\{X_i\}$ are independent and identically distributed. This is the topic of Question 6.1.2 and is covered in Stroock (2010).

Q1.2 Questions – Markov Processes

Q1.2.1* – Random walk on the integers

Introduction. A simple random walk on the integers is one of the most basic stochastic processes: at each time point, the process moves $+1$ or -1 with equal probability. This question studies the properties of the simple random walk, particularly the first time it returns to 0 if the process starts at 0. Regardless of where the process starts, it will eventually reach 0; this phenomenon is known as the gambler's ruin. A gambler with a finite amount of money will eventually lose when playing a fair game against a bank with an infinite amount of money. The gambler's money will perform a random walk, and it will reach zero at some point.

Question. A simple random walk $\{S_n\}_{n=0}^{\infty}$ is a Markov process such that $S_0 = 0$ and

$$\mathrm{P}(S_{n+1} = S_n + 1) = \mathrm{P}(S_{n+1} = S_n - 1) = \tfrac{1}{2}.$$

Alternatively, one can describe S_n via a sequence of independent Bernoulli random variables $\{X_n\}$ with probabilities $\mathrm{P}(X_n = +1) = \mathrm{P}(X_n = -1) = \frac{1}{2}$ and

$$S_n = \sum_{i=1}^{n} X_i.$$

(i) Explain why

$$\mathrm{P}(S_n = 0) = \begin{cases} \binom{n}{n/2}\left(\frac{1}{2}\right)^n, & n \text{ is even} \\ 0, & n \text{ is odd} \end{cases}.$$

(ii) Define $T_0 = \min_{n \geq 1}\{S_n = 0\}$, the time at which the process first returns to 0. Show that it is possible to write

$$\mathrm{P}(S_n = 0) = \sum_{k=1}^{n} \mathrm{P}(S_{n-k} = 0)\,\mathrm{P}(T_0 = k)$$

for $n \geq 1$.

(iii) By considering $G(t) = \sum_{n=0}^{\infty} t^n\,\mathrm{P}(S_n = 0)$ for $0 < t < 1$, prove that the generating function for T_0, defined by

$$G_0(t) = \sum_{k=1}^{\infty} t^k\,\mathrm{P}(T_0 = k),$$

is equal to $G_0(t) = 1 - \sqrt{1 - t^2}$. Note that $G(t) = (1 - t^2)^{-1/2}$ for $0 < t < 1$.

(iv) What is the probability that the process returns to 0 at some point?

(v) Now, suppose that $\mathrm{P}(X_n = +1) = p$ and $\mathrm{P}(X_n = -1) = q = 1 - p$, where $p \neq \frac{1}{2}$. What changes?

Q1.2.2 – Properties of random sums

Introduction. A random sum of independent random variables arises naturally in many contexts, such as branching processes. Working with sums of random variables is difficult even if the variables are independent because the distribution of the sum often involves multiple–fold convolutions. Fortunately, generating functions may be used to learn about the distribution of the sum. This question uses generating functions to obtain the expectation, variance, and other properties of a random sum.

Question. Let $X_1, X_2, \ldots$ be non–negative independent and identically distributed random variables with mean μ and variance σ^2. For the positive integer-valued random variable $N \in \{1, 2, \ldots\}$, let $\mathrm{E}(N) = m$ and $\mathrm{Var}(N) = \tau^2$. Define $S = \sum_{i=1}^{N} X_i$.

(i) Find $\mathrm{E}(S)$ and $\mathrm{Var}(S)$.

(ii) Suppose $\mathrm{E}\left(\phi^N\right) = H(\phi)$ for $0 < \phi < 1$. Show that $\mathrm{E}\left(e^{-\theta S}\right) = H(G(\theta))$, where $G(\theta) = \mathrm{E}\left(e^{-\theta X}\right)$.

(iii) Suppose $Z_0 = 1$ and

$$Z_n = X_1 + \cdots + X_{Z_{n-1}} \text{for } n \geq 1.$$

If each X_i is an independent Bernoulli random variable with $\mathrm{P}(X_i = 1) = p$, prove that $G_n(s) = G_{n-1}(1 - p + ps)$ for $0 < s < 1$, where $G_n(s) = \mathrm{E}\left(s^{Z_n}\right)$.

(iv) Prove that $G_n(s) = 1 - p^n + p^n s$.

(v) Use part (iv) to show that $\mathrm{P}(Z_n = 0) \to 1$.

Q1.2.3 – Properties of a two-state discrete Markov chain

Introduction. Linear algebra is used to analyze the matrices that describe the transition probabilities between states in a Markov chain; these matrices are known as transition or stochastic matrices. In particular, matrix algebra can be used to understand the behavior of the chain over time, which involves the spectrum of the transition matrix. This question uses linear algebra to explore results for a two-state discrete chain, the most simple example of a Markov process.

Question. Suppose $\{X_n\}_{n \geq 0}$ is a time–homogeneous Markov chain on $\{0, 1\}$ with transition matrix

$$P = \begin{pmatrix} \alpha & 1 - \alpha \\ 1 - \beta & \beta \end{pmatrix},$$

where $0 \leq \alpha, \beta \leq 1$. Thus, for all n, $\mathrm{P}(X_{n+1} = 0 \mid X_n = 0) = \alpha$ and $\mathrm{P}(X_{n+1} = 1 \mid X_n = 1) = \beta$.

(i) Without calculating the eigenvalues of P, demonstrate why $\lambda = 1$ is an eigenvalue.

(ii) Find the value of the smallest eigenvalue and explain the special cases that arise. What does the process look like for these special cases?

(iii) If it exists, find the probability vector π satisfying $\pi = \pi P$. What is the interpretation of π?

(iv) Suppose that $v = (v_1, v_2)'$ and $v P = \lambda' v$, where $\lambda' < 1$. Prove that $v_1 + v_2 = 0$.

(v) Define the probability vector $q^{(n)} = (\mathrm{P}(X_n = 0), \mathrm{P}(X_n = 1))'$ for $n \geq 0$, and let $-1 < \lambda' < 1$. Show that $q^{(n)} \to \pi$ as $n \to \infty$. Consider starting with

$$q^{(0)} = a\pi + b\,v$$

for some scalars a and b. Discuss the implications with respect to the special cases in part (ii).

Q1.2.4* – Properties of discrete Markov chains

Introduction. Markov chains have become an important tool for Bayesian analysis. For example, sampling from the posterior distribution is a popular method for Bayesian inference given an intractable posterior distribution. Posterior sampling requires a Markov chain to have the posterior as the stationary distribution, which is usually not difficult to achieve. It then becomes necessary to determine convergence aspects, such as irreducibility and ergodicity; ergodicity is a consequence of aperiodicity and irreducibility when the state space is finite. Further, how fast the chain converges to the stationary distribution is a matter of great importance. This question studies all of these phenomenon in a finite state space and shows that all the key ideas are related to the eigenvalues and corresponding eigenvectors of the transition probability matrix.

Question. Let $P = (p_{ij})$ be a $M \times M$ transition matrix for a discrete Markov chain $\{X_n\}$ on $\{1, \ldots, M\}$ for $M > 1$. Hence, for all $n \geq 1$,

$$\mathrm{P}(X_n = j \mid X_{n-1} = i) = p_{ij}.$$

Let $1 = \lambda_1 > \lambda_2 > \cdots > \lambda_M$ be the eigenvalues of P.

(i) Show that all the eigenvalues of P lie between -1 and $+1$. Additionally, show that 1 is the largest eigenvalue. Hint: For any vector $v = (v_1, \ldots, v_M)'$ of dimension M,

$$v_{(1)} \leq \sum_{j=1}^{M} p_{ij}\, v_j \leq v_{(M)},$$

where $v_{(1)}$ and $v_{(M)}$ are the minimum and maximum of v, respectively.

(ii) If P is reducible, why is the second largest eigenvalue of P also 1?

(iii) Now, suppose that π is the unique stationary probability for P (i.e., $\pi P = \pi$), P is irreducible, and the smallest eigenvalue of P is greater than -1. Let $(\pi, w_2, \ldots, w_M)$ be the left eigenvectors of P. Prove that $w_j'\mathbf{1} = 0$, where $\mathbf{1}$ is the M–dimensional vector of 1's.

(iv) Denote the start of the chain as $\boldsymbol{q}^{(0)}$, where $\mathrm{P}(X_0 = j) = q_j^{(0)}$. Indicate when

$$\boldsymbol{q}^{(0)} = \alpha\, \boldsymbol{\pi} + \sum_{j=2}^{M} \alpha_j\, \boldsymbol{w}_j,$$

and show that $\alpha = 1$ using part (iii).

(v) Show that the probability mass function for X_n is

$$\boldsymbol{q}^{(n)} = \boldsymbol{\pi} + \sum_{j=2}^{M} \alpha_j\, \lambda_j^n\, \boldsymbol{w}_j,$$

and explain why $\boldsymbol{q}^{(n)} \to \boldsymbol{\pi}$.

Q1.2.5* – The Poisson process from infinitesimal probabilities

Introduction. This question concerns the homogeneous Poisson process from an infinitesimal perspective. It is well-known that the Poisson process has jumps that are Poisson distributed; the number of jumps in intervals

$$(0, t_1), (t_1, t_2), \ldots, (t_{m-1}, t_m)$$

are independent Poisson random variables with means

$$\lambda\, t_1, \lambda\, (t_2 - t_1), \ldots, \lambda\, (t_m - t_{m-1})$$

for some $\lambda > 0$. From a fundamental point of view, the existence of such a process is due to the following notion: the sum of two independent Poisson random variables with means λ and μ is a Poisson random variable with mean $\lambda + \mu$.

Question. Suppose N_t, for $t \geq 0$, is a discrete-valued non-decreasing Markov process in continuous time with $N_0 = 0$. For any $n \geq 0$ and small h,

$$\mathrm{P}(N_{t+h} = n + 1 \mid N_t = n) = h\lambda + o(h),$$
$$\mathrm{P}(N_{t+h} = n \mid N_t = n) = 1 - h\lambda + o(h),$$
$$\mathrm{P}(N_{t+h} > n + 1 \mid N_t = n) = o(h).$$

Define $p_t(n) = \mathrm{P}(N_t = n)$.

(i) Show that $p_t'(0) = -\lambda\, p_t(0)$ and $p_t'(n) = -\lambda\, p_t(n) + \lambda\, p_t(n-1)$ for $n \geq 1$, where $'$ denotes differentiation with respect to t.

(ii) Find the matrix $G = (g_{ij})_{0 \leq i,j < \infty}$ such that $P_t' = P_t\, G$, where $P_t' = (p_t'(0), p_t'(1), \ldots)$ and $P_t = (p_t(0), p_t(1), \ldots)$.

(iii) Show that the solution to the differential equation in parts (i) and (ii) is

$$p_t(n) = \frac{(t\lambda)^n}{n!}\, e^{-t\lambda},$$

and explain why $p_0(0) = 1$.

(iv) Show that the solution in part (iii) can be written as $p_t(n) = [\exp(tG)]_n$, where $\exp(tG)$ is the matrix

$$\sum_{l=0}^{\infty} \frac{t^l G^l}{l!}$$

and $[\cdot]_n$ denotes the $(1, n)$ element of the matrix. Pay special attention to the meaning of G^0.

(v) How long does the Poisson process stay at state 0? Work this out from the infinitesimal probabilities provided at the start of the question and then let $h \to 0$.

Q1.2.6 – The Poisson process from exponential random variables

Introduction. The Poisson process can be constructed from a sequence of independent exponential random variables. The process remains at each height for an independent exponential time before increasing by one. This question establishes that the height at a particular point in time is Poisson-distributed and that, for the process to be Markov, the exponential times spent at each height are necessary.

Question. Suppose the $\{X_i\}_{i\geq 1}$ are independent exponential random variables with mean $1/\lambda$. For some random integer $N \geq 0$, define

$$S_N = X_1 + \cdots + X_N$$

with $S_0 = 0$.

(i) For a fixed $n \geq 1$, show that

$$\mathrm{E}\left(e^{-\theta S_n}\right) = \left(\frac{\lambda}{\lambda + \theta}\right)^n$$

for any $\theta > 0$.

(ii) Show that S_n has a gamma distribution. Use $\mathrm{E}\left(e^{-\theta G}\right)$ to find the parameters for the gamma distribution, where G is a gamma random variable with mean a/b and variance a/b^2.

(iii) Define the integer-valued random variable $N \geq 0$ such that $S_N < t$ and $S_{N+1} > t$ for some $t > 0$. Show that

$$\mathrm{P}(N = n) = \frac{(\lambda t)^n}{n!} e^{-\lambda t}.$$

(iv) Using the $\{X_i\}$, define a Markov stochastic process on time $t > 0$ that has a Poisson variable as the height gained between arbitrary time points s and t, where $s < t$.

(v) Using the memoryless property of an exponential random variable, demonstrate the Markov property of the process; do this by finding the time remaining for the process to stay at height n if it is known to be there at time s.

S1.2 Solutions & Further Reading – Markov Processes

S1.2.1* – Random walk on the integers

(i) Firstly, $S_n = 0$ is only possible if there are enough positive 1's to cancel out the negative 1's, which can only happen if n is even. Thus, $P(S_n = 0) = 0$ for all odd n. For even n, there must be $n/2$ positive 1's and $n/2$ negative 1's. Consider the binomial distribution with n trials, probability parameter $p = \frac{1}{2}$, and $n/2$ successes, i.e. $Y \sim \text{Binom}(n, \frac{1}{2})$. Then,

$$\mathrm{P}(S_n = 0) = \mathrm{P}(Y = n/2)$$

and, for all even n,

$$\mathrm{P}(S_n = 0) = \binom{n}{n/2}(\tfrac{1}{2})^{n/2}(\tfrac{1}{2})^{n-n/2} = \binom{n}{n/2}(\tfrac{1}{2})^{n}.$$

(ii) Using the law of total probability,

$$\mathrm{P}(S_n = 0) = \sum_{k=1}^{n} \mathrm{P}(S_n = 0 \mid T_0 = k)\,\mathrm{P}(T_0 = k).$$

Now, $\mathrm{P}(S_n = 0 \mid T_0 = k) = \mathrm{P}(S_{n-k} = 0)$ from the homogeneity of the process. Note that $\mathrm{P}(S_n = 0 \mid T_0 > n) = 0$ for all n.

(iii) Starting with $G(t)$,

$$\begin{aligned}
G(t) &= \sum_{n=0}^{\infty} t^n \mathrm{P}(S_n = 0) \\
&= 1 + \sum_{n=1}^{\infty} t^n \mathrm{P}(S_n = 0) \quad \text{(because } \mathrm{P}(S_0 = 0) = 1\text{)} \\
&= 1 + \sum_{n=1}^{\infty}\sum_{k=1}^{n} t^n \mathrm{P}(S_{n-k} = 0)P(T_0 = k) \quad \text{(using part (ii) to get } \mathrm{P}(T_0 = k) \text{ involved)} \\
&= 1 + \sum_{k=1}^{\infty}\sum_{n=k}^{\infty} t^n \mathrm{P}(S_{n-k} = 0)P(T_0 = k) \quad \text{(triangular sum property)} \\
&= 1 + \sum_{k=1}^{\infty} \mathrm{P}(T_0 = k) \sum_{m=0}^{\infty} t^{m+k} \mathrm{P}(S_m = 0) \quad (m = n - k) \\
&= 1 + \sum_{k=1}^{\infty} t^k \mathrm{P}(T_0 = k) \sum_{m=0}^{\infty} t^m \mathrm{P}(S_m = 0) \\
&= 1 + \sum_{k=1}^{\infty} t^k \mathrm{P}(T_0 = k)G(t) \\
&= 1 + G_0(t)G(t).
\end{aligned}$$

Because $G(t) = 1/\sqrt{1-t^2}$, it must be that $G_0(t) = 1 - \sqrt{1-t^2}$.

(iv) Because $G_0(1) = 1$, the outcome space for T_0 is precisely the even numbers from 2 onwards, so $P(T_0 \in \{2, 4, 6, \cdots\}) = 1$. Thus, the probability that the process returns to 0 is 1.

(v) The difference now is that $G(t) = 1 - (1 - 4pqt^2)^{-1/2}$, which is left as an exercise to the reader. Thus,

$$G_0(t) = 1 - \sqrt{1 - 4pqt^2}.$$

Now, $G_0(1) = 1 - \sqrt{1 - 4pq} \neq 1$. Hence, T_0 exists with probability $G_0(1)$. Thus, the probability that the path does not return to 0 is $\sqrt{1 - 4pq}$.

Further Reading

Historical Background. The first random walks were introduced by Pearson, with the simple random walk being the most basic (Pearson, 1905). Grimmett and Stirzaker (1982) and Durrett (1999) are good books on the simple one–dimensional random walk on the integers. Modern extensions allow for random walks in all environments, most notably on graphs (Lawler and Limic, 2010).

Points of Interest.

1. There are a number of properties available to be discovered for the simple random walk. For example, the random time T_a at which the process first visits $a > 0$ is the sum of a independent copies of T_1. Additionally, $G_1(t)$ is the generating function for T_1, which is available from $G_0(t)$ via the law of total probability: for $n \geq 1$,

$$P(T_0 = n) = \tfrac{1}{2} P(T_0 = n \mid S_1 = 1) + \tfrac{1}{2} P(T_0 = n \mid S_1 = -1)$$

and $P(T_0 = n \mid S_1 = 1) = P(S_n = 0 \mid S_1 = -1) = P(T_1 = n - 1)$. Therefore, $P(T_0 = n) = P(T_1 = n - 1)$ for $n \geq 1$, and $G_0(t) = t\, G_1(t)$ provides the generating function for all T_a.

2. Other properties of the simple random walk can be found from the reflection principle. If $N_n^0(a, v)$ are the number of paths from $a > 0$ to $v > 0$ in a time n and which pass through 0, then

$$N_n^0(a, v) = N_n(-a, v) = N_n(0, a + v).$$

The last term is easily available because, in the time interval n, there must be u moves up and d moves down for each path with $u + d = n$ and $u - d = a + v$. Therefore, there will be

$$C(n, u) = \binom{n}{\frac{1}{2}(n + a + v)}$$

such paths. This reflection principle assists with the finding of $P(M_n \geq r)$, where $M_n = \max\{S_0, S_1, \ldots, S_n\}$.

3. Direct access to the probabilities for T_0 via combinatorics is difficult. However, they can be obtained from the generating function

$$G_0(t) = \sum_{n=1}^{\infty} P(T_0 = 2n)\, t^{2n}.$$

To find $P(T_0 = 2n)$, consider the Taylor expansion for $\sqrt{1-x}$:

$$\sqrt{1-x} = 1 - \tfrac{1}{2}x - \tfrac{1}{2}(\tfrac{1}{2}-1)x^2/2! - \cdots - \tfrac{1}{2}(\tfrac{1}{2}-1)\ldots(\tfrac{1}{2}-n+1)x^n/n! - \cdots,$$

which implies that

$$1 - \sqrt{1-x} = \sum_{n=1}^{\infty} \tfrac{1}{2}^n \frac{1}{n!} 1.3\ldots(2n-3)\,x^n.$$

Therefore,

$$P(T_0 = 2n) = \tfrac{1}{2}^n \frac{1}{n!} 1.3\ldots(2n-3).$$

Note that the product of odd numbers can be expressed in terms of factorials.

Miscellaneous. Random walks have been used in many disciplines to describe physical and biological phenomena. Codling et al. (2008) provides a brief overview of random walk processes in biology. Kempe (2003) lists several uses of random walks in physics. Finally, Weiss (1983) provides a historical glimpse into the adoption of the random walk in various scientific disciplines.

S1.2.2 – Properties of random sums

(i) The key is to notice that the expectation of $\sum_{i=1}^{N} X_i$ acts on all random variables: N and the $\{X_i\}_{i=1}^{N}$. Recall the law of total probability in the form of conditional expectations:

$$E(S) = E\left(\sum_{i=1}^{N} X_i\right) = E\left\{E\left(\sum_{i=1}^{N} X_i \mid N\right)\right\} = E(N)\,\mu = m\,\mu.$$

The variance is based on the conditional variance formula:

$$\begin{aligned}\mathrm{Var}(S) &= \mathrm{Var}\left(\sum_{i=1}^{N} X_i\right)\\ &= E\left\{\mathrm{Var}\left(\sum_{i=1}^{N} X_i \mid N\right)\right\} + \mathrm{Var}\left\{E\left(\sum_{i=1}^{N} X_i \mid N\right)\right\}\\ &= E(N)\,\sigma^2 + \mathrm{Var}(N)\mu^2 = m\sigma^2 + \mu^2\tau^2.\end{aligned}$$

(ii) Reemploying conditional expectations,

$$\begin{aligned}E\left(e^{-\theta S}\right) &= E\left(e^{-\theta\sum_{i=1}^{N} X_i}\right) = E\left\{E\left(e^{-\theta\sum_{i=1}^{N} X_i} \mid N\right)\right\}\\ &= E\left[\left\{E\left(e^{-\theta X_1}\right)\right\}^N\right] \quad \text{(by independence of the } X\text{'s)}\\ &= E\left[\{G(\theta)\}^N\right] = H(G(\theta)).\end{aligned}$$

(iii) Using the answer to part (ii),

$$\begin{aligned} G_n(s) &= \mathrm{E}(s^{Z_n}) \\ &= \mathrm{E}\left\{\mathrm{E}\left(s^{X_1+\cdots+X_{Z_{n-1}}}|Z_{n-1}\right)\right\} \quad \text{(expectation is conditioned on } Z_{n-1}) \\ &= \mathrm{E}\left\{E(s^{X_1})\cdots E(s^{X_{Z_{n-1}}})\right\} \\ &= \mathrm{E}\left\{(1-p+ps)^{Z_{n-1}}\right\} \quad \text{(because } G(s)=\mathrm{E}(s^X)=(1-p)+ps) \\ &= G_{n-1}(1-p+ps). \end{aligned}$$

In the general case, $G_n(s) = G_{n-1}(G_X(s))$.

(iv) An important insight now is to see that $G_n(s) = G_X(G_{n-1}(s))$. The solution can be obtained using induction. Clearly, the hypothesis is true for $n = 1$. Assume true for n and note that

$$G_{n+1}(s) = 1 - p + sG_n(s) = 1 - p + p(1 - p^n + sp^n) = 1 - p^{n+1} + p^{n+1}s$$

as required.

(v) In part (iv), it is shown that the probability generating function of the distribution of Z_n is $G_n(s) = 1 - p^n + p^n s$. Therefore,

$$\mathrm{P}(Z_n = k) = \frac{1}{k!}\frac{d^k}{ds^k}G_n(s)\bigg|_{s=0}.$$

Because $k = 0$,

$$\mathrm{P}(Z_n = 0) = \frac{1}{0!}\frac{d^0}{ds^0}G_n(s)\bigg|_{s=0} = G_n(s)\bigg|_{s=0} = 1 - p^n \to 1$$

as $n \to \infty$. So, $\mathrm{P}(Z_n = 0) \to 1$ as $n \to \infty$. Note that $Z_n \sim \mathrm{Bern}(p^n)$.

Further Reading

Historical Background. The simple branching process, also known as the simple birth process or the Galton–Watson process, is the most basic population model. Each individual in a given generation gives birth to a random number of offspring, the sum of which forms the population size of the next generation. The question of interest is whether the population goes extinct or not. It is an interesting exercise to show that the probability of extinction is the smallest root of the equation $\eta = G_X(\eta)$. Harris (1964) and Grimmett and Stirzaker (1982) gave a comprehensive survey of branching processes. Finally, Athreya and Ney (2004) serves as a comprehensive monograph on branching processes; see chapter 1 for their discussion on the Galton–Watson process, which is more flexible than the simple branching process considered here.

Points of Interest.

1. Because G_X is a probability generating function, it is that $G_X(1) = 1$, so 1 is always a root of the equation. Whether there is a smaller root depends on the gradient of $G_X(t)$ at 1. Because G_X is a convex function and $G_X(0) = \mathrm{P}(X = 0) > 0$, there is at most one other root, which exists when $G_X'(1) > 1$. The observant reader will know that $G_X'(1)$ is the first moment $\mathrm{E}(X)$. In short, if $\mathrm{E}(X) > 1$, then the probability of extinction is greater than 0 but less than 1; if $\mathrm{E}(X) \leq 1$, the probability of extinction is 1.

2. In this question, X is a Bernoulli random variable, so each generation has either a population of 0 or 1. If 0, the population has gone extinct. To survive, a positive success is needed at each generation. The population size at generation n is Bernoulli with probability of success p^n. It is impossible to keep getting a success with independent Bernoulli experiments; eventually a 0 must occur, so the population goes extinct with probability one. Also, note that $E(X) = p < 1$.

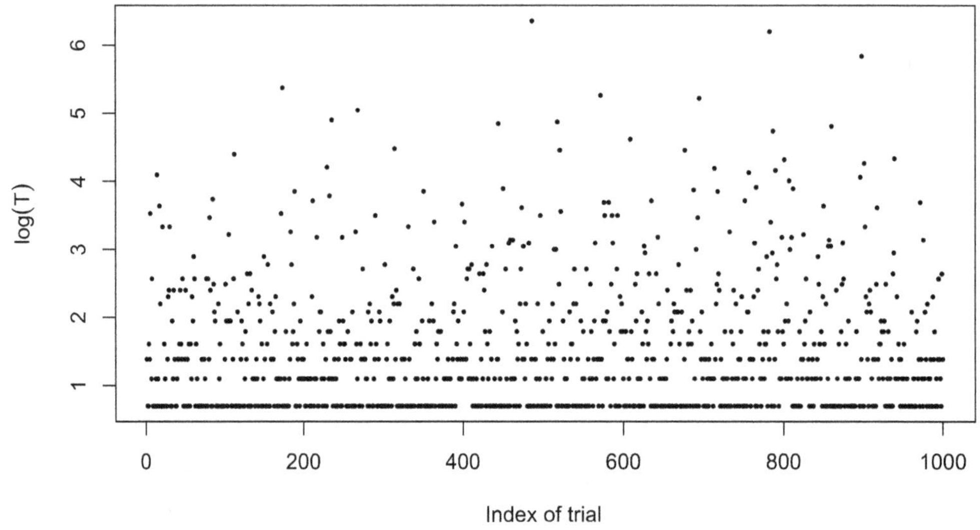

FIGURE 1.2
Plot of 1000 log-extinction times for Poisson-distributed offspring with mean 1.

Demonstration. If the random number of offspring is Poisson-distributed with mean 1, then the population will go extinct. Extinction may take some time to occur. A plot of the log-extinction times are provided in Fig. 1.2 for 1000 processes; some of the times are very large. The mean extinction time is 13.1, and the variance is approximately 5796.1.

S1.2.3 – Properties of a two-state discrete Markov chain

(i) Because P is a stochastic matrix, $P\mathbf{1} = \mathbf{1}$, where $\mathbf{1}$ is a vector of ones. Thus, $\lambda = 1$ is an eigenvalue of P.

(ii) Consider the characteristic equation $\det(P - \lambda I) = 0$, which can be expanded to obtain

$$(\alpha - \lambda)(\beta - \lambda) - (1 - \alpha)(1 - \beta) = 0.$$

This is equivalent to $\lambda^2 - (\alpha+\beta)\lambda + \alpha + \beta - 1 = 0$, which has roots 1 and $\alpha+\beta-1$; the latter root is smallest because $0 \le \alpha, \beta \le 1$. One special case arises when $\alpha = \beta = 1$, which implies that the process stays at the starting point and may not transition between states. The other special case is when $\alpha = \beta = 0$, which implies that the process must oscillate between the two states.

(iii) The vector $\pi = (\pi_1, \pi_2)'$ must satisfy $\pi = \pi P$ and $\pi_1 + \pi_2 = 1$, which is equivalent to satisfying

$$\begin{aligned} \alpha\pi_1 + (1-\beta)\pi_2 &= \pi_1, \\ (1-\alpha)\pi_1 + \beta\pi_2 &= \pi_2, \\ \pi_1 + \pi_2 &= 1. \end{aligned}$$

The first two of these equations are actually the same. Substituting the third equation into the first, one has $\pi_1 = (\beta-1)/(\alpha+\beta-2)$. By substituting this back into $\pi_1 + \pi_2 = 1$, one has $\pi_2 = (\alpha-1)/(\alpha+\beta-2)$. Because $\pi = \pi P$, π is the stationary distribution of the chain and only exists when $\alpha+\beta-2 \neq 0$ (i.e., the second eigenvalue is strictly smaller than 1).

(iv) First, multiply the $\mathbf{1}$ vector: $v\,P\mathbf{1} = \lambda'\,v\,\mathbf{1}$. Because P is a transition matrix, $P\mathbf{1} = \mathbf{1}$ and $v\,\mathbf{1} = \lambda'\,v\,\mathbf{1}$, which may be written as $v_1 + v_2 = \lambda'(v_1+v_2)$. Given that $\lambda' < 1$ and $v \neq \mathbf{0}$, the only solution to the equation is $v_1 + v_2 = 0$.

(v) Because $q^{(0)}$ is a probability vector and $v\,\mathbf{1} = 0$, then $q^{(0)}\,\mathbf{1} = 1$ and $a = 1$. The nth step of the process may be written as

$$q^{(n)} = q^{(0)}\,P^n = (\pi + bv)P^n = \pi + b\,\lambda'^n\,v,$$

where $\pi P^n = \pi$ and $vP^n = \lambda'^n v$. Because $|\lambda'| < 1, q^{(n)} \to \pi$ as $n \to \infty$. If $\lambda' = 1$, then π does not exist and the chain is reducible. If $\lambda' = -1$, then $q^{(n)}$ does not converge and the chain is periodic; $q^{(n)}$ is either $\pi + b\,v$ or $\pi - b\,v$.

Further Reading

Historical Background. Since the advent of Markov chain Monte Carlo (MCMC) methods, a substantial amount of work has gone into demonstrating rates of convergence for discrete-space Markov chains, such as the Metropolis algorithm. Convergence rates are determined by the second largest and smallest eigenvalues of the transition matrix. The exercise explored this relationship in the most simple space possible: a two-state chain. In many cases, it is not possible to evaluate the eigenvalues explicitly. Thus, much research focuses on obtaining bounds on the eigenvalues. The most well-known bounds are Cheeger bounds and, more recently, geometric bounds (Diaconis and Stroock, 1991).

Points of Interest.

1. To look at the convergence of a chain, one can express any probability vector as the linear combination of the left eigenvectors of P. This eigenbasis is problematic if eigenvalues other than the largest are repeated. In this case, a similar procedure to that described in this question can be performed using Jordan forms (Gallager, 1997).

2. If the smallest eigenvalue is -1, then $|\lambda_2| = 1$. Using part (v) and assuming all other eigenvalues are between -1 and 1, it is seen that, for large n, the $q^{(n)}$ will be approaching $\pi + (-1)^n \alpha_2 w_2$. This chain will not converge because it will be $\pi + \alpha_2 w_2$ for even n and $\pi - \alpha_2 w_2$ for odd n. This oscillating behavior can easily be illustrated for a two-state chain: the matrix P with $\lambda_2 = -1$ is given by

$$P = \begin{pmatrix} 0 & 1 \\ 1 & 0 \end{pmatrix},$$

which has $\pi = (\frac{1}{2}, \frac{1}{2})$. This chain is irreducible and flips between the two states at each iteration.

3. Studying the spectral behavior of the transition kernels of a Markov chain is a valuable technique when analyzing their long runtime behavior; the analysis of such behavior is essential for correctly implementing MCMC algorithms (Hastings, 1970; Robert and Casella, 2004). Assuming that all the left eigenvalues of P (i.e. $\pi, w_2, \ldots, w_M$) are distinct, one can take

$$q^{(0)} = \pi + \sum_{j=2}^{M} \alpha_j w_j,$$

for some $\{\alpha_j\}_{j=2}^{M}$. Then,

$$\begin{aligned} q^{(n)} = q^{(0)} P^n &= \left(\pi + \sum_{j=1}^{M} \alpha_j w_j \right) P^n \\ &= \pi P^n + \sum_{j=1}^{M} \alpha_j w_j P^n \\ &= \pi + \sum_{j=1}^{M} \alpha_j \lambda_j^n w_j. \end{aligned}$$

The leading eigenvalue of a Markov kernel is always 1, and the contribution of the other eigenvalues vanish as $n \to \infty$. Indeed, the other eigenvalues need to be strictly between -1 and 1 for convergence. Aperiodicity arises when the smallest eigenvalue is -1, which leads to oscillations, and irreducibility arises when the second largest eigenvalue is 1. The slowest decaying eigenvalue is either the second largest or the smallest eigenvalue; the smaller the maximum absolute value of these two eigenvalues, the faster the convergence to the stationary probability.

4. The analysis of time-inhomogenous chains are well-studied in the time series and econometrics literature (Shiryaev, 1999; Shumway and Stoffer, 2000). Such an extension to this simple example is natural in many applications where processes experience temporal dependence. The arguments used throughout this problem may be extended for chains with more than two states, such as in the next problem.

Demonstration. To demonstrate the relationship between the convergence rate and the maximum absolute value of the second largest and the smallest eigenvalues, set

$$P = \begin{pmatrix} 0.3 & 0.7 \\ 0.2 & 0.8 \end{pmatrix}.$$

The stationary probability is $\pi = (\pi_1, \pi_2)' = (0.222, 0.778)'$, and the second eigenvalue is $\lambda_2 = 0.1$. The convergence of $\boldsymbol{q}^{(n)}$ should be geometric with the speed of convergence to 0 determined by 0.1^n, which is quite fast. Fig. 1.3 illustrates the convergence of $q_1^{(n)}$ to $\pi_1 = 0.222$. Because convergence occurs quickly, even $q_1^{(5)}$ is equal to 0.222.

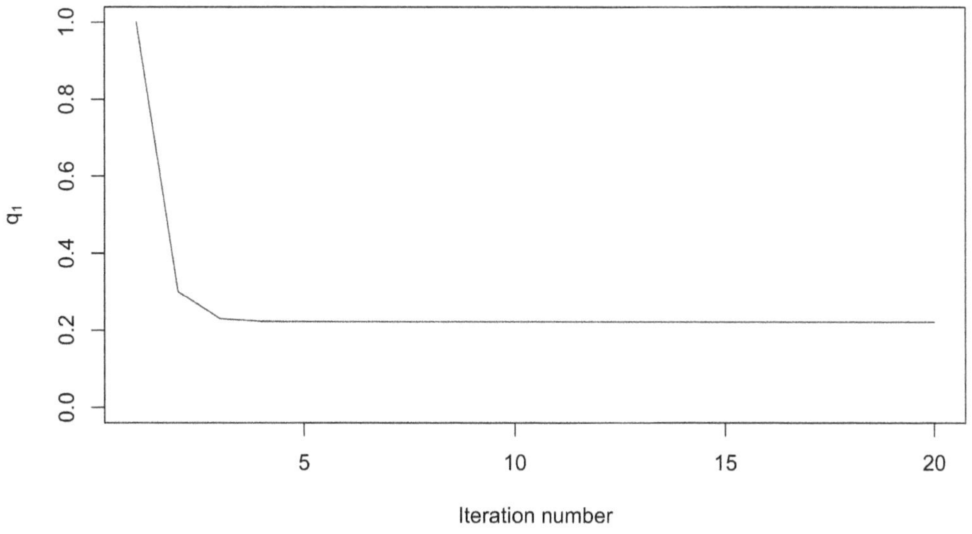

FIGURE 1.3
Convergence of $q_1^{(n)}$ to the stationary probability π_1.

S1.2.4* – Properties of discrete Markov chains

(i) Note that $\mathbf{1} = (1,\dots,1)'$ corresponds to an eigenvalue of 1: $P\mathbf{1} = \mathbf{1}$. Now, let $\lambda \neq 1$ be any other eigenvalue and v the corresponding left eigenvector of P, which may be written as $v'P = \lambda v'$. Then, $v'P\mathbf{1} = \lambda v'\mathbf{1}$ and, because $v \neq 0$ and $\lambda \neq 1$, it must be true that $v'\mathbf{1} = 0$. Therefore, $v_{(M)} > 0$ and $v_{(1)} < 0$. For all $i \in \{1,\dots,M\}$,

$$\sum_{j=1}^{M} p_{ij}\, v_j = \lambda\, v_i,$$

where λ is an eigenvalue and $v = (v_1,\dots,v_M)'$ is the corresponding eigenvector. Because the (p_{ij}) form a set of weights, it is true that $\lambda\, v_{(M)} \leq v_{(M)}$ and $\lambda \leq 1$. Further, $\lambda\, v_{(M)} \geq v_{(1)}$ and $\lambda\, v_{(1)} \leq v_{(M)}$ implies $|\lambda| \leq 1$.

(ii) If P is reducible, it can be represented in the form

$$P = \begin{pmatrix} P_1 & 0 \\ 0 & P_2 \end{pmatrix},$$

where P_1 and P_2 are probability matrices. Therefore, 1 is an eigenvalue associated with both P_1 and P_2, and there are two eigenvalues of value 1 for P.

(iii) This has been done in part (i); also, see part (iv) of Question 1.2.3. Note that v stands for an arbitrary left eigenvector.

(iv) If the left eigenvectors form a basis, then writing any M–dimensional vector $q^{(0)}$ in this way is possible. This occurs when the eigenvalues of P are distinct. Because $w_j'\mathbf{1} = 0$ for all $j = 2,\dots,M$, it is true that $1 = q^{(0)\prime}\mathbf{1} = \alpha\pi'\mathbf{1} = \alpha$.

(v) By repeatedly applying the P operator, it is seen that $q^{(1)} = q^{(0)}P$ and, in general, $q^{(n)} = q^{(0)}P^n$. Using the form of part (iv) yields $q^{(n)}$. If all the $\{|\lambda_j| < 1\}_{j\geq 2}$, then $q^{(n)} \to \pi$ as $n \to \infty$.

Further Reading

Historical Background. This question is a more general version of the previous one. It cements the idea that, for a finite discrete chain, the existence and convergence of the chain relies solely on the second largest eigenvalue not being 1 (i.e., the chain is irreducible) and the smallest eigenvalue not being -1 (i.e., the chain is aperiodic). In the first case, the stationary probability vector exists and is unique. In the second case, the chain converges to stationary probability. If there is an infinite amount of states, more conditions for convergence are needed. In particular, the states must be persistent and non-null.

Points of Interest.

1. The ideas behind this question are concerned with the convergence of Markov chains. The existence of 1 as the largest eigenvalue and that the corresponding eigenvector having non–negative components arises as a special case of the Perron–Frobenius theory; this theory is about square non-negative matrices and provides results on the largest eigenvalue and properties of the corresponding eigenvector.

2. The rate of convergence in part (v) will be determined by the second largest absolute value of the eigenvalues. If this value, say $|\lambda_2|$, is close to 1, then convergence can be slow. Finding bounds for the second largest eigenvalue has been an active area of research; see Diaconis and Stroock (1991) for a key paper on the topic. Additionally, Polson (1996) considered the more traditional Cheeger-type bounds.

3. A general article on convergence rates of Markov chains is contained in lecture notes provided by Lalley (2009). Additionally, Grimmett and Stirzaker (1982) has an excellent chapter on Markov chains.

S1.2.5* – The Poisson process from infinitesimal probabilities

(i) First, consider the law of total probability:

$$\mathrm{P}(N_{t+h} = n) = \sum_{m=0}^{n} \mathrm{P}(N_{t+h} = n \mid N_t = m)\,\mathrm{P}(N_t = m).$$

The conditional probabilities can be separated according to the given conditional probabilities:

$$\begin{aligned}\mathrm{P}(N_{t+h} = n) &= \mathrm{P}(N_{t+h} = n \mid N_t = n)\,\mathrm{P}(N_t = n)\\ &\quad + \mathrm{P}(N_{t+h} = n \mid N_t = n-1)\,\mathrm{P}(N_t = n-1)\\ &\quad + \sum_{m<n-1} \mathrm{P}(N_{t+h} = n \mid N_t = m)\,\mathrm{P}(N_t = m).\end{aligned}$$

Hence,

$$\mathrm{P}(N_{t+h} = n) = (1 - \lambda h)\,\mathrm{P}(N_t = n) + \lambda h \mathrm{P}(N_t = n-1) + o(h).$$

Take $h \to 0$ to get

$$p_t'(n) = -\lambda p_t(n) + \lambda\, p_t(n-1),$$

which is valid for all $n \geq 1$. Setting up the differential equation with $n = 0$ follows similarly.

(ii) From part (i), it is seen that $g_{i,i} = -\lambda$ and $g_{i,i+1} = \lambda$ for all $i \geq 1$. All other elements are 0.

(iii) Differentiate the $p_t(n)$ given in the question to get

$$p_t'(n) = \frac{t^{n-1}\lambda^n}{(n-1)!} e^{-t\lambda} - \lambda p_t(n),$$

and the answer follows. Note that $p_0(0) = 1$ because the process starts at $N_0 = 0$.

(iv) The solution to the equation $P_t' = P_t\, G$ is

$$P_t = I + \sum_{l=1}^{\infty} \frac{t^l\, G^l}{l!}$$

because

$$P_t' = \sum_{l=1}^{\infty} \frac{t^{l-1}G^{l-1}}{(l-1)!} G = P_t G.$$

Further, the starting conditions are $P_0(0) = I$, where I is the identity matrix. Then, $p_t(n)$ is the probability of going from 0 to n in time t and is the $(1,n)$ element of P_t. The probability of going from m to n in time t would be the $(m+1,n)$ element of P_t.

(v) For the process defined on intervals h, the probability the process has not moved by time t is

$$\{1 - \lambda h + o(h)\}^{t/h}.$$

This is seen to converge to $\exp(-\lambda t)$ as $h \to 0$, demonstrating the connection with the exponential random variables.

Further Reading

Historical Background. When defined on the real line, the Poisson process has several definitions and interpretations, all of which imply that the number of events in a period of time is Poisson distributed and independent of the numbers of events in non–overlapping time periods. This exercise explores the definition of the Poisson process in terms of events occurring in infinitesimal periods of time. Calculus ensues.

Points of Interest.

1. Grimmett and Stirzaker (1982) is an excellent book that covers Poisson processes. It considers the more general starting equations

$$\mathrm{P}(N_{t+h} = n+1 \mid N_t = n) = h\lambda + o(h),$$
$$\mathrm{P}(N_{t+h} = n \mid N_t = n) = 1 - h\lambda + o(h),$$
$$\mathrm{P}(N_{t+h} > n+1 \mid N_t = n) = o(h),$$

where the holding times are state dependent. Note that $h\lambda$ is the (approximate) probability of a single jump in an interval of time h. It becomes more accurate as $h \to 0$. The probability of no jump is $1 - h\lambda$, and the probability of two or more jumps is $o(h)$. If time is discretized into chunks of time h and $t = Mh$, then the probability of no jump in $(0,t)$ is

$$(1 - \lambda h)^M = (1 - \lambda h)^{t/h} \to \exp(-\lambda t)$$

as $h \to 0$. Additionally, the probability of one jump is

$$M\,(1 - \lambda h)^{M-1}\,\lambda h \to \lambda\, t \exp(-\lambda t)$$

as $h \to 0$. A further exploration of such Poisson processes can be found in Cox and Isham (1980).

2. Infinitesimal probabilities are useful for modeling continuous-time phenomena. However, if one is looking to understand the probability of a change over a large interval of time, the length of time might be problematic; this is in contrast to the discrete time setting, where one-step probabilities are sufficient. In many instances, it is ideal to study changes over smaller intervals of time. Then, one could derive the probabilities of change accumulated over a larger time interval comprised of the smaller intervals. This approach is widely used in finance, where fundamental changes over small intervals of time are represented by stochastic differential equations.

3. The Poisson process can also be understood from random exponential holding times. The number of jumps N in time $(0, t)$ is the maximum n for which $T_1 + \cdots + T_n < t$, where $\{T_i\}_{i=1}^n$ are independent copies of an exponential random variable with parameter λ. Thus, $\mathrm{P}(N = n) = \mathrm{P}(Z < t, Z + S > t)$, where $Z \sim \mathrm{Ga}(n, \lambda)$ and $S \sim \mathrm{Exp}(\lambda)$. Specifically,

$$\mathrm{P}(N = n) = \int\int_{z<t,s>t-z} \frac{\lambda^n z^{n-1}}{\Gamma(n)} e^{-\lambda z} \lambda e^{-\lambda s} \, dz \, ds.$$

   For the process $\{N_t\}$ to be Markov, it is necessary that

$$\mathrm{P}(N_t = m \mid N_{r \leq s}) = \mathrm{P}(N_t = m \mid N_s), \quad s < t.$$

   Suppose $N_s = m$. From the Markov property, the time in which the process remains at state m does not depend on the time the process has spent at state m prior to time s. Therefore, the time spent in a state must have a "lack of memory property," which is satisfied by the exponential class of random variables.

4. Last and Penrose (2017) discussed various characteristics of Poisson processes, stationary point processes, Poisson integrals, perturbation analysis, and more. In general, their book serves as an excellent foundation in statistics and measure theory related to the Poisson process.

5. If there are two states, the adapted version of the Poisson process is provided by the generator matrix

$$G = \lambda \begin{pmatrix} -1 & 1 \\ 1 & -1 \end{pmatrix}.$$

   For $n \geq 1$, it is possible to find G^n:

$$G^n = \lambda^n 2^{n-1} (-1)^{n-1} \begin{pmatrix} -1 & 1 \\ 1 & -1 \end{pmatrix}.$$

   It is also possible to find P_t, which has

$$\tfrac{1}{2} + \tfrac{1}{2} e^{-2\lambda t} \quad \text{and} \quad \tfrac{1}{2} - \tfrac{1}{2} e^{-2\lambda t}$$

   as the diagonal elements and off-diagonal elements, respectively.

6. There is a continuous-time version of the discrete infinitesimal probabilities discussed in the first point of interest. For a continuous-time transition probability of a Markov process, let $p_t(y)$ denote the density of the height of the process at time t, which started at 0 at time 0. The small interval conditions are

$$\int p_h(y) \, dy = 1, \quad \int y \, p_h(y) \, dy = 0, \quad \int y^2 \, p_h(y) \, dy = h.$$

   For $m \geq 3$, it is that $\int y^m p_h(y) \, dy = o(h) \equiv$ negligible. We have negligible$/h \to 0$ as $h \to 0$.

   To ensure the existence of a process, the law of total probability yields

$$p_{t+h}(y) = \int p_h(x) \, p_t(y - x) \, dx.$$

   To exploit the small interval conditions, use a Taylor expansion for $p_t(y - x)$:

$$p_t(y - x) = p_t(y) - x \, dp_t(y)/dy + \tfrac{1}{2} x^2 d^2 p_t(y)/dy^2 + \text{higher order terms},$$

where the higher order terms can be ignored because they will disappear when integrated with the $p_h(x)$. Put this back into the law of total probability to get

$$p_{t+h}(y) = p_t(y) + \tfrac{1}{2}\, h\, d^2 p_t(y)/dy^2 + o(h).$$

Now, rearrange and let $h \to 0$ to get $dp_t(y)/dt = \frac{1}{2} d^2 p_t(y)/dy^2$. This has the form of a heat equation, and a solution is $p_t(y) = (2t\pi)^{-1/2} \exp(-\frac{1}{2}y^2/t)$. This particular process is known as Brownian motion.

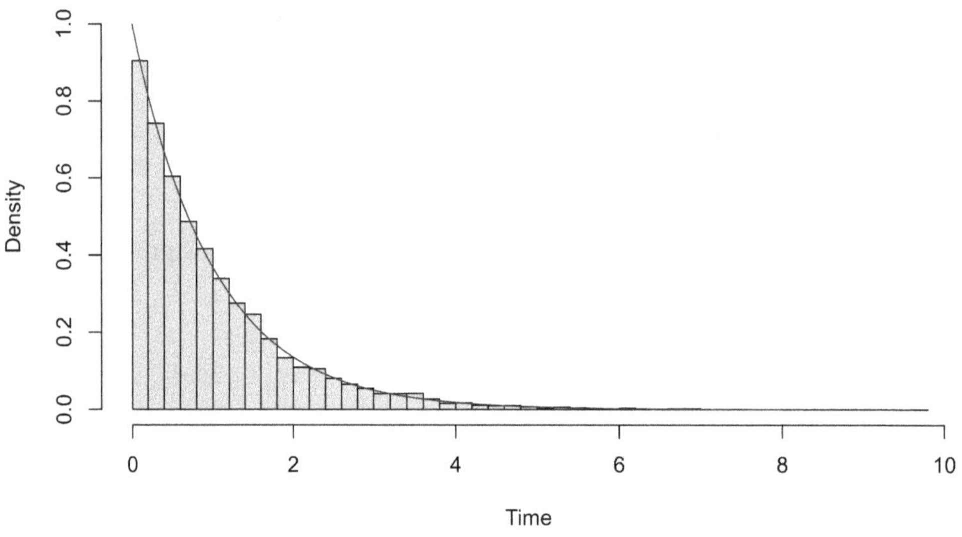

FIGURE 1.4
Histogram of random times constructed from Bernoulli random variables with the true exponential density overlaid.

Demonstration. A key feature of the infinitesimal construction is as follows: for a sequence of independent Bernoulli random variables with probability h, the integer of the first success multiplied by h approximates an exponential random variable increasingly well as $h \to 0$. This phenomenon is illustrated in Fig. 1.4, where the Bernoulli-generated random times are represented by the histogram and the overlaid line is the true exponential density function. Note that the times generated from the Bernoulli random variables are actually geometric random variables.

S1.2.6 – The Poisson process from exponential random variables

(i) Because the $\{X_i\}$ are independent and identically distributed,

$$\mathrm{E}\left(e^{-\theta S_n}\right) = \mathrm{E}\left(e^{-\theta(X_1+\cdots+X_n)}\right) = \prod_{i=1}^{n} \mathrm{E}\left(e^{-\theta X_i}\right) = \left\{\mathrm{E}\left(e^{-\theta X_i}\right)\right\}^n,$$

where

$$\mathrm{E}\left(e^{-\theta X_i}\right) = \int_0^\infty e^{-\theta x}\lambda e^{-\lambda x}\,\mathrm{d}x = \int_0^\infty e^{-(\lambda+\theta)x}\lambda\,\mathrm{d}x = \frac{\lambda}{\lambda+\theta}.$$

Therefore,

$$\mathrm{E}\left(e^{-\theta S_n}\right) = \left(\frac{\lambda}{\lambda+\theta}\right)^n.$$

(ii) Finding the Laplace transform of G,

$$\begin{aligned}\mathrm{E}\left(e^{-\theta G}\right) &= \int_0^\infty e^{-\theta g}\frac{b^a}{\Gamma(a)}g^{a-1}e^{-bg}\,\mathrm{d}g\\ &= \frac{b^a}{\Gamma(a)}\int_0^\infty e^{-(b+\theta)g}g^{a-1}\,\mathrm{d}gb\\ &= \left(\frac{b}{b+\theta}\right)^a.\end{aligned}$$

By the uniqueness of Laplace transforms, part (i) implies that $S_n \sim \mathrm{Ga}(n,\lambda)$.

(iii) Because $S_{n+1} = S_n + X_{n+1}$,

$$\mathrm{P}(N=n) = \mathrm{P}(S_n < t, S_n + X_{n+1} > t),$$

where S_n and X_{n+1} are independent. Thus,

$$\mathrm{P}(N=n) = \int_{s<t,x>t-s} f_{S_n}(s)\,f_X(x)\,ds\,dx.$$

Solving the x integral first,

$$\mathrm{P}(N=n) = \int_0^t \frac{\lambda^n}{\Gamma(n)}s^{n-1}e^{-\lambda s}\,e^{-\lambda(t-s)}\,ds,$$

from which the answer follows.

(iv) Consider the construction of the Poisson process using exponential random variables and their memoryless property. The process can be constructed by taking, over any interval (s,t), the change in height to be a Poisson random variable with parameter $\lambda(t-s)$; this is independent of any change outside of the interval (s,t). Such a process exists by a construction of independent exponential random variables. The process also satisfies the Chapman–Kolmogorov equation because the sum of two independent Poisson random variables with intensity parameters λ_1 and λ_2 is also Poisson-distributed with intensity $\lambda_1+\lambda_2$.

(v) The process defined in part (iv) is Markov due to the memoryless property of the exponential random variable. For example, if the height is at $N(s)$ at time s, then the time remaining at state $N(s)$ is an exponential random variable that does not depend on previous heights.

Further Reading

Historical Background. The Poisson process may be defined using independent exponential random variables to model the times between counts; compare this to the infinitesimal construction of Question 1.2.5. The memoryless property of the exponential distribution plays a crucial role in this new construction. The uniqueness of the exponential distribution holding the memoryless property is attributed to the following fact: if $g(t+s) = g(t)g(s)$ for all non–negative s and t, then g must be the exponential function.

Points of Interest.

1. The Poisson process is the simplest Markov process that is continuous in time and has a discrete state space. However, more general Markov processes may be considered. Typically, the rate parameter of the exponential random variable is used to construct such processes because it determines the length of stay in any state. For example, denote the rate for state j as $-g_{jj}$ with $g_{jj} < 0$. The probability that the process moves from state j to state k is $-g_{jk}/g_{kk}$. For the Poisson process, $g_{jj} = -\lambda$ and $g_{k-1,k} = \lambda$ with all other $g_{jk} = 0$.

2. In general, the probability that the process moves from state j to state k during the interval (s, t) is

$$P(N(t) = k \mid N(s) = j) = \left[e^{(t-s)\,G}\right]_{jk},$$

where $G = (g_{jk})$ and $[\cdot]_{jk}$ evaluates the expression at the (j, k) component of the matrix. Recall the definition of the matrix e^A:

$$e^A = I + \sum_{n=1}^{\infty} \frac{A^n}{n!}.$$

One can use calculus to obtain the above conditional probability. For an arbitrarily small h,

$$P(N(t+h) = k \mid N(t) = j) = \begin{cases} 1 + g_{jj}\,h & j = k \\ g_{jk}\,h & j \neq k \end{cases}.$$

Using the law of total probability and letting $h \to 0$, one arrives at

$$p'_{jk}(t) = \sum_l g_{lk}\, p_{jl}(t),$$

where $p_{jk}(t) = P(N(t) = k \mid N(0) = j)$ and $'$ denotes differentiation with respect to t; this equation can be written in matrix form as $P'(t) = P(t)\,G$, where $P(t) = e^{tG}$. The Markov solutions for $p_{jk}(t)$ need to satisfy the law of total probability for each interim time point $0 < s < t$:

$$P(N(t) = k) = \sum_l P(N(t) = k \mid N(s) = l)\, P(N(s) = l \mid N(0) = j).$$

This follows from the set up of the infinitesimal starting point and the law of total probability. Because $P(t) = e^{tG}$, the law of total probability is equivalent to the result $e^{tG} = e^{(t-s)G}\,e^{sG}$.

3. Consider a two state process with

$$G = \begin{pmatrix} -\lambda_1 & \lambda_1 \\ \lambda_2 & -\lambda_2 \end{pmatrix}$$

for some $\lambda_1, \lambda_2 > 0$. In this case, $P(t)$ can be found exactly. However in general, it is not possible to find the analytical solution for $P(t)$, though numerically there are algorithms and software that can compute e^{tG} to a high degree of accuracy.

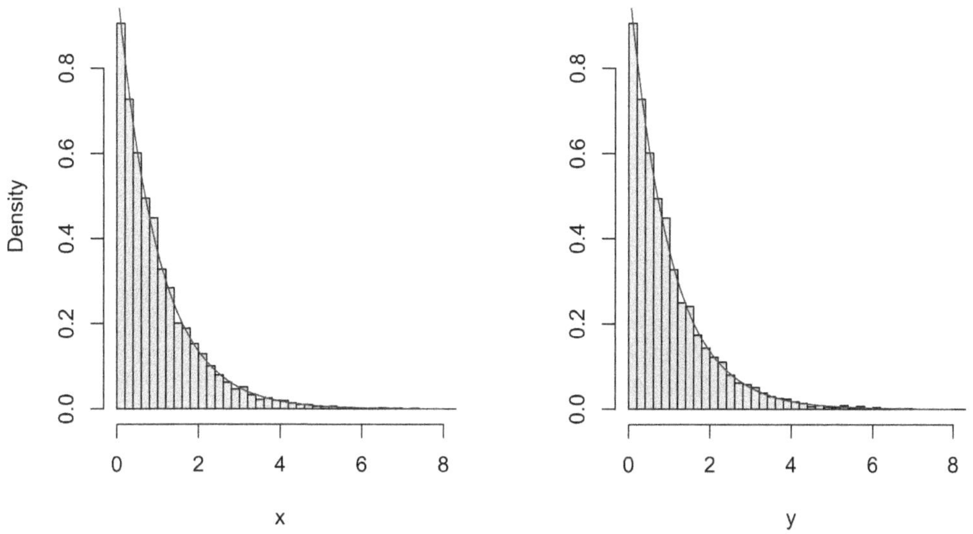

FIGURE 1.5
The histograms of the $\{x_i\}$ and $\{y_i\}$ samples with the standard exponential density overlaid.

Demonstration. The memoryless property of the exponential distribution is often mentioned. However, its implications usually are not well understood, even though it can be expressed mathematically as $\mathrm{P}(T > s + t \mid T > s) = \mathrm{P}(T > t)$. Consider the sample $x_{1:n}$ of independent standard exponential random variables with $n = 10{,}000$. Then, define

$$y_i = \begin{cases} x_i, & x_i < 1 \\ 1 + \epsilon_i, & x_i > 1 \end{cases},$$

where the $\{\epsilon_i\}$ are independent standard exponential random variables. The memoryless property asserts that the $\{y_i\}$ are also a set of independent standard exponential random variables. For a visual representation, see Fig. 1.5.

Miscellaneous. The Poisson process is often constructed with infinitesimal probabilities because probabilistic changes over smaller intervals of time may be more easily specified than for larger intervals of time; see Question 1.2.5 for more details. The idea is that the probabilistic properties of the process over a large interval of time can be found by solving an infinitesimal version, which necessarily involves calculus. Two good references for the Poisson process are Norris (1997) and Asmussen (2003). A more general treatment for the construction of Markov processes can be found in Rogers and Williams (2000).

2

Likelihood Estimation

The use of probability models as likelihoods to answer data-driven questions has a long history in science dating back to the pioneering efforts of Fisher (Fisher, 1922; Aldrich, 1997). An all-encompassing history of the likelihood and maximum likelihood is provided by Stigler (2007). With the adoption of likelihoods, many statistical questions that were difficult or unsolved were within reach. For example, likelihoods enabled scientists to quantify the uncertainty of a given estimator and to test for the mean of a population by taking a particular value.

Likelihoods quickly became a popular statistical tool because they provided an easy means to obtain estimators and uncertainty intervals from the data. Naturally, there was interest in the properties of these estimators and intervals; the investigation of these properties was possible once estimators were recognized as random variables. Their key basic properties are mean, variance, and covariance (when there are multiple parameters).

The field of decision theory was born to measure the appropriateness of estimators and model specification. It is widely acknowledged that a good criterion for an estimator is to minimize mean squared error (potentially subject to other constraints). The mean squared error measures the average squared difference between the estimated values and the actual values, and it is often expressed as the sum of the variance and squared bias. Due to the bias-variance trade-off, an unbiased estimator can not have an arbitrarily small variance. There are a number of results that provide lower bounds for variances of both biased and unbiased estimators. The most popular lower bound on the variance of an unbiased estimator is the Cramér–Rao lower bound. A less well-known corresponding result for biased estimators exists and is explored in Question 2.1.1.

The maximum likelihood estimator (MLE) is the most popular estimator derived from the likelihood function (Fisher, 1922). However, finding the MLE is not always a trivial exercise. Question 2.1.2 presents a Poisson regression model, where the MLE of the regression coefficient is estimated via the Newton–Raphson algorithm. Question 2.1.3 considers maximum likelihood estimation for random or mixed effects models using the Expectation-Maximization (EM) algorithm. Though this question is analytically tractable by integrating out the random effects, it is instructive to see how the EM algorithm works.

Properties of the MLE are mostly concerned with consistency–the notion that the estimator converges with probability one to the true value. Question 2.1.4 explores the case when the support of the observations depends on unknown parameters; determining this support may also be considered an estimation problem. The usual techniques for establishing consistency no long apply in this scenario.

Question 2.1.5 concerns ordered data and asks the following: if all the data are observed, independent, and identically distributed, then is there a need to consider their ordering when estimating model parameters? Additionally, when the largest observation is missing, how should one proceed to estimate the unknown parameters?

Questions 2.1.6 and 2.1.7 look at models for which the normalizing constant is intractable. In these cases, the normalizing constants involve integrals that do not have closed-form solutions. Models with intractable normalizing constants are often estimated

DOI: 10.1201/9781003493471-2

using MLE or Bayesian methods that circumvent the intractability problem. Alternatively, an estimator based solely on the score function (the first derivative of the log-likelihood) will avoid the problem; this is the idea behind using the Fisher information distance in these two questions. Essentially, the aim is to minimize the Fisher distance between the empirical distribution and the family of distributions with respect to the parameter. The MLE arises by using Kullback–Leibler divergence in place of Fisher distance.

There are two questions on nonparametric maximum likelihood estimation (NPMLE). Maximizing a fully nonparametric model often has no solution. For example, no solution exists when maximizing a likelihood of independent and identically distributed observations over all possible density functions. However, there are models for which NPMLE is possible, such as certain mixing distributions and discrete data models. Question 2.1.8 constructs a nonparametric density estimator using kernels, and Question 2.1.9 uses NPMLE for a survival model with discrete observations.

Q2.1 Questions – Likelihood Estimation

Q2.1.1 – Lower bound of variance for unbiased/biased estimator

Introduction. The Cramér–Rao lower bound provides a lower bound for the variance of an unbiased estimator. In likelihood analyses, an estimator that achieves this lower bound is considered efficient, which essentially means that the estimator achieves the lowest possible mean squared error (MSE) among all unbiased estimators. The estimator that achieves the Cramér–Rao lower bound is also known as the minimum variance unbiased estimator (MVUE). This question reviews the Cramér–Rao lower bound for the Gaussian distribution and extends the bound to work for biased estimators. The proof for the extended Cramér–Rao bound is obtained through intermediate steps and is used to compute the lower bound for both biased and unbiased estimators.

Question. Suppose that observations $\{y_i\}$ for $i = 1, \ldots, n$ follow a normal distribution with unknown mean parameter θ and known variance σ^2. Consider an estimator for θ of the form $\widehat{\theta} = c\,\overline{y}$, where $\overline{y}$ is the sample mean.

(i) Find the mean squared error for $\widehat{\theta}$ in terms of the true value θ^*. What is the optimal choice for c in terms of θ^* under the mean squared error criterion?

(ii) For a non-zero θ^*, why is it necessary for $c \to 1$ as $n \to \infty$ to ensure that $\widehat{\theta}$ converges to θ^*?

(iii) If $\widetilde{\theta}$ is an unbiased estimator of θ (i.e., $\mathrm{E}(\widetilde{\theta}) = \overline{y}$), then the Cramér–Rao bound yields

$$\mathrm{Var}\,(\widetilde{\theta}) \geq \sigma^2/n. \tag{2.1}$$

Alternatively, if $\widehat{\theta}$ is a biased estimator of θ with bias $b(\theta^*)$, then

$$\mathrm{Var}\,(\widehat{\theta}) \geq \{1 + b'(\theta^*)\}^2\,\sigma^2/n. \tag{2.2}$$

Confirm that the variance of $\widehat{\theta} = c\overline{y}$ equals the lower bound given in (2.2).

(iv) Prove the lower bound in part (iii). That is, if $\widehat{\theta}$ has bias $b(\theta^*)$, then (2.2) holds.

(v) Explain why choosing $c < 0$ or $c > 1$ is inadmissible with respect to mean squared error.

Q2.1.2 – Newton–Raphson algorithm for MLE

Introduction. The maximum likelihood estimator is one of the most intuitive likelihood-based estimators. However, the MLE may not always be available in closed form because the analytical maximization of the log-likelihood may be intractable. In such cases, several numerical algorithms may be used to estimate the parameters of interest, including the popular Newton–Raphson method. This question shows that the Newton–Raphson method is based on basic results, yet it requires some conditions to hold for the algorithm to work efficiently.

Question. Consider the Poisson regression model

$$y_i = \text{Pois}\left(e^{x_i\beta}\right)$$

for $i = 1, \ldots, n$. Here, x_i is a scalar and β is a one-dimensional parameter.

(i) Write down the log–likelihood function $L(\beta)$ for β.

(ii) Find the first and second derivatives of the log-likelihood function $L(\beta)$, and show that the expected value of the first derivative is 0. Also, confirm that the log–likelihood is concave (i.e., the second derivative is non-positive).

(iii) Describe a Newton–Raphson algorithm for finding the maximum likelihood estimator $\widehat{\beta}$.

(iv) Using a first-order Taylor expansion of $L'(\widehat{\beta})$ about β, describe why

$$\widehat{\beta} - \beta \approx -L'(\beta)/L''(\beta).$$

(v) Explain the approximation

$$\widehat{\beta} \approx \text{N}\Big(\beta, -1/L''(\beta)\Big).$$

Indicate how this could be used to test the hypothesis $H_0 : \beta = 0$.

Q2.1.3 – EM algorithm for a random effects model

Introduction. The Expectation–Maximization (EM) algorithm is often used to find the MLE for a model when the complexity of the model precludes the use of gradient descent or Newton–Raphson algorithms. For example, the EM algorithm is often used when the likelihood involves an integral over latent variables. This question presents a simple random effects model in which the parameters may be estimated analytically; however, an EM algorithm is used for illustration.

Question. Consider the random effects model

$$y_i = a + x_i b_i + \sigma\epsilon_i, \quad b_i = \mathrm{N}(\mu, \tau^2), \quad i = 1, \dots, n,$$

where the (σ, τ) are known and the $\{\epsilon_i\}$ are independent standard normal random variables. The likelihood function for $\boldsymbol{\theta} = (a, \mu)'$ is given by

$$l(\boldsymbol{\theta}) = \prod_{i=1}^{n} \int p(y_i \mid b_i, a)\, p(b_i \mid \mu)\, db_i.$$

Let $L(\boldsymbol{\theta}) = \log l(\boldsymbol{\theta})$ denote the log-likelihood for $\boldsymbol{\theta}$. The maximum likelihood estimator $\widehat{\boldsymbol{\theta}}$ can be found using the EM algorithm.

(i) The conditional distribution of b_i given y_i and $\boldsymbol{\theta}$ is a normal distribution with mean ξ_i and variance ψ_i^2. Find ξ_i and ψ_i.

(ii) Let $\boldsymbol{y} = (y_1, \dots, y_n)'$ and $\boldsymbol{b} = (b_1, \dots, b_n)'$ and define

$$Q(\boldsymbol{\theta}' || \boldsymbol{\theta}) = \int \log\{p(\boldsymbol{y}, \boldsymbol{b} \mid \boldsymbol{\theta}')\}\, p(\boldsymbol{b} \mid \boldsymbol{y}, \boldsymbol{\theta})\, d\boldsymbol{b}.$$

Show that $Q(\boldsymbol{\theta}' || \boldsymbol{\theta})$ can be written as

$$L(\boldsymbol{\theta}') + \int \log p(\boldsymbol{b} \mid \boldsymbol{y}, \boldsymbol{\theta}')\, p(\boldsymbol{b} \mid \boldsymbol{y}, \boldsymbol{\theta})\, d\boldsymbol{b}.$$

(iii) If $\widetilde{\boldsymbol{\theta}} = \arg\max_{\boldsymbol{\theta}'} Q(\boldsymbol{\theta}' || \boldsymbol{\theta})$, explain why $Q(\widetilde{\boldsymbol{\theta}} || \boldsymbol{\theta}) \geq Q(\boldsymbol{\theta} || \boldsymbol{\theta})$, and show that

$$L(\widetilde{\boldsymbol{\theta}}) - L(\boldsymbol{\theta}) \geq \int \log \frac{p(\boldsymbol{b} \mid \boldsymbol{y}, \boldsymbol{\theta})}{p(\boldsymbol{b} \mid \boldsymbol{y}, \widetilde{\boldsymbol{\theta}})}\, p(\boldsymbol{b} \mid \boldsymbol{y}, \boldsymbol{\theta})\, d\boldsymbol{b}.$$

(iv) Using the inequality $\log z \geq 1 - 1/z$ for $z > 0$, show that the right side of part (iii) is non-negative and hence $L(\widetilde{\boldsymbol{\theta}}) \geq L(\boldsymbol{\theta})$. Explain the significance of these findings.

(v) Show that the $(\widetilde{a}, \widetilde{\mu})$ are given by

$$\widetilde{a} = \overline{y} - n^{-1} \sum_{i=1}^{n} x_i \xi_i \quad \text{and} \quad \widetilde{\mu} = n^{-1} \sum_{i=1}^{n} \xi_i.$$

Q2.1.4* – Asymptotic behavior of MLE estimators

Introduction. Maximum likelihood estimation is the most popular approach to parametric estimation for a specified probability model. As the name implies, the maximum likelihood estimator is obtained by finding the value of the parameter that maximizes the likelihood function. When observations are discrete, the MLE will obtain the parameter value that maximizes the probability of what was observed. This question explores the relationship between a parameter and its MLE through the derivation of a density function for a transformed random variable.

Question. Suppose $x_1, \ldots, x_n$ are independent and identically distributed from a uniform distribution on the interval $[0, \theta]$, where $\theta > 0$ is unknown. That is,

$$f_X(x) = \begin{cases} 1/\theta, & 0 \le x \le \theta \\ 0, & \text{otherwise.} \end{cases}$$

Let θ^* denote the correct value of θ.

(i) Find the maximum likelihood estimator $\widehat{\theta}$ of θ^*.

(ii) Show that the distribution function of $\widehat{\theta}$ is given by

$$F_{\widehat{\theta}}(z) = \begin{cases} 0, & z < 0 \\ (z/\theta^*)^n, & 0 \le z \le \theta^* \\ 1, & z > \theta^*. \end{cases}$$

(iii) Find the density function for the transformed variable $y = n(\theta^* - \widehat{\theta})$, and explain why $y \ge 0$.

(iv) Show that the density function from part (iii) becomes

$$f_Y(y) = (1/\theta^*) \exp(-y/\theta^*), \quad y \ge 0$$

as $n \to \infty$.

(v) Find an approximate $100(1-\alpha)\%$ confidence interval for θ^*.

Q2.1.5* – Maximum ordered statistics and the likelihood principle

Introduction. The ordered statistics $x_{(1)}, \ldots, x_{(n)}$ are random variables arising from the sorting of realizations of random variables $x_1, \ldots, x_n$ in increasing order. They are a fundamental tool in nonparametric statistics and density estimation, and they are often used as estimators for the quantiles of an underlying distribution. This question explores the maximum ordered statistic, prompting the computation of its distribution and density functions. These results are then applied to investigate characteristics of the maximum ordered statistic for the exponential distribution. Additionally, some intuition is formed regarding the relationship between ordered statistics, outlier removal, and the likelihood principle.

Question. Suppose $x_1, \ldots, x_n$ are independent and identically distributed from the density function $f_X(x)$ with corresponding distribution function $F_X(x)$. Let

$$x_{(1)} < \cdots < x_{(n)}$$

be the ordered statistics.

(i) Find $\mathrm{P}(X_{(n)} \le x)$ in terms of $F_X(x)$ and n. Then, find the density function of $x_{(n)}$.

(ii) Using the likelihood principle, explain why the joint density function of $\{x_{(1)}, \ldots, x_{(n)}\}$ is of the form

$$f_{x_{(1)},\ldots,x_{(n)}}(x_1, \ldots, x_n) \propto 1(x_{(1)} < \cdots < x_{(n)}) \prod_{i=1}^{n} f_X(x_i),$$

where $n!$ is the constant of proportionality.

(iii) If $f_X(x) = (1/\theta)\exp(-x/\theta)$ for some $\theta > 0$ and $x > 0$, find the expected value of $x_{(n)}$.

(iv) If the estimator for θ in part (iii) is the sample mean (the maximum likelihood estimator), what is the bias of the estimator if $x_{(n)}$ is missing? That is, what is the bias of the estimator

$$\widehat{\theta}_{(-n)} = \frac{1}{n-1} \sum_{i=1}^{n-1} x_{(i)}?$$

(v) Find the joint likelihood of $\{x_{(1)}, \ldots, x_{(n-1)}\}$ and the corresponding maximum likelihood estimator for θ. Does this new estimator have a smaller bias than the estimator in part (iv)?

Q2.1.6 – Estimation via Fisher information distance

Introduction. The Fisher information distance is a metric on the space of probability density functions. Similar to the Kullback–Leibler divergence, the Fisher information distance can be used to approximate the divergence between the true density and the model density via Monte Carlo estimation. As shown in this question, minimizing the Fisher information distance yields a consistent estimator for the parameter of interest.

Question. Suppose $p_\theta(x)$ is a density function for each $\theta \in \Theta$, $p_0(x)$ is a member of the family for a particular θ_0, and observations $\{x_1, \ldots, x_n\}$ are independent and identically distributed from p_0. The aim is to estimate θ by minimizing the Fisher information distance

$$F(p_0, p_\theta) = \int p_0(x) \left\{ \frac{p_0'(x)}{p_0(x)} - \frac{p_\theta'(x)}{p_\theta(x)} \right\}^2 dx,$$

where $p_0'(x)$ denotes the derivative of $p_0(x)$ with respect to x.

(i) Under regularity conditions (which should be stated) show that it is possible to write

$$F(p_0, p_\theta) = \kappa + \int p_0(x) \left[\left\{ \frac{p'_\theta(x)}{p_\theta(x)} \right\}^2 + 2\frac{d}{dx} \left\{ \frac{p'_\theta(x)}{p_\theta(x)} \right\} \right] dx,$$

where κ does not depend on θ.

(ii) Define

$$p_\theta(x) = \frac{\exp\{w(x)/\theta\}}{\int \exp\{w(x)/\theta\}\, dx}$$

for some suitable function w. Find

$$l(\theta, x) = \left\{ \frac{p'_\theta(x)}{p_\theta(x)} \right\}^2 + 2\frac{d}{dx} \left\{ \frac{p'_\theta(x)}{p_\theta(x)} \right\}.$$

(iii) Explain how θ may be estimated using an approximation to $F(p_0, p_\theta)$.

(iv) Estimate θ using the framework described in part (iii) for the model in part (ii).

(v) What are the asymptotic properties of the estimator found in part (iv)? That is, show that it converges to the true parameter value.

Q2.1.7 – Estimation with intractable normalizing constants

Introduction. When distributions have intractable normalizing constants, their parameters may not be estimated via traditional estimation procedures, such as maximum likelihood estimation. Alternative estimation approaches often minimize a distance between the empirical distribution function and the distribution of interest. This question concerns the estimation of the concentration parameter in the von Mises distribution, which has an intractable normalizing constant.

Question. Consider the family of von Mises density functions given by

$$f_\theta(x) = f(x \mid \theta) = \frac{\exp(\theta \cos x)}{Z(\theta)}, \quad 0 < x < 2\pi,$$

where $\theta \in \mathbb{R}_+$ and $Z(\theta) = \int_0^{2\pi} \exp(\theta \cos x)\, dx$. The Fisher distance between f_θ and f_{θ^*} is given by

$$F(\theta^*, \theta) = \int f_{\theta^*}(x) \left\{ f'_{\theta^*}(x)/f_{\theta^*}(x) - f'_\theta(x)/f_\theta(x) \right\}^2 dx,$$

where $'$ denotes differentiation with respect to x.

(i) Show that minimizing the Fisher distance is equivalent to minimizing

$$\int f_{\theta^*}(x) \left[\{(f'_\theta(x)/f_\theta(x)\}^2 + 2\, \{f'_\theta(x)/f_\theta(x)\}' \right] dx$$

with respect to θ.

(ii) Explain why the minimization solution to part (i) does not involve Z.

(iii) Find $\widehat{\theta}$, the solution to the minimization problem in part (i), when f_{θ^*} is replaced by the empirical density mass function based on an independent and identically distributed sample $\{x_1, \ldots, x_n\}$ coming from $f(\cdot \mid \theta^*)$.

(iv) Show that $\mathrm{E}\{\cos(X)\} = \theta^* \, \mathrm{E}\{\sin^2(X)\}$, where X is from the true density function.

(v) Show that the estimator is consistent: $\widehat{\theta} \overset{a.s.}{\to} \theta^*$ as $n \to \infty$.

Q2.1.8* – A sin kernel for density estimation

Introduction. Kernel density estimation is a nonparametric approach to estimate density functions using kernels. Although the Gaussian kernel is the most commonly used, many kernels exist and may be more appropriate for certain criteria. This question investigates the bias of a density estimator arising from a sin kernel.

Question. Suppose $x_1, \ldots, x_n$ are independent and identically distributed from the density f^*. A kernel density estimator using the sin function is given by

$$f_R(x) = (n\pi)^{-1} \sum_{i=1}^{n} \frac{\sin\{R(x - x_i)\}}{x - x_i},$$

where R is a positive smoothing parameter for the kernel. The aim is to find the bias of $f_R(x)$ when the true density is standard normal.

(i) Show that

$$\int_{-\infty}^{\infty} \cos(Rx)\, \phi(x)\, dx = \exp(-\tfrac{1}{2}R^2),$$

where $\phi(x)$ is the standard normal density.

(ii) Show that

$$\int_{-\infty}^{\infty} \cos(Rx)\, \sigma^{-1}\, \phi((x-\mu)/\sigma)\, dx = \cos(R\mu)\, \exp(-\tfrac{1}{2}\sigma^2 R^2).$$

(iii) Show that

$$\int_{-\infty}^{\infty} \cos\{R(y-x)\}\, \sigma^{-1}\, \phi((x-\mu)/\sigma)\, dx = \cos(R(y-\mu))\, \exp(-\tfrac{1}{2}\sigma^2 R^2).$$

(iv) Find $\mathrm{E}\{f_R(x)\}$.

(v) Show that $\pi \left|\mathrm{E}\{f_R(x)\} - f^*(x)\right| \le R^{-1} \exp(-R^2/2)$.

Q2.1.9 – Survival analysis with censored data

Introduction. The field of survival analysis focuses on modeling time-to-event data. The outcome of interest in such data is whether the events of interest occurred and, if so, when they occurred. One of the most prevalent challenges in survival analysis is censoring, which occurs when observations do not experience the events of interest by the end of the study period; thus, censoring may be seen as a missing data problem. Time-to-event data arise in many fields, ranging from deaths in biological organisms to failures in mechanical systems. This question concerns the probabilistic modeling of a survival random variable in the presence of censored data.

Question. Consider a sample $\{t_i, \delta_i\}_{i=1}^n$ from n individuals with potentially right-censored observations, where δ_i is an indicator for whether the event was observed for individual i during the study and t_i is the time of event for individual i (if observed). The event of interest is the death of an individual, so it is assumed that the event inevitably occurs. A censored observation $(k, \delta = 0)$ means the event was not observed during the study, so $T \geq k$ for this individual; here, T is a survival random variable modeled discretely by $\mathrm{P}(T = k \mid T \geq k) = h_k$ for $k \in \{1, 2, \ldots\}$. Note that $0 < h_k < 1$ is an unknown probability for all k.

(i) Derive $\mathrm{P}(T = k)$ and $\mathrm{P}(T > k)$ in terms of $\{h_1, \ldots, h_k\}$.

(ii) Write the likelihood function for $\{h_1, h_2, \ldots\}$ in terms of the sample $\{t_i, \delta_i\}_{i=1}^n$, and find the maximum likelihood estimator for h_k.

(iii) Find the estimator for $\mathrm{P}(T = k)$.

(iv) Detail any particular anomaly if the largest observation is censored.

(v) If there is no censoring, show that the estimator for $\mathrm{P}(T = k)$ becomes n_k/n, where n_k is the number of events at time k.

S2.1 Solutions – Likelihood Estimation

S2.1.1 – Lower bound of variance for unbiased/biased estimator

(i) The MSE for $\widehat{\theta}$ with a quadratic loss function is given by

$$\begin{aligned}\mathrm{MSE}(\widehat{\theta}) &= \mathrm{E}\{(\widehat{\theta}-\theta^*)^2\}\\ &= \mathrm{E}\left\{\left(\frac{c}{n}\sum_{i=1}^{n} y_i - \theta^*\right)^2\right\}\\ &= \frac{c^2}{n^2}\mathrm{E}\left\{\left(\sum_{i=1}^{n} y_i\right)^2\right\} - \frac{2c\theta^*}{n}\sum_{i=1}^{n}\mathrm{E}(y_i) + (\theta^*)^2\\ &= c^2\sigma^2/n + (\theta^*)^2(c-1)^2,\end{aligned}$$

which is in the form variance + bias^2. To find the optimal choice for c in terms of θ^*, the MSE is minimized with respect to c:

$$\frac{\partial}{\partial c}\mathrm{MSE}(\widehat{\theta}) \;=\frac{\partial}{\partial c}\left[c^2\sigma^2/n + c^2(\theta^*)^2 - 2c(\theta^*)^2 + (\theta^*)^2\right] = 2c\sigma^2/n + 2c(\theta^*)^2 - 2(\theta^*)^2.$$

Therefore, $c = n(\theta^*)^2/(\sigma^2 + n(\theta^*)^2)$.

(ii) The bias is given by $\theta^*(c-1)$. Therefore, for the estimator to converge to θ^*, it must be that $c \to 1$. Convergence will occur according to the MSE criterion because the variance also goes to 0 as $n \to \infty$.

(iii) The bias is $b(\theta^*) = \theta^*(c-1)$, so $b'(\theta^*) + 1 = c$. Thus, (2.2) may be written as $\mathrm{Var}\,(\widehat{\theta}) \geq c^2\sigma^2/n$. In this case, when $\hat{\theta} = c\bar{y}$, $\mathrm{Var}\,(\widehat{\theta}) = c^2\sigma^2/n$, which equals the extended Cramér–Rao lower bound in (2.2).

(iv) Start with $\mathrm{E}(T) = \theta + b(\theta)$ and $Z = \partial \log p(y \mid \theta)/\partial\theta$, and show that $\mathrm{E}(Z) = 0$ using

$$\int p(y \mid \theta)\, dy = 1 \quad\Longrightarrow\quad \int \frac{\partial}{\partial\theta}\, p(y \mid \theta)\, dy = 0,$$

where the order of differentiation and integration is switched. Further,

$$\mathrm{E}(Z^2) = \int \left\{\frac{\partial}{\partial\theta}\log p(y \mid \theta)\right\}^2 p(y \mid \theta)\, dy = n\, I(\theta),$$

where $I(\theta)$ is the Fisher information. For the Gaussian model with mean θ and variance σ^2, $I(\theta) = 1/\sigma^2$. Now,

$$E(T\,Z) = \int T\,\frac{\partial}{\partial\theta}\, p(y \mid \theta)\, dy,$$

which can be obtained from $(\partial/\partial\theta)\int T\, p(y \mid \theta)\, dy$. Hence, $E(T\,Z) = \partial\mathrm{E}(T)/\partial\theta = 1 + b'(\theta)$ because $\mathrm{E}(T) = \theta + b(\theta)$.

Putting everything together and using the upper bound of a covariance in terms of variances,

$$\mathrm{Cov}(T, Z) = \mathrm{E}(T\,Z) \leq \sqrt{\mathrm{Var}(T)\,\mathrm{Var}(Z)},$$

where $\mathrm{E}(Z) = 0$. Hence,

$$\mathrm{Var}(T) \geq \frac{\{1 + b'(\theta)\}^2}{E(Z^2)}$$

and the result is completed by noting that $\mathrm{E}(Z^2) = n\,I(\theta) = n/\sigma^2$.

(v) If $c < 0$, $-c\overline{y}$ is a better estimator because it will have the same variance as $c\overline{y}$ with a smaller bias squared. If $c > 1$, then both bias squared and variance are larger than with $c = 1$.

Further Reading

Historical Background. The Cramér–Rao lower bound for the variance of an unbiased estimator was developed in the mid-1940s. The key inequality that makes it work is that a correlation coefficient lies between -1 and $+1$. Understanding the proof of the Cramér–Rao lower bound readily leads to a corresponding lower bound for biased estimators. For further reading on the technical details of the Cramér–Rao bound, see Joshi (1976). For generalizations of the Cramér–Rao bound, see Hammersley (1950) and Chapman and Robbins (1951).

Points of Interest.

1. While the Cramér–Rao lower bound for unbiased estimators is well-known, less well-known is the corresponding lower bound for biased estimators. Biased estimators arise naturally as shrinkage estimators, which are obtained when attempting to reduce an estimator's variance. In this problem, shrinkage occurs when one is interested in the estimator $c\overline{y}$ with $c < 1$. The Chapman–Robbins bound is a generalization of the Cramér–Rao bound that provides a lower bound for biased estimators.

2. Fisher information is a measure of how much information a random variable contains about an unknown parameter. In one dimension, the Cramér–Rao bound is the reciprocal of the Fisher information. In multiple dimensions, the Cramér–Rao bound is the inverse of the Fisher information matrix.

3. For the Cramér–Rao bound to exist, two regularity conditions must be satisfied. First, the Fisher information must be defined for all θ: for all y such that $p(y \mid \theta) > 0$, $\partial/\partial\theta \log p(y \mid \theta)$ is finite and exists. Second, define the estimator $\widehat{\theta} = T(y)$. Then, the operations of integration with respect to y and differentiation with respect to θ must be interchangeable in the expectation of T. If these two weak regularity conditions are met, then the Cramér–Rao bound exists.

S2.1.2 – Newton–Raphson algorithm for MLE

(i) Take the logarithm of the likelihood:

$$\begin{aligned} L(\beta) &= \log\left\{\prod_{i=1}^{n} \frac{\exp\left(-e^{x_i\beta}\right) e^{x_i y_i \beta}}{y_i!}\right\} \\ &= \log\left\{\prod_{i=1}^{n}\left(\frac{1}{y_i!}\right)\exp\left(-\sum_{i=1}^{n} e^{x_i\beta}\right) e^{\sum_{i=1}^{n} x_i y_i \beta}\right\} \\ &= -\sum_{i=1}^{n}\log(y_i!) - \sum_{i=1}^{n} e^{x_i\beta} + \sum_{i=1}^{n} x_i y_i \beta. \end{aligned}$$

(ii) Find the first derivative:

$$\begin{aligned} \frac{\partial}{\partial\beta}L(\beta) &= \frac{\partial}{\partial\beta}\left\{-\sum_{i=1}^{n}\log(y_i!) - \sum_{i=1}^{n} e^{x_i\beta} + \sum_{i=1}^{n} x_i y_i \beta\right\} \\ &= -\sum_{i=1}^{n} x_i e^{x_i\beta} + \sum_{i=1}^{n} x_i y_i. \end{aligned}$$

The expected value of the first derivative is

$$\begin{aligned} \mathrm{E}\left\{\frac{\partial}{\partial\beta}L(\beta)\right\} &= \mathrm{E}\left(-\sum_{i=1}^{n} x_i e^{x_i\beta} + \sum_{i=1}^{n} x_i Y_i\right) \\ &= -\sum_{i=1}^{n} x_i e^{x_i\beta} + \sum_{i=1}^{n} x_i \mathrm{E}(Y_i) \\ &= -\sum_{i=1}^{n} x_i e^{x_i\beta} + \sum_{i=1}^{n} x_i e^{x_i\beta} \\ &= 0. \end{aligned}$$

Then, obtain the second derivative:

$$\frac{\partial^2}{\partial\beta^2}L(\beta) = \frac{\partial}{\partial\beta}\left(-\sum_{i=1}^{n} x_i e^{x_i\beta} + \sum_{i=1}^{n} x_i y_i\right) = -\sum_{i=1}^{n} x_i^2 e^{x_i\beta},$$

which is negative. Thus, the log-likelihood is a concave function.

(iii) The Newton–Raphson algorithm estimates $\widehat{\beta}$ by iteratively updating the sequence $\{\beta^{(k)}\}$, where

$$\beta^{(k)} = \beta^{(k-1)} - L'(\beta)/L''(\beta).$$

Convergence is reached when $\beta^{(k+1)} = \beta^{(k)}$, where it is true that $L'(\beta^{(k)}) = 0$. Therefore, $\widehat{\beta}$ would characterize the MLE.

(iv) In general, the Taylor expansion for a function $f(x)$ about scalar a is

$$f(x) = \sum_{n=0}^{\infty} \frac{f^{(n)}(a)}{n!}(x-a)^n.$$

The first-order Taylor expansion of $L'(\widehat{\beta})$ about β is

$$L'(\widehat{\beta}) \approx L'(\beta) + L''(\beta) \cdot (\widehat{\beta} - \beta).$$

Because $L'(\widehat{\beta}) = 0$, it is that $L'(\beta) + L''(\beta) \cdot (\widehat{\beta} - \beta) \approx 0$, which implies that $\widehat{\beta} - \beta \approx -L'(\beta)/L''(\beta)$.

(v) From part (iv), $\widehat{\beta} \approx \beta - L'(\beta)/L''(\beta)$; the goal becomes finding the approximate distribution for $-L'(\beta)/L''(\beta)$. First, it is known that $\mathrm{E}\left\{L'(\beta)\right\} = 0$ and

$$\begin{aligned} L'(\beta) &= -\sum_{i=1}^{n} x_i e^{x_i\beta} + \sum_{i=1}^{n} x_i y_i \\ &= \sum_{i=1}^{n} x_i \left(y_i - e^{x_i\beta}\right). \end{aligned}$$

The central limit theorem indicates that $L'(\beta)/n$ will be approximately normal. Regarding variance, note that

$$\mathrm{Var}\{L'(\beta)\} = \sum_{i=1}^{n} x_i^2 \mathrm{Var}(Y_i) = \sum_{i=1}^{n} x_i^2 e^{x_i\beta} = -L''(\beta).$$

Hence, $\mathrm{Var}(\widehat{\beta}) = -1/L''(\beta)$ and

$$\widehat{\beta} \approx \mathrm{N}\left(\beta, -1/L''(\beta)\right).$$

In short, $-L'(\beta)/L''(\beta)$ is approximately a normal random variable with mean 0 and variance $-1/L''(\beta)$. Thus, it is possible to test $H_0: \beta = 0$ with test statistic

$$T = \widehat{\beta}\sqrt{-L''(\widehat{\beta})},$$

which will be approximately standard normal under the null hypothesis. Then, the standard normal test may be applied as usual.

Further Reading

Historical Background. The origins of the Newton–Raphson algorithm, named after English mathematicians Sir Isaac Newton and Joseph Raphson, date back to the 17th century. However, it was not until the middle of the 20th century that proofs of convergence, both local linear and local quadratic, were achieved in a series of papers by Soviet mathematician Leonid Kantorovich.

In statistics, the Newton–Raphson algorithm is often used to maximize the log-likelihood function $L(x)$ and requires the second derivative of $L(x)$. However, the original version of the method aimed to find a zero of the function $f(x)$ (i.e., find the x for which $f(x) = 0$). The idea is quite simple: if $f'(x) > 0$ locally about the zero, then the algorithm takes $x \to x'$ given by $x' = x - f(x)/f'(x)$. Thus, the method ensures that $x' < x$ if $f(x) > 0$ or $x' > x$ if $f(x) < 0$.

Points of Interest.

1. The terms in part (v) (i.e., $L'(\beta)$ and $L''(\beta)$) depend on the sample size. An approximation can be used based on the Fisher information. For example, $-L''(\beta)/n$ converges to what can be written as

$$I(\beta) = \lim_{n\to\infty} n^{-1} \sum_{i=1}^{n} x_i^2 e^{x_i\beta}.$$

Hence, $\mathrm{Var}(\widehat{\beta}) \approx \{nI(\beta)\}^{-1}$ and

$$\widehat{\beta} \approx \mathrm{N}\left(\beta, \{nI(\beta)\}^{-1}\right).$$

2. This exercise uses the Newton–Raphson algorithm to compute the MLE for a Poisson regression model. As part (iv) shows, the MLE for β can be obtained by iteratively using the ratio of the first and second derivatives of the log-likelihood function. The algorithm assumes that the function of interest is twice differentiable and uses a second-order Taylor approximation. In high-dimensional regression models, the first and second derivatives become vectors and (Hessian) matrices, respectively.

3. If the evaluation of the second derivative is infeasible, gradient descent methods may be used (Curry, 1994). Although gradient descent methods are slower, the size of their steps can be better controlled. For an introductory comparison between these two approaches, see Battiti (1992).

Demonstration. Consider a multivariate $\boldsymbol{\beta} = (0.5, -0.5, 0.3)'$ with covariates $\mathbf{x}_i \sim \mathrm{N}(\mathbf{0}_3, \mathbf{I}_3)$ for $i = 1, \ldots, n$, and take $n = 100,000$ samples from the Poisson regression model. The Newton–Raphson algorithm estimates $\boldsymbol{\beta}$ very well by the fifth iteration, depicted in Fig. 2.1 (upper). Additionally, the log-likelihood appears to be maximized by the fifth iteration, which is seen in Fig. 2.1 (lower).

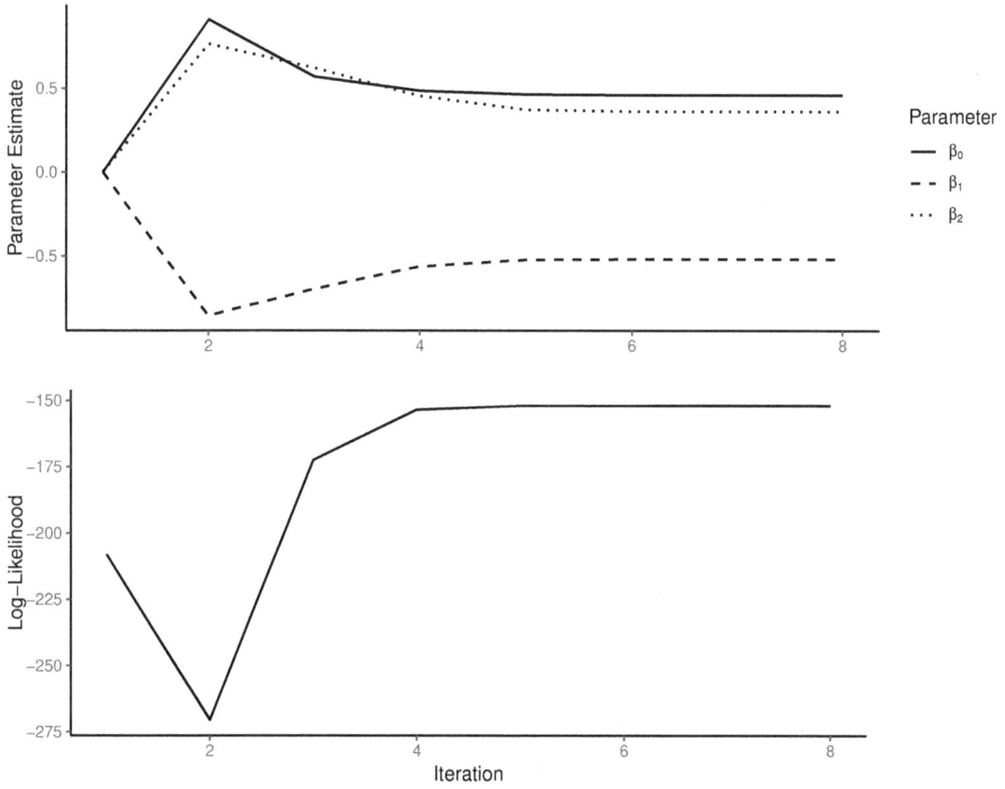

FIGURE 2.1
Upper: The convergence of $\boldsymbol{\beta}$ using the Newton–Raphson method for the Poisson regression model. *Lower*: The log-likelihood evaluated at each iteration.

S2.1.3 – EM algorithm for a random effects model

(i) The conditional distribution $p(b_i \mid y_i, \boldsymbol{\theta})$ can be written as

$$\begin{aligned}
p(b_i \mid y_i, \boldsymbol{\theta}) &\propto p(y_i \mid b_i, \boldsymbol{\theta}) p(b_i \mid \boldsymbol{\theta}) \\
&\propto \exp\left\{-\frac{1}{2\sigma^2}(y_i - a - x_i b_i)^2\right\} \exp\left\{-\frac{1}{2\tau^2}(b_i - \mu)^2\right\} \\
&\propto \exp\left\{-\frac{1}{2\sigma^2}(-2y_i x_i b_i + 2a x_i b_i + x_i^2 b_i^2) - \frac{1}{2\tau^2}(b_i^2 - 2b_i \mu)\right\} \\
&= \exp\left[-\frac{1}{2}\left\{b_i^2\left(\frac{x_i^2}{\sigma^2} + \frac{1}{\tau^2}\right) - 2\left(\frac{y_i x_i}{\sigma^2} - \frac{a x_i}{\sigma^2} + \frac{\mu}{\tau^2}\right) b_i\right\}\right],
\end{aligned}$$

which indicates a normal distribution with mean ξ_i and variance ψ_i^2 given by

$$\psi_i^2 = \left(x_i^2/\sigma^2 + 1/\tau^2\right)^{-1} \quad \text{and} \quad \xi_i = \psi_i^2 \cdot \left((y_i - a)x_i/\sigma^2 + \mu/\tau^2\right).$$

(ii) Now,

$$\begin{aligned}
Q(\boldsymbol{\theta}' \| \boldsymbol{\theta}) &= \int \log\{p(\boldsymbol{y}, \boldsymbol{b} \mid \boldsymbol{\theta}')\} p(\boldsymbol{b} \mid \boldsymbol{y}, \boldsymbol{\theta}) d\boldsymbol{b} \\
&= \int \log\{p(\boldsymbol{b} \mid \boldsymbol{\theta}', \boldsymbol{y}) p(\boldsymbol{y} \mid \boldsymbol{\theta}')\} p(\boldsymbol{b} \mid \boldsymbol{y}, \boldsymbol{\theta}) d\boldsymbol{b} \\
&= \int \{\log p(\boldsymbol{b} \mid \boldsymbol{\theta}', \boldsymbol{y}) + \log p(\boldsymbol{y} \mid \boldsymbol{\theta}')\}\, p(\boldsymbol{b} \mid \boldsymbol{y}, \boldsymbol{\theta}) d\boldsymbol{b} \\
&= \int \log\{p(\boldsymbol{b} \mid \boldsymbol{\theta}', \boldsymbol{y})\} p(\boldsymbol{b} \mid \boldsymbol{y}, \boldsymbol{\theta}) d\boldsymbol{b} + L(\boldsymbol{\theta}'),
\end{aligned}$$

where the second equality follows as $p(\boldsymbol{y}, \boldsymbol{b} \mid \boldsymbol{\theta}') = p(\boldsymbol{b} \mid \boldsymbol{\theta}', \boldsymbol{y}) p(\boldsymbol{y} \mid \boldsymbol{\theta}')$.

(iii) If $\widetilde{\boldsymbol{\theta}} = \arg\max_{\boldsymbol{\theta}'} Q(\boldsymbol{\theta}' \| \boldsymbol{\theta})$, then by definition $Q(\widetilde{\boldsymbol{\theta}} \| \boldsymbol{\theta}) \geq Q(\boldsymbol{\theta} \| \boldsymbol{\theta})$. Now, use the result from part (ii) to get

$$\begin{aligned}
L(\widetilde{\boldsymbol{\theta}}) - L(\boldsymbol{\theta}) &\geq \int \log\{p(\boldsymbol{b} \mid \boldsymbol{\theta}, \boldsymbol{y})\} p(\boldsymbol{b} \mid \boldsymbol{y}, \boldsymbol{\theta}) d\boldsymbol{b} - \int \log\{p(\boldsymbol{b} \mid \widetilde{\boldsymbol{\theta}}, \boldsymbol{y})\} p(\boldsymbol{b} \mid \boldsymbol{y}, \boldsymbol{\theta}) d\boldsymbol{b} \\
&= \int \log\left\{\frac{p(\boldsymbol{b} \mid \boldsymbol{\theta}, \boldsymbol{y})}{p(\boldsymbol{b} \mid \widetilde{\boldsymbol{\theta}}, \boldsymbol{y})}\right\} p(\boldsymbol{b} \mid \boldsymbol{y}, \boldsymbol{\theta}) d\boldsymbol{b}.
\end{aligned}$$

(iv) Using the given inequality,

$$\log\left\{\frac{p(\boldsymbol{b} \mid \boldsymbol{\theta}, \boldsymbol{y})}{p(\boldsymbol{b} \mid \widetilde{\boldsymbol{\theta}}, \boldsymbol{y})}\right\} \geq 1 - \frac{p(\boldsymbol{b} \mid \widetilde{\boldsymbol{\theta}}, \boldsymbol{y})}{p(\boldsymbol{b} \mid \boldsymbol{\theta}, \boldsymbol{y})},$$

so

$$\begin{aligned}
L(\widetilde{\boldsymbol{\theta}}) - L(\boldsymbol{\theta}) &\geq \int \left\{1 - \frac{p(\boldsymbol{b} \mid \widetilde{\boldsymbol{\theta}}, \boldsymbol{y})}{p(\boldsymbol{b} \mid \boldsymbol{\theta}, \boldsymbol{y})}\right\} p(\boldsymbol{b} \mid \boldsymbol{y}, \boldsymbol{\theta}) d\boldsymbol{b} \\
&= \int \{p(\boldsymbol{b} \mid \boldsymbol{y}, \boldsymbol{\theta}) - p(\boldsymbol{b} \mid \boldsymbol{y}, \widetilde{\boldsymbol{\theta}})\} d\boldsymbol{b} \\
&= 0,
\end{aligned}$$

implying that the log-likelihood (and likelihood) at $\tilde{\theta}$ is larger than or equal to the log-likelihood at θ. The significance is that the value of the log-likelihood increases after each iteration.

(v) The aim is to find the $\widetilde{\theta} = (\widetilde{a}, \widetilde{\mu})'$ that maximizes $Q(\widetilde{\theta}||\theta)$. In other words, find $\theta' = (a', \mu')$ that maximizes

$$\int \log\{p(y, b \mid \theta')\} p(b \mid y, \theta) db \propto \int \{\log p(y \mid b, \theta') + \log p(b \mid \theta')\}\, p(b \mid y, \theta) db$$

$$\propto \int \left[-\frac{1}{2\sigma^2} \sum_{i=1}^{n} (y_i - a' - x_i b_i)^2 - \frac{1}{2\tau^2} \sum_{i=1}^{n} (\mu' - b_i)^2 \right] \mathrm{N}(b_i \mid \xi_i, \psi_i^2) db.$$

Take the derivatives and set to 0:

$$\int \left\{ na' - \sum_{i=1}^{n} (y_i - x_i b_i) \right\} \mathrm{N}(b_i \mid \xi_i, \psi_i^2) db = 0$$

so that

$$n\widetilde{a} = \sum_{i=1}^{n} y_i - \int \sum_{i=1}^{n} x_i b_i \mathrm{N}(b_i \mid \xi_i, \psi_i^2) db$$
$$= \sum_{i=1}^{n} y_i - \sum_{i=1}^{n} x_i \xi_i,$$

which implies $\widetilde{a} = \overline{y} - n^{-1} \sum_{i=1}^{n} x_i \xi_i$ and

$$n\widetilde{\mu} - \sum_{i=1}^{n} \xi_i = 0 \quad \Rightarrow \quad \widetilde{\mu} = \frac{1}{n} \sum_{i=1}^{n} \xi_i.$$

Further Reading

Historical Background. The EM algorithm was introduced to mainstream statistical analyses with the classic paper of Dempster et al. (1977). Although specific versions of the algorithm had appeared previously, Dempster's paper demonstrated a general use for problems involving likelihood functions with missing data.

Points of Interest.

1. The main point of this problem is to derive an EM algorithm, which increases the value of the likelihood at each iteration. As such, it would converge at least to a local mode. A number of different starting points could be used to check that a global maximum has been found. For this problem, maximizing the log-likelihood directly is possible because the normal distributions can be integrated out. However, for nonlinear problems, the EM algorithm would be required.

2. There are many variants of the EM algorithm. For example, the maximization step can be done conditionally when the number of parameters is large and a single maximization over all parameters simultaneously is infeasible. Such extensions are the ECM algorithm (Meng and Rubin, 1993) and the ECME algorithm (Liu and Rubin, 1994). Another variant addresses the problem that arises when the expectation component can not be computed; solutions include numerical and stochastic approaches, such as Monte Carlo integration in the MCEM algorithm (Wei and Tanner, 1990).

3. Because the EM algorithm is frequently used for parameter estimation of mixture models, it has been widely adopted in various scientific disciplines, including psychometrics (Bock and Aitkin, 1981), quantitative genetics (Diffey et al., 2017), and neural networks (Greff et al., 2017). Recent extensions of the EM algorithm are detailed in Mclachlan and Krishnan (2007).

Demonstration. Consider the following settings to generate 1,000 observations: $\sigma = 1$, $\tau = 1$, $a = 2$, and $\mu = 0.5$. The EM algorithm only required a few iterations to adequately estimate the parameters because convergence was quite fast. Fig. 2.2 shows the convergence of a and μ (left) and the log-likelihood (right).

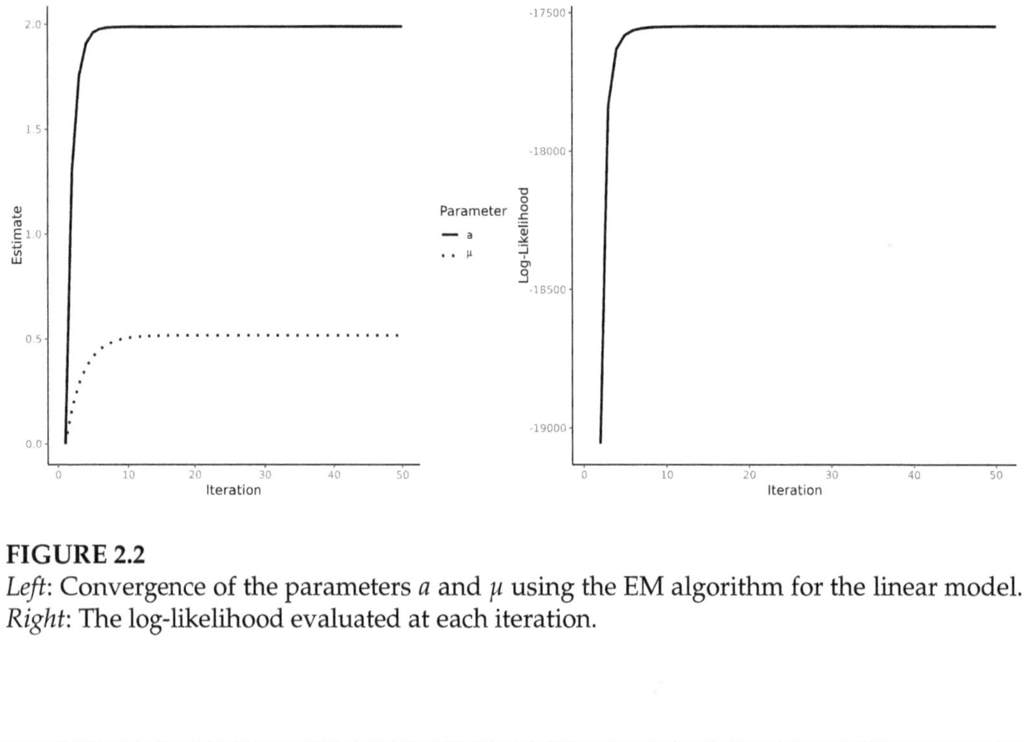

FIGURE 2.2
Left: Convergence of the parameters a and μ using the EM algorithm for the linear model. *Right*: The log-likelihood evaluated at each iteration.

S2.1.4* – Asymptotic behavior of MLE estimators

(i) The likelihood function $l(\theta)$ can be written as

$$l(\theta) = \prod_{i=1}^{n} \theta^{-1} 1(x_i \leq \theta) \propto \theta^{-n} 1(x_{(n)} \leq \theta),$$

where $x_{(n)}$ is the maximum ordered statistic and $1(A)$ is an indicator function that evaluates to 1 if A is true and 0 otherwise. If $x_{(n)} > \theta$, then $l(\theta) = 0$. If $x_{(n)} \leq \theta$, then $l(\theta) \propto \theta^{-n}$ is a decreasing function of θ. Thus, $l(\theta)$ is maximized by choosing the smallest possible θ given the constraint $x_{(n)} \leq \theta$. Therefore, $\widehat{\theta} = x_{(n)}$ is the MLE.

(ii) The distribution function of $\widehat{\theta}$ can be obtained through direct computation:

$$F_{\widehat{\theta}}(z) = \mathrm{P}(\widehat{\theta} \le z) = \mathrm{P}(X_{(n)} \le z) = \mathrm{P}(X_1 \le z, \ldots, X_n \le z) = \{\mathrm{P}(X_1 \le z)\}^n,$$

where the independence of the $\{x_i\}$ is used to obtain the final equality. For the true value θ^*,

$$\mathrm{P}(X_1 \le z) = \int_0^z (\theta^*)^{-1} dx = z/\theta^*.$$

Note that the density function is positive for $0 \le z \le \theta$ and, if $z > \theta^*$, the value is 1. Thus,

$$F_{\widehat{\theta}} = \begin{cases} 0 & z < 0 \\ (z/\theta^*)^n & 0 \le z \le \theta^* \\ 1 & z > \theta^*. \end{cases}$$

(iii) From part (ii), the density function for $\widehat{\theta}$ is

$$f_{\widehat{\theta}}(z) = n\, z^{n-1}/(\theta^*)^n, \quad 0 \le z \le \theta^*. \tag{2.3}$$

To obtain the density function of a transformed random variable, use

$$f_Y(y) = f_Z(g^{-1}(y)) \left| \frac{d}{dy} g^{-1}(y) \right|, \tag{2.4}$$

where $z = \widehat{\theta}$ and $y = g(z) = n(\theta^* - z)$. Note that $g^{-1}(y) = \theta^* - y/n$ and $\left| \frac{d}{dy} g^{-1}(y) \right| = 1/n$. Thus,

$$\begin{aligned} f_Y(y) &= \frac{n}{(\theta^*)^n}(\theta^* - y/n)^{n-1} \cdot \frac{1}{n} = \frac{(\theta^* - y/n)^n}{(\theta^* - y/n)(\theta^*)^n} \\ &= \frac{\{1 - y/(n\theta^*)\}^n (\theta^*)^n}{(\theta^* - y/n)(\theta^*)^n} = \frac{\{1 - y/(n\theta^*)\}^n}{\theta^* - y/n}, \end{aligned}$$

where $z = \theta^* - y/n$ is plugged into (2.3), and (2.4) is used to obtain the first equality. Recall from part (i) that $\widehat{\theta} = x_{(n)}$, which must be less than or equal to θ^*. Therefore, $\mathrm{P}(Z \le \theta^*) = 1$ and $\mathrm{P}(Y \ge 0) = 1$.

(iv) Recall that $\lim_{n \to \infty}(1 + x/n)^n = e^x$. Hence, as $n \to \infty$,

$$\{1 - y/(n\theta^*)\}^n \to \exp(-y/\theta^*),$$

and the density in part (iii) becomes $f_Y(y) = \exp(-y/\theta^*)/\theta^*$. The formal result is a consequence of Scheffé's theorem.

(v) In part (iv), y is asymptotically exponentially distributed with scale parameter θ^*. The $100(1 - \alpha)\%$ confidence interval for the scale parameter is

$$\frac{2n\widehat{\theta}}{\chi^2_{1-\alpha/2,2n}} < \theta^* < \frac{2n\widehat{\theta}}{\chi^2_{\alpha/2,2n}}. \tag{2.5}$$

However, one must use an approximate confidence interval because the exponential distribution holds as $n \to \infty$. Fortunately, the exponential distribution is in the exponential family, so a normal approximation to the $\chi^2_{p,\nu}$ distribution can be used. Thus, the approximate $100(1-\alpha)\%$ confidence interval for θ^* is

$$\left(X_{(n)} - z_{\alpha/2}\frac{X_{(n)}}{\sqrt{n}}, X_{(n)} + z_{\alpha/2}\frac{X_{(n)}}{\sqrt{n}} \right),$$

using the asymptotic result of $\mathrm{E}(Y) = 1/\widehat{\theta}$, $\mathrm{Var}(Y) = 1/\widehat{\theta}^2$, and $\widehat{\theta} = x_{(n)}$.

Further Reading

Historical Background. A standard condition for the convergence of a MLE is that the space of observations must not depend on the parameter because the support of the density $f(\cdot \mid \theta)$ must be the same for all θ. The Kullback–Leibler divergence between densities with different supports is not guaranteed to exist. Thus, without identical supports, each case needs a self-contained proof.

Points of Interest.

1. The challenging aspect of this problem is that the support depends on an unknown parameter θ, so the usual method of differentiating a likelihood function is not available. It is seen that the maximum data point acts as a sufficient statistic (Mukhopadhyay, 2014).

2. In part (iii), a transformation is defined to evaluate the penalized distance between θ^* and its MLE $\widehat{\theta}$. If $\widehat{\theta} = \theta^*$, then $y = 0$. As the distance between θ^* and $\widehat{\theta}$ increases, y increases. The relationship between θ^* and $\widehat{\theta}$ is explored in part (iv), where the asymptotic distribution of the penalized distance between θ^* and $\widehat{\theta}$ is found. In most cases, the asymptotic distribution is a normal distribution; however, in this exercise, it was not.

3. A more complicated version of this exercise where some data are right-censored is provided in Yu (2021).

4. This exercise comes across Scheffé's theorem, which is the result that, if a sequence of density functions $f_n(x)$ converges to $f(x)$ point-wise for each x, then $f_n(x)$ converges to $f(x)$ with respect to the L_1 distance; i.e.,

$$\int |f_n(x) - f(x)|\, dx \to 0.$$

To see this, write

$$|f_n(x) - f(x)| = f_n(x) - f(x) + 2\max\{f(x) - f_n(x), 0\},$$

so

$$\int |f_n(x) - f(x)|\, dx = 2\int \max\{f(x) - f_n(x), 0\}\, dx.$$

The term on the right is upper bounded by $2\int \max\{f(x), 0\}\, dx$, which is finite. The result now follows from the dominated convergence theorem.

S2.1.5* – Maximum ordered statistics and the likelihood principle

(i) Let A denote the event $\{X_{(n)} \leq x\}$ and B denote the event $\{X_i \leq x : i = 1, \ldots, n\}$. By definition, $P(X_i \leq X_{(n)}) = 1$ for all i. Thus, events A and B are equivalent and

$$\begin{aligned} P(X_{(n)} \leq x) &= P(X_1 \leq x, \ldots, X_n \leq x) \\ &\overset{\text{ind}}{=} P(X_1 \leq x) \cdots P(X_n \leq x) \\ &\overset{\text{i.d.}}{=} \{P(X_1 \leq x)\}^n \\ &= F_X(x)^n. \end{aligned}$$

Differentiate $F_{X_{(n)}}(x) = P(X_{(n)} \leq x)$ to obtain the density function

$$f_{X_{(n)}}(x) = \frac{d}{dx}\{F_X(x)^n\} = nF_X(x)^{n-1} f_X(x).$$

(ii) The likelihood principle argues that all information in a sample relevant to the model parameters is contained in the likelihood function. The set of ordered statistics $\{x_{(i)}\}_{i=1}^n$ contains the same information as the set of random variables $\{x_i\}_{i=1}^n$. Therefore, the likelihood function for these sets of random variables are equivalent. Thus, by the likelihood principle,

$$l(f \mid x_1, \ldots, x_n) = f_{X_{(1)}, \ldots, X_{(n)}}(x_1, \ldots, x_n) \propto \prod_{i=1}^n f_X(x_i).$$

In particular, there are $n!$ permutations of the values $\{X_i\}_{i=1}^n$, so

$$f_{X_{(1)}, \ldots, X_{(n)}}(x_1, \ldots, x_n) = n! \prod_{i=1}^n f_X(x_i).$$

(iii) From part (i), the density function of $x_{(n)}$ requires $F_X(x)$:

$$\begin{aligned} F_X(x) &= \int_0^x \frac{1}{\theta} \exp(-t/\theta) dt \\ &= -\int_0^{-x/\theta} \exp(u)\, du \\ &= 1 - \exp(-x/\theta), \end{aligned}$$

where a u-substitution of $u = -t/\theta$ is used to obtain the second line. Then, use part (i) to obtain the density function of $x_{(n)}$ to compute $E(X_{(n)})$:

$$f_{X_{(n)}}(x) = n\{1 - \exp(-x/\theta)\}^{n-1} \theta^{-1} \exp(-x/\theta).$$

Unfortunately, the integral $E(X_{(n)}) = \int_0^\infty x f_{X_{(n)}}(x) dx$ is difficult to compute analytically. Because $x_{(n)}$ is a non-negative random variable, its expectation may instead be computed using its cumulative distribution function:

$$E(X_{(n)}) = \int_0^\infty \{1 - (1 - e^{-x/\theta})^n\} dx.$$

However, this integral is also difficult to compute analytically. Rather, a telescoping method is employed in which analytically tractable summands are recursively computed. Define $I_n = \text{E}(X_{(n)})$, which may be treated as a function of n. Then, write $I_n = I_0 + (I_1 - I_0) + (I_2 - I_1) + \cdots + (I_n - I_{n-1})$. The nth telescoped term is

$$\begin{aligned} I_n - I_{n-1} &= \int_0^\infty \left\{1 - (1 - e^{-x/\theta})^n\right\} - \left\{1 - (1 - e^{-x/\theta})^{n-1}\right\} dx \\ &= \int_0^\infty \left(1 - e^{-x/\theta}\right)^{n-1} e^{-x/\theta} dx \\ &= \theta \int_0^1 u^{n-1} du = \theta / n \,, \end{aligned}$$

where a u-substitution of $u = 1 - e^{-x/\theta}$ is used to obtain the third equality. Note that $I_n - I_{n-1}$ is more tractable than just working with I_n because the first "1-" term cancels out. Starting with $I_0 = \int_0^\infty \{1 - (1 - e^{-x/\theta})^0\} dx = 0$, recursively compute all the terms in the telescoping sum for I_n:

$$I_n = 0 + \theta + \theta/2 + \cdots + \theta/n = \theta \sum_{k=1}^{n} \frac{1}{k} = \theta\, H(n),$$

where $H(n)$ denotes the nth harmonic number. Therefore, the expected value of the maximum ordered statistic for a sample of size n from an exponential distribution with scale parameter θ is $\text{E}(X_{(n)}) = \theta\, H(n)$.

(iv) Using the definition of bias,

$$\begin{aligned} \text{bias}_\theta(\widehat{\theta}_{(-n)}) = \text{E}\left(\widehat{\theta}_{(-n)} - \theta\right) &= \text{E}\left(\frac{1}{n-1} \sum_{i=1}^{n-1} X_{(i)}\right) - \theta \\ &= \frac{1}{n-1} \text{E}\left(\sum_{i=1}^{n} X_{(i)} - X_{(n)}\right) - \theta \\ &= \frac{1}{n-1} \left\{\text{E}\left(\sum_{i=1}^{n} X_i\right) - \text{E}(X_{(n)})\right\} - \theta \\ &= \frac{1}{n-1} \left\{\sum_{i=1}^{n} \text{E}(X_i) - \theta H(n)\right\} - \theta \\ &= \frac{n\theta - \theta H(n)}{n-1} - \theta \\ &= \frac{\theta\{1 - H(n)\}}{n-1}, \end{aligned}$$

where the fourth equality follows because $\sum_{i=1}^{n} x_{(i)} = \sum_{i=1}^{n} x_i$. Note that $\text{E}(X_i) = \theta$ using the mean of an exponential distribution.

(v) Obtain the joint distribution of the ordered statistics excluding the maximum ordered statistic by integrating over the maximum observed value in the joint distribution of the ordered statistics:

$$\begin{aligned}
f_{X_{(1)},\dots,X_{(n-1)}}(x_{(1)},\dots,x_{(n-1)}) &= \int_{x_{(n-1)}}^{\infty} f_{X_{(1)},\dots,X_{(n)}}(x_{(1)},\dots,x_{(n)})dx_{(n)} \\
&= \int_{x_{(n-1)}}^{\infty} n!\prod_{i=1}^{n} f_X(x_{(i)})dx_{(n)} \\
&= n!\prod_{i=1}^{n-1} f_X(x_{(i)}) \int_{x_{(n-1)}}^{\infty} f_X(x_{(n)})dx_{(n)} \\
&= n!\prod_{i=1}^{n-1} \frac{1}{\theta}\exp(-x_{(i)}/\theta) \int_{x_{(n-1)}}^{\infty} \frac{1}{\theta}\exp(-x_{(n)}/\theta)dx_{(n)} \\
&= n!\prod_{i=1}^{n-1} \frac{1}{\theta}\exp(-x_{(i)}/\theta)\cdot\exp(-x_{(n-1)}/\theta).
\end{aligned}$$

Now, maximize the log-likelihood to obtain the MLE:

$$\begin{aligned}
\frac{\partial l(\theta)}{\partial\theta} &= \frac{\partial}{\partial\theta}\left[\log(n!) + \sum_{i=1}^{n-1}\log\left\{\frac{1}{\theta}\exp(-x_{(i)}/\theta)\right\} - x_{(n-1)}/\theta\right] \\
&= \frac{\partial}{\partial\theta}\left\{(n-1)\log(1/\theta) - \sum_{i=1}^{n-1} x_{(i)}/\theta - x_{(n-1)}\theta^{-1}\right\} \\
&= -(n-1)/\theta + \sum_{i=1}^{n-1} x_{(i)}/\theta^2 + x_{(n-1)}/\theta^2 \overset{\text{set}}{=} 0,
\end{aligned}$$

which implies

$$\widehat{\theta}^* = \frac{\sum_{i=1}^{n-1} x_{(i)} + x_{(n-1)}}{n-1}.$$

Note that the MLE under this updated joint likelihood can be expressed as a function of the estimator in part (iv): $\widehat{\theta}^* = \widehat{\theta}_{(-n)} + x_{(n-1)}/(n-1)$. Because $\mathrm{E}(X_{(n-1)}) > 0$, the bias for $\widehat{\theta}^*$ is greater than for $\widehat{\theta}_{(-n)}$.

Further Reading

Historical Background. A common problem in data collection is sampling bias, which can manifest itself in many ways. The underlying theme is that the observations are not a genuine random sample from the relevant population. In short, some members of the population are more likely to be sampled than others. This question explored sampling bias when the largest of n samples was not available or ignored. Analysis proceeded by attempting to correct for the bias. The standard model representation of this problem is

$$g(x) \propto b(x)f(x),$$

where $f(x)$ is the target density function to sample, yet samples are coming from $g(x)$ due to the bias function $b(x)$. The sample from $g(x)$ is required to learn about $f(x)$.

Points of Interest.

1. This question started with fundamental results regarding ordered statistics. In the second part of the question, a heuristic argument was used to construct the joint density function of the ordered statistics; notably, the likelihood principle was used to argue that it was proportional to the product of the marginals. Several other key results for ordered statistics may be worth exploring. For example, there is often interest in the density function for the ith ordered statistic $x_{(i)}$ or the joint distribution of $x_{(i)}$ and $x_{(j)}$ for $i \neq j$; to obtain these results, one can either create a heuristic argument as in part (ii) or complete a non-trivial amount of algebra (Casella and Berger, 2021). Typically, the former is more rewarding and time-efficient.

2. The next two parts of the question concerned the maximum ordered statistic for the exponential distribution. In part (iv), $x_{(n)}$ was not available to estimate θ. The bias of the new estimator helped quantify the effect that dropping the highest value had on inference; such a case arises when removing outliers from data sets, a common preprocessing step in many statistical frameworks. Thus, ordered statistics provide a way to explore the impacts that the removal of outliers may have on resulting inference.

3. In the final part of the question, the bias between $\widehat{\theta}_{(-n)}$ and $\widehat{\theta}^*$ was compared. The first estimator assumed that ignoring the maximum ordered statistic would not significantly affect the average of the observations, which was a faulty assumption – especially because the sample was exponentially distributed. The second estimator dropped this assumption by explicitly computing the MLE for the joint distribution of all observations excluding the maximum ordered statistic. In doing so, the first $n-1$ ordered statistics' distributional forms were updated to reflect the removal of the maximum ordered statistic, and, as a result, the second estimator had higher bias and lower variance.

4. Conventional applications of ordered statistics include nonparametric density estimation and inference, where assumptions about the underlying distribution are minimized (Arnold et al., 2008). For example, rank-based tests, such as the Wilcoxon signed-rank test or Mann–Whitney U test, utilize ordered or rank statistics. Additionally, ordered statistics are essential to rank correlation coefficients, including the Spearman's rank correlation coefficient and Kendall's τ coefficient, which quantify the degree of monotonicity between two variables without assuming a specific functional relationship. Ordered statistics find application in survival analysis, where they are used to analyze time-to-event data. The survival function, essential to survival analysis, may be estimated using ordered statistics. More recently, ordered statistics have been used in extreme value-based methods to model risk assessment, extreme weather events, and phenology (Jäntschi, 2020). Finally, ordered statistics can be used in mathematical statistics for distributional shape analysis, quantile estimation, and percentile calculations, providing insights into the relative standing of a particular data point within a data set (David and Nagaraja, 2004).

Demonstration. Consider $n = 25$ samples from an exponential distribution with scale parameter $\theta = 1/5$. The following three estimators for θ are computed and compared: the MLE assuming all samples are available ($\widehat{\theta} = \overline{x}$), the estimator in part (iv), and the estimator in part (v). To compare these estimators, the resulting densities are plotted in Fig 2.3. The density for the MLE is included as a proxy for how well the other estimated densities are expected to perform for such a small sample size. Note that the density using the third estimator better approximates the true density than the density using the second estimator; this is because the third estimator updates the joint distribution of all samples, then computes the MLE, rather than computing the MLE on the original joint distribution without having the maximum observed value.

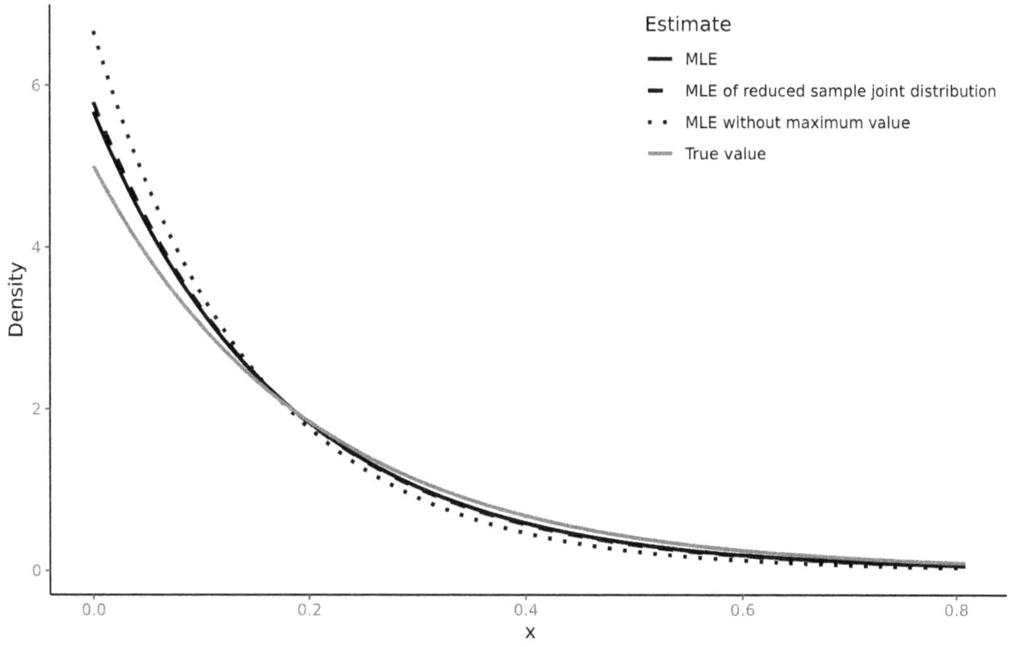

FIGURE 2.3
The estimated densities resulting from different estimates of θ. The estimator from part (v) (dashed line) better approximates the true density (solid gray line) than the estimator from part (iv) (dotted line). The density using the standard MLE (solid black line) is shown as a proxy for how well the estimated densities may be expected to perform for such a small sample size.

S2.1.6 – Estimation via Fisher information distance

(i) Start by expanding the squared expression:

$$\begin{aligned} F(p_0, p_\theta) &= \int p_0(x) \left\{ \frac{p_0'(x)}{p_0(x)} - \frac{p_\theta'(x)}{p_\theta(x)} \right\}^2 dx \\ &= \int p_0(x) \left\{ \frac{p_0'(x)}{p_0(x)} \right\}^2 dx + \int p_0(x) \left\{ \frac{p_\theta'(x)}{p_\theta(x)} \right\}^2 dx - 2 \int p_0'(x) \frac{p_\theta'(x)}{p_\theta(x)} dx. \end{aligned}$$

Use integration by parts on the final term to obtain

$$\int p_0'(x) \frac{p_\theta'(x)}{p_\theta(x)} dx = p_0(x) \frac{p_\theta'(x)}{p_\theta(x)} \bigg|_{-\infty}^{\infty} - \int p_0(x) \frac{d}{dx} \left\{ \frac{p_\theta'(x)}{p_\theta(x)} \right\} dx,$$

where $u = p'_\theta(x)/p_\theta(x)$ and $dv = p'_0(x)dx$. The first term on the right side is 0 because the end points are assumed to vanish to 0. Thus,

$$F(p_0, p_\theta) = \underbrace{\int p_0(x) \left\{ \frac{p'_0(x)}{p_0(x)} \right\}^2 dx}_{\kappa} + \int p_0(x) \left\{ \frac{p'_\theta(x)}{p_\theta(x)} \right\}^2 dx$$
$$+ \int 2p_0(x) \frac{d}{dx} \left\{ \frac{p'_\theta(x)}{p_\theta(x)} \right\} dx,$$

which gives the desired form for $F(p_0, p_\theta)$.

(ii) Write $p_\theta(x) = c\exp\{w(x)/\theta\}$ for some normalizing constant c, and take the derivative of $p_\theta(x)$ with respect to x: $p'_\theta(x) = p_\theta(x)\, w'(x)/\theta$, where $w'(x) = dw/dx$. Thus,

$$\frac{p'_\theta(x)}{p_\theta(x)} = \frac{w'(x)}{\theta},$$
$$\left\{ \frac{p'_\theta(x)}{p_\theta(x)} \right\}^2 = \left\{ \frac{w'(x)}{\theta} \right\}^2,$$
$$\frac{d}{dx} \left(\frac{p'_\theta(x)}{p_\theta(x)} \right) = \frac{w''(x)}{\theta}.$$

Putting the above together, $l(\theta, x) = \{w'(x)/\theta\}^2 + 2\, w''(x)/\theta$.

(iii) An estimator of θ can be obtained by minimizing $F(p_0, p_\theta)$, which is equivalent to minimizing $\int p_0(x) l(\theta, x) dx$. Substitute p_0 with a sample of size n to obtain a Monte Carlo estimator:

$$\widehat{\theta} = \arg\min_{\theta \in \Theta} \frac{1}{n} \sum_{i=1}^{n} l(\theta, x_i).$$

(iv) Using part (ii) and part (iii), the minimum point of the Monte Carlo estimator is

$$\widehat{\theta} = -\frac{\sum_{i=1}^{n} w'(x_i)^2}{\sum_{i=1}^{n} w''(x_i)}.$$

(v) Dividing the numerator and denominator by n and assuming suitable regularity conditions for the function $w(x)$, the ratio for $\widehat{\theta}$ will converge to

$$-\frac{\int w'(x)^2 p_{\theta^*}(x)\, dx}{\int w''(x)\, p_{\theta^*}(x)\, dx},$$

where θ^* is the true parameter value. The aim is to show that

$$\int p_{\theta^*}(x) \{w'(x)^2 + \theta^*\, w''(x)\}\, dx = 0,$$

demonstrating that the estimator converges to θ^*. This can be shown using integration by parts: set $u = p_{\theta^*(x)}$, $v' = w''(x)$, so $u' = p_{\theta^*}(x)\, w'(x)/\theta^*$ and $v = w'(x)$.

Further Reading

Historical Background. The relative Fisher information distance is not as well-known as the standard Fisher information distance. The former is the distance defined in the question, which is connected to Fisher information for a density function given by $I(f) = \int f'(x)^2/f(x)\,dx$. This is in contrast to the Fisher information for a parameter given by $I(\theta) = \int \{\partial f(x \mid \theta)/\partial\theta\}^2/f(x \mid \theta)\,dx$. The Fisher distance is heavily utilized in some functional analysis problems.

Hyvarinen (2005) introduced the use of the relative Fisher information distance for parameter estimation and pointed out two ideas. First, the distance can be approximated using a Monte Carlo sample after an integration by parts, which requires some regularity conditions. Second, the distance is based on the score function of the family of density functions, so any intractable normalizing constant depending on unknown parameters conveniently disappears.

For further reading on Fisher information distance, see Costa et al. (2015). Some estimation problems using the Fisher information distance have been studied by Hyvarinen (2005). Finally, on the topic of obtaining estimators by minimizing divergences, see Broniatowski (2021).

Points of Interest.

1. The standard maximum likelihood estimator arises by exchanging the Fisher information distance with the Kullback–Leibler divergence

$$D(p_0, p_\theta) = \int p_0(x)\,\log\{p_0(x)/p_\theta(x)\}\,dx.$$

   In particular, the MLE is obtained by maximizing $\int p_0(x)\,\log p_\theta(x)\,dx$, which is equivalent to maximizing the log-likelihood for the observed sample: $\sum_{i=1}^n \log p_\theta(x_i)$. However, not every measure of distance or divergence yields a Monte Carlo estimation procedure as in part (iii). For example, half the squared Hellinger distance

$$\tfrac{1}{2}\,d_H^2(p_0, p_\theta) = 1 - \int \sqrt{p_0(x)\,p_\theta(x)}\,\mathrm{d}x$$

   can not be written as a Monte Carlo estimator of the form $\sum_{i=1}^n l(\theta, x_i)$.

2. If the discrete empirical distribution does not render a Monte Carlo estimator $\widehat{\theta}$, one may instead use a continuous nonparametric estimator of the true density, denoted $\widehat{f}_n$, based on an independent and identically distributed sample $\{x_i\}_{i=1}^n$. Then, one must estimate $\theta \in \Theta$ from a parametric model $f_\theta(\cdot) \equiv f(\cdot \mid \theta)$ by minimizing $d(\widehat{f}_n, f_\theta)$ for some distance d between density functions. For example, Beran (1977) uses the Hellinger distance.

Demonstration. When $w(x) = -\frac{1}{2}x^2$, the MLE and the estimator from part (iv) are the same: $\sum_{i=1}^n x_i^2/n$. When $w(x) = -\frac{1}{2}x^4$, the estimators are different; the MLE is $2\sum_{i=1}^n x_i^4/n$, and the Fisher estimator from part (iv) is $(2/3)\,\sum_{i=1}^n x_i^6/\sum_{i=1}^n x_i^2$. Fig. 2.4 presents a histogram of the MLE (left) and the Fisher estimator (right) based on a sample size of $n = 100$. They are very similar; the means are 1.00 and 0.99, respectively, and the variances are 0.04

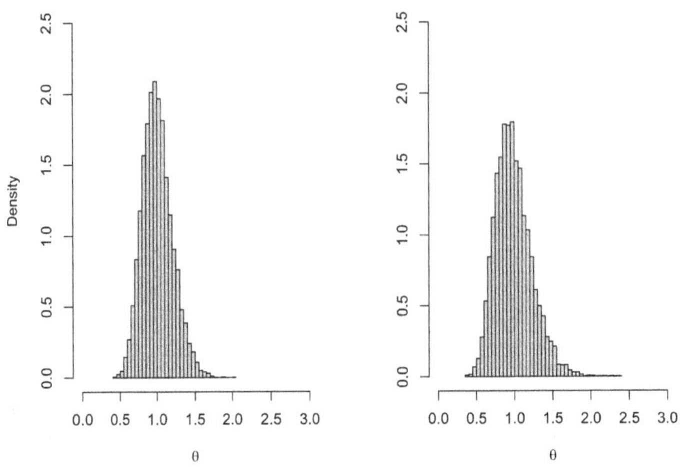

FIGURE 2.4
MLEs (left panel) and Fisher estimators (right panel) from 10,000 experiments, each with a sample size of 100.

and 0.06, respectively. Based on this demonstration, if the MLE is difficult to find, then the Fisher estimator may provide an excellent alternative.

S2.1.7 – Estimation with intractable normalizing constants

(i) Minimizing the Fisher distance between f_{θ^*} and f_θ with respect to θ is equivalent to minimizing

$$\int f_{\theta^*}(x)\{f'_\theta(x)/f_\theta(x)\}^2\,dx - 2\int \{f'_{\theta^*}(x)f'_\theta(x)/f_\theta(x)\}\,dx.$$

The second term can be rearranged using integration by parts provided that $f_{\theta^*}(x)\,f'_\theta(x)/f_\theta(x)$ disappears at the boundary for all θ; this is true for the von Mises family because $f'_\theta(x)/f_\theta(x) = -\theta\sin(x)$, which is 0 when $x = 0$ and $x = 2\pi$. Then,

$$\int \{f'_{\theta^*}(x)\,f'_\theta(x)/f_\theta(x)\}\,dx = -\int f_{\theta^*}(x)\,(f'_\theta(x)/f_\theta(x))'\,dx,$$

and the result follows.

(ii) It does not involve Z because $f'_\theta(x)/f_\theta(x) = -\theta\sin(x)$ does not depend on Z.

(iii) A Monte Carlo estimator for θ can be derived from part (i) by minimizing

$$\sum_{i=1}^{n}\left\{\alpha(x_i,\theta)^2 + 2\alpha'(x_i,\theta)\right\},$$

where $\alpha(x,\theta) = f'_\theta(x)/f_\theta(x)$. Thus, the Monte Carlo estimator is

$$\widehat{\theta} = \frac{\sum_{i=1}^{n} \cos(x_i)}{\sum_{i=1}^{n} \sin^2(x_i)}.$$

(iv) $\mathrm{E}\left\{\sin^2(X)\right\}$ can be written as

$$\int_0^{2\pi} \sin(x)\, \sin(x)\, \exp\{\theta \cos(x)\}/Z(\theta)\, dx,$$

which can be simplified to $\theta^{-1}\,\mathrm{E}\left\{\cos(X)\right\}$ using integration by parts, where $u = \sin(x)$ and $v' = \sin(x)\exp\{\theta\cos(x)\}$.

(v) By the law of large numbers, the numerator of $\widehat{\theta}$ divided by n converges almost surely to $\mathrm{E}\left\{\cos(X)\right\}$. Similarly, the denominator divided by n converges almost surely to $\mathrm{E}\left\{\sin^2(X)\right\}$. Therefore, $\widehat{\theta}$ converges almost surely to θ^*.

Further Reading

Historical Background. In Question 2.1.6, it was stated that the Fisher distance can be used for parameter estimation, even in the presence of an intractable normalizing constant. This question highlights how the Fisher distance can be used with the von Mises family of density functions. The original work in this direction was provided by Hyvarinen (2005). In Bayesian statistics, Markov chain Monte Carlo sampling may be used for models with intractable normalizing constants (Gelman and Meng, 1998). More recent ideas appear in Murray et al. (2006).

Points of Interest.

1. The von Mises distribution can be thought of as a bivariate normal distribution constrained to lie on a circle. It is similar to the normal in that it is unimodal and the "variance" is determined by the concentration parameter. In this question, the mode is assumed to be 0 and θ is the concentration parameter. The von Mises distribution is used predominantly for modeling circular or directional data (Mardia and Jupp, 1999).

2. Estimation of the von Mises concentration parameter has received much attention due to the intractability of the normalizing constant. For example, Marrelec and Giron (2024) explored twelve estimators for θ; compared to other distributions in the exponential family, this might seem excessive. However, the log-likelihood function is given by

$$l(\theta) = -n \log Z(\theta) + \theta \sum_{i=1}^{n} \cos(x_i),$$

yielding a nonlinear estimator for the ratio

$$Z'(\widehat{\theta})/Z(\widehat{\theta}) = n^{-1} \sum_{i=1}^{n} \cos(x_i).$$

Thus, numerical methods are needed to estimate $\widehat{\theta}$.

3. The Fisher information approach provides a simple estimator for all members of the exponential family. Consider the form of the exponential family

$$f(x \mid \theta) = c(x)\exp\{\theta g(x) - b(\theta)\}.$$

Then, $f'(x \mid \theta)/f(x \mid \theta) = c'(x)/c(x) + \theta\, g'(x)$. To estimate θ using this approach, one needs to minimize

$$\theta^2 \sum_{i=1}^{n} g'(x_i)^2 + 2\theta \sum_{i=1}^{n} \{g''(x_i) + g'(x_i)\, c'(x_i)/c(x_i)\}$$

with respect to θ. The solution is

$$\widehat{\theta} = -\frac{\sum_{i=1}^{n}\{g''(x_i) + g'(x_i)\, c'(x_i)/c(x_i)\}}{\sum_{i=1}^{n} g'(x_i)^2},$$

and it can be shown that $\widehat{\theta}$ is consistent:

$$\mathrm{E}\left\{g''(X) + g'(X)\, c'(X)/c(X) + \theta^*\, g'(X)^2\right\} = 0.$$

As an illustration, consider a normal model with mean θ and variance 1. Then, $g(x) = x$ and $c(x) = \exp(-\frac{1}{2}x^2)$, which yields $\widehat{\theta} = \overline{x}$.

S2.1.8* – A sin kernel for density estimation

(i) Define $I(R) = \int_{-\infty}^{\infty} \cos(Rx)\,\phi(x)\,dx$ as a function of R. Then, differentiate $I(R)$ with respect to R: $I'(R) = -\int_{-\infty}^{\infty} x\sin(Rx)\,\phi(x)\,dx$. Additionally, note that $\phi'(x) = -x\phi(x)$. Using integration by parts with $u = \sin(Rx)$ and $dv = -x\phi(x)dx$, it is that $I'(R) = -R\,I(R)$. Thus,

$$I(R) = c\,\exp(-\tfrac{1}{2}R^2)$$

for some constant c. Because $I(0) = \int_{-\infty}^{\infty}\phi(x)dx = 1$, it is that $c = 1$.

(ii) Write the integral as $J(R)$. First, use the transform $z = x - \mu$ and note that $\cos\{R(z + \mu)\} = \cos(Rz)\,\cos(R\mu) - \sin(Rz)\sin(R\mu)$. The sin integral disappears because sin is an odd function. Therefore,

$$J(R) = \cos(R\mu)\int_{-\infty}^{\infty}\frac{1}{\sigma}\cos(Rz)\,\phi(z/\sigma)\,dz.$$

Finally, use the transformation $y = z/\sigma$ to show that $J(R) = \cos(R\mu)\,I(R\sigma)$.

(iii) The result follows by using the transformation $x \to z = y - x$.

(iv) The result from part (i) gives

$$\int_0^{\infty}\cos(sx)\,e^{-\frac{1}{2}s^2}\,ds = \pi\phi(x).$$

Writing $\mathrm{E}\{f_R(x)\} = m_R(x)$, one obtains $\pi\,dm_R/dx = \cos(Rx)\,e^{-\frac{1}{2}R^2}$. Therefore, $\pi\,m_R(x) = \int_0^R \cos(sx)\,e^{-\frac{1}{2}s^2}\,ds$ because $m_0(x) = 0$.

(v) Because $\int_0^\infty \cos(sx)\, e^{-\frac{1}{2}s^2}\, ds = \pi\phi(x)$,

$$\pi|m_R(x) - \phi(x)| = \left| \int_R^\infty \cos(sx)\, e^{-\frac{1}{2}s^2}\, ds \right|,$$

which is upper bounded by

$$\int_R^\infty e^{-\frac{1}{2}s^2}\, ds \leq R^{-1}\, e^{-\frac{1}{2}R^2}.$$

Further Reading

Historical Background. The nonparametric estimator of a density function was born from an attempt to obtain gradients from the empirical distribution function (Parzen, 1962). It was found that kernel density estimators could be used to estimate empirical distributions; the most common kernel used to smooth empirical distributions is the Gaussian. There was an immediate interest in the properties of different kernels, and the Epanechnikov kernel was found to be the most efficient (Moraes et al., 2021). However, other kernels may be more appropriate for certain criteria. For example, kernels that allow for negative values can have significantly smaller MSE.

Points of Interest.

1. When $f^*(x)$ is the standard normal density, the bias of the density estimator is smaller than $\exp(-R^2/2)$. The Gaussian kernel density estimator

$$\widehat{f}(x) = \frac{1}{nh} \sum_{i=1}^{n} \phi\left(\frac{x - x_i}{h}\right)$$

has a bias of $O(h^2)$, where the bandwidth h is the standard deviation of the normal kernel. The variance for $\widehat{f}(x)$ is of size $1/(nh)$, whereas the sin kernel yields a variance of size R/n. Matching the variances by taking $R = 1/h$, the corresponding bias of the sin kernel is significantly smaller than that for the normal kernel (i.e., $\exp\{-R^2/2\}$ compared to $1/R^2$). This result also holds for any normal model f^* and for a mixture of normal models. In the latter case, the reduced bias will hold for a large class of true models.

2. The normal and sin kernels behave quite differently. As $h \to 0$, the function $\phi(x/h)/h$ converges to a point mass at 0. On the other hand, the sin kernel $\sin(Rx)/(x\pi)$ also spikes, but at multiple points around 0.

3. In multi-dimensions, the bandwidth for the normal kernel needs to be a covariance matrix, which is not trivial to define. The benefit of the sin kernel is that the bivariate estimator

$$f_R(x, y) = n^{-1}\pi^{-2} \sum_{i=1}^{n} \frac{\sin\{R(x - x_i)\}}{x - x_i} \frac{\sin\{R(y - y_i)\}}{y - y_i}$$

is based on a bivariate sample $\{x_i, y_i\}_{i=1:n}$. Thus, the sin kernel does not require a covariance estimator because the correctly estimated covariance structure is already present within $f_R(x, y)$.

4. The drawback of using $f_R(x)$ is that it can take negative values. However, the sin kernel still integrates to 1. Additionally, if there are negative values, they will occur in the tails with low probability. The estimator $f_R(x)$ can be constrained to be non-negative, for example, by forcing the corresponding cumulative distribution function to be non-decreasing.

Demonstration. The plan here is to show that the MSE for estimating a point on the density

$$f(x) = 0.3\,\mathrm{N}(x \mid 0,1) + 0.7\,\mathrm{N}(x \mid 3,1/4)$$

is smaller when one uses the sin kernel rather than the Gaussian kernel. One of the more difficult points to estimate is the sharp mode at $x = 3$ for which the true value is 0.5598. For the Gaussian kernel, the value of the density at $x = 3$ is estimated with the rule-of-thumb bandwidth $1.06\widehat{\sigma}\,n^{-1/5}$, where $\widehat{\sigma}$ is the sample standard deviation. The value of n is set to 100. The mean value of the estimator over 1000 simulations of the data is 0.3484. For the sin kernel with a value of $R = 1/h^2 = 5.90$, the mean value is 0.5536. The respective MSEs are 0.0159 and 0.0005. The choice of R is based on the result that the variances are of order $1/(nh^2)$ and R/n. Therefore, when the variances are matched, the bias for the sin kernel estimator is substantially smaller than that for the Gaussian kernel estimator.

Now, consider a new data set of size 1000 and estimate the entire density using both kernels. Fig. 2.5 depicts the true density (full line), the sin kernel density estimator (dashed line), and the Gaussian kernel density estimator (dotted line). It is clear that the sin kernel density estimator is the better one.

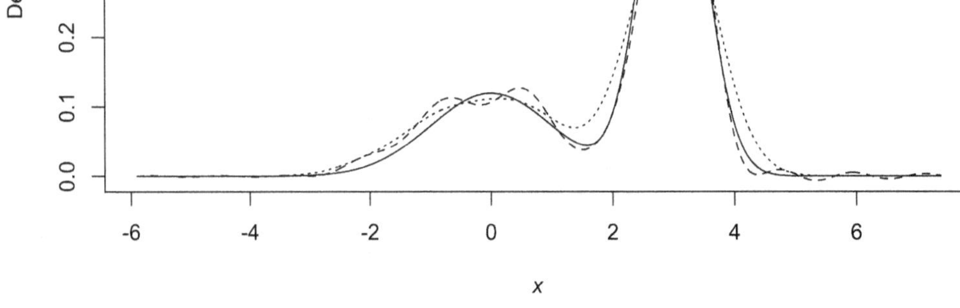

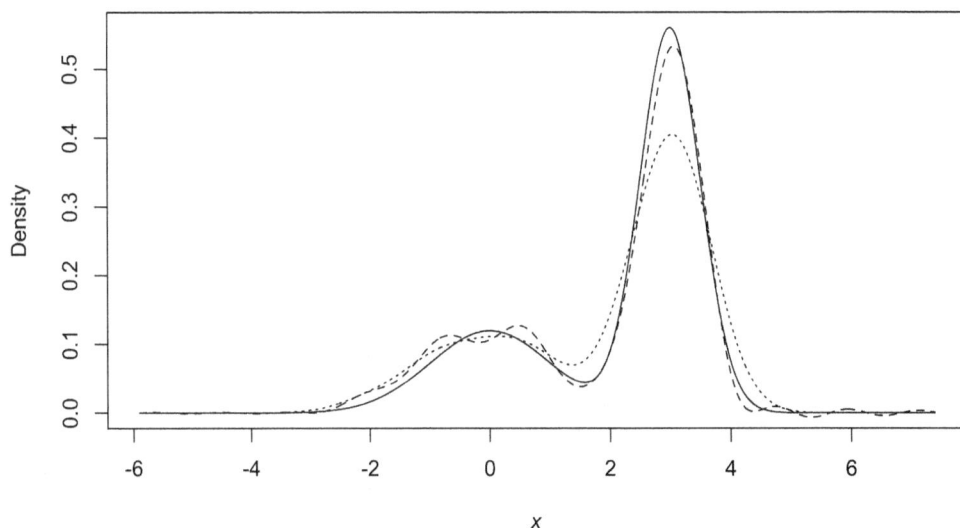

FIGURE 2.5
A comparison of sin and Gaussian kernel density estimators depicted as dashed and dotted lines, respectively. The true density is depicted as the full line.

S2.1.9 – Survival analysis with censored data

(i) Start with the simplest case: $k = 1$. Relying on the law of total probability, it is possible to see a pattern emerge:

$$\begin{aligned}
&P(T = 1) = P(T = 1 \mid T \geq 1) = h_1 \\
&P(T > 1) = P(T > 1 \mid T \geq 1) = 1 - h_1 \\
&P(T = 2) = P(T = 2 \mid T \geq 2)P(T \geq 2) = h_2(1 - h_1) \\
&P(T > 2) = P(T > 2 \mid T \geq 2)P(T \geq 2) = (1 - h_2)(1 - h_1) \\
&\quad\vdots \\
&P(T = k) = P(T = k \mid T \geq k)P(T \geq k) = h_k \prod_{1 \leq j < k} (1 - h_j) \\
&P(T > k) = P(T > k \mid T \geq k)P(T \geq k) = \prod_{1 \leq j \leq k} (1 - h_j).
\end{aligned}$$

For $\sum_{k=1}^{\infty} P(T = k) = 1$, it is that $\prod_{k=1}^{\infty}(1 - h_k) = 0$.

(ii) The likelihood function can be written as

$$\begin{aligned}
L &= \prod_{i=1}^{n} \left(h_{t_i} \prod_{1 \leq j < t_i} (1 - h_j) \right)^{1(\delta_i = 1)} \left(\prod_{1 \leq j < t_i} (1 - h_j) \right)^{1(\delta_i = 0)} \\
&= \prod_{k=1}^{\infty} h_k^{\sum_i 1(\delta_i = 1, t_i = k)} (1 - h_k)^{\sum_i [1(\delta_i = 1)1(t_i > k) + 1(\delta_i = 0)1(t_i > k)]} \\
&= \prod_{k=1}^{\infty} h_k^{\sum_i 1(\delta_i = 1, t_i = k)} (1 - h_k)^{\sum_i 1(t_i > k)} \\
&= \prod_{k=1}^{\infty} h_k^{A_k} (1 - h_k)^{B_k},
\end{aligned}$$

where

$$A_k = \sum_{i=1}^{n} 1(\delta_i = 1,\ t_i = k) \quad \text{and} \quad B_k = \sum_{i=1}^{n} 1(t_i > k).$$

To calculate the MLE of h_k, note that the log-likelihood is

$$l = \sum_k A_k \log h_k + B_k \log(1 - h_k).$$

Taking the derivative with respect to h_k,

$$\frac{\partial l}{\partial h_k} = \frac{A_k}{h_k} - \frac{B_k}{1 - h_k} \overset{\text{set}}{=} 0.$$

Therefore, the MLE is $\widehat{h}_k = A_k / (A_k + B_k)$.

(iii) Using the form of $P(T = k)$ in part (i) and $\widehat{h}_k$ from part (ii), the estimator for $P(T = k)$ is given by

$$\widehat{p}_k = \widehat{h}_k \prod_{1 \leq j < k} (1 - \widehat{h}_j) = \frac{A_k}{A_k + B_k} \prod_{1 \leq j < k} \frac{B_j}{A_j + B_j}.$$

(iv) If the largest observation is censored, it is not known exactly when the individual concerned actually dies. Hence, the estimator of the survival function will not reach zero, and the height will be $1/n$ at the largest observed censoring time.

(v) Denote the number of deaths at time k by n_k. Then,

$$A_k = n_k \quad \text{and} \quad B_k = \sum_{j>k} n_j,$$

so $A_k + B_k = \sum_i 1(t_i \geq k) = \sum_{j \geq k} n_j$. Therefore,

$$\widehat{p}_k = \frac{A_k}{A_k + B_k} \prod_{1 \leq j < k} \frac{B_j}{A_j + B_j} = \frac{n_k}{\sum_{j \geq k} n_j} \frac{\sum_{j>k-1} n_j}{\sum_{j \geq k-1} n_j} \cdots \frac{\sum_{j>1} n_j}{\sum_{j \geq 1} n_j}$$

$$= \frac{n_k}{\sum_{j \geq 1} n_j} = \frac{n_k}{n},$$

where many terms cancel (similar to a telescoping product).

Further Reading

Historical Background. This question is based on a discrete version of the Kaplan–Meier estimator that estimates a survival function in the presence of right-censored observations (Kaplan and Meier, 1958). The estimator is piece-wise constant and drops at exact failure times. If the final observation is censored, the estimator does not drop to 0.

The discrete-time setting is quite illustrative for the estimator. In this setting, survival is taken as a sequence of independent Bernoulli outcomes that determine whether the individual survives to the next time point. If censoring occurs, outcomes are observed up to the end of the study. The Kaplan–Meier estimator is similar, except the time intervals collapse to 0. See Machin et al. (2006) for a comprehensive overview of survival analysis and Ibrahim et al. (2001) for a Bayesian treatment.

Points of Interest.

1. One of the most common challenges in survival analysis is censoring. For example, survival analysis is often used in medical statistics as a means to study how well a drug is able to prevent a disease from occurring. In this pharmaceutical example, censoring may occur if individuals contract the disease after the study period ends.

2. Inference techniques for survival data can trace their roots to three key publications. The first is Kaplan and Meier (1958), which developed a nonparametric estimator of the survival function; this "Kaplan–Meier" estimator is found in part (iii) of this question and is a modern version of estimators arising from the life tables used in actuarial science that were developed by Edmond Halley in 1693. The second work is David Cox's seminal paper, which proposed the most popular regression model for survival data to date (Cox, 1972). The third work is the development of the theory of counting processes and their use in survival analysis (Aalen, 1978). In particular, the counting process techniques of Aalen (1978) can be used to derive the properties of the Cox model. The Cox model can be extended quite easily to model time-dependent covariates, multiple states, or random effects (Klein et al., 2014).

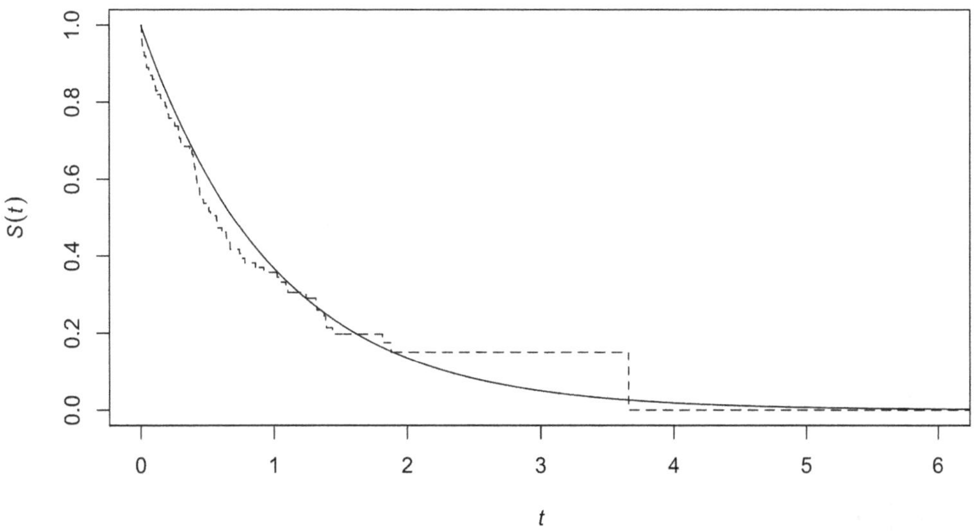

FIGURE 2.6
Kaplan–Meier estimator of a survival function with random right censoring (dashed line) and the true survival function (full line).

Demonstration The Kaplan–Meier estimator is the continuous-time version of the survival estimator given in this question and is specified as

$$\widehat{S}(t) = \prod_{t_i \le t, d_i = 1} (1 - 1/m_i),$$

where it is assumed that the failure times are continuous and distinct and $m_i = \#\{t_j \ge t_i\}$ is the number of observations that have failure times greater than or equal to t_i. Here, $d_i = 1$ indicates an exact failure time of t_i. Fig. 2.6 presents the Kaplan–Meier estimator when the failure times $\{y_i\}$ were taken to be standard exponential random variables and the censoring times $\{c_i\}$ were independent $\mathrm{Ga}(2, 1)$ random variables. Thus, the observed data was $\{t_i, d_i\}$, where $t_i = \min(y_i, c_i)$ and $d_i = 1(t_i = y_i)$. The sample size was 100, and there were 30 censored observations. The survival estimator in Fig. 2.6 (dashed line) drops at observed failure times. The true survival function $S(t) = \exp(-t)$ appears in Fig. 2.6 as a full line. The estimator becomes more unreliable as time progresses due to the diminishing sample size.

3

Exponential Family

The exponential family is a set of probability distributions important to data analysis. The canonical form for the family is given by

$$f(y \mid \theta) = c(y) \exp\{\theta\, t(y) - b(\theta)\},$$

where $c(y)$ must be non-negative. Many principles in statistics are demonstrated using the exponential family, such as asymptotics, completeness, minimum variance, and sufficiency. Such concepts are demonstrated with the exponential family because the parameters and data interact in a single term. However, the exponential family is used for much more than pedagogy; many distributions that find common use across science belong to the family, including the beta, exponential, gamma, normal, and Poisson distributions. Indeed, there are several books written solely about the exponential family and the role it plays in statistical modeling and inference (Sundberg, 2019; Efron, 2022).

Questions 3.1.1 and 3.1.2 concern the asymptotics and estimation of maximum likelihood estimators for parameters belonging to the exponential family. If the model is specified correctly, the MLE converges to the true parametric value. However, under the inevitable model misspecification, how does the MLE behave? Ideally, the MLE converges to the parametric value that minimizes some distance or divergence between the model and the true density function. This is the case when the model belongs to the exponential family and the divergence metric is the Kullback–Leibler divergence.

In the exponential family, properties of the MLE are easily established due to the convexity of the likelihood function. For example, an asymptotic distribution for the MLE is available, which is an improvement over the normal approximation. Questions 3.1.3 and 3.1.4 look at likelihood ratio tests for the exponential family. In particular, likelihood ratios for the exponential and Poisson distributions are considered.

Question 3.1.5 shows that, for the exponential family, the Cramér–Rao lower bound for the variance of an unbiased estimator is attained. Question 3.1.6 explores alternative estimators to the MLE, which may not be available if the normalizing constant is intractable. In particular, the parameters are estimated by minimizing a Monte Carlo approximation to the Fisher information distance between the correct model and the exponential family. Finally, Question 3.1.7 provides a Bayesian treatment of the exponential family. In particular, the exercise investigates the predictive density when a conjugate prior is chosen.

DOI: 10.1201/9781003493471-3

Q3.1 Questions – Exponential Family

Q3.1.1 – MLE for the exponential family

Introduction. The exponential family is widely used in statistics due to its convenient properties. Distributions belonging to the family include the Bernoulli, beta, exponential, gamma, Gaussian, geometric, and Poisson distributions. This question focuses on properties of the maximum likelihood estimator that are applicable for all distributions in the exponential family.

Question. Consider the canonical exponential family model

$$f_Y(y \mid \theta) = c(y)\, \exp\left\{y\theta - b(\theta)\right\},$$

where $c(y)$ is a known function and $b(\theta)$ is part of the normalizing constant.

(i) Show that $\mathrm{E}\,(Y) = b'(\theta)$, where the derivative $b'(\theta)$ is assumed to exist.

(ii) Prove directly that $b(\theta)$ is a convex function, assuming it exists (i.e., $b''(\theta) \geq 0$).

(iii) Show that $\mathrm{Var}\,(Y) = b''(\theta)$.

(iv) Denote the MLE as $\widehat{\theta}$, which is given by $b'(\widehat{\theta}) = \overline{y}$, where $\overline{y}$ is the sample mean. If the true model is $f^*(y)$, show that $\widehat{\theta}$ converges to

$$\theta^* = \arg\max_{\theta} \int f^*(y)\, \log f(y \mid \theta)\, dy.$$

(v) Interpret the result in part (iv).

Q3.1.2 – Properties of the MLE for the exponential family

Introduction. This question builds upon Question 3.1.1, where it was shown that the MLE converges to the best possible parameter that minimizes the Kullback–Leibler divergence from the true density. Now, more properties of the MLE for the exponential family are explored, including its convergence in probability.

Question. Consider the density function with exponential family form

$$f(y \mid \theta) = c(y)\, \exp\{y\theta - b(\theta)\},$$

where $c \geq 0$.

(i) If $L(\theta, y) = \log f(y \mid \theta)$, find the form of the Fisher information

$$I(\theta) = -\mathrm{E}\left\{\frac{\partial^2}{\partial\theta^2} L(\theta, Y)\right\},$$

and show that $I(\theta)$ is the same as

$$\mathrm{E}\left[\left\{\frac{\partial}{\partial\theta} L(\theta, Y)\right\}^2\right].$$

Here, expectation is with respect to $f(\cdot \mid \theta)$.

(ii) Show that $b(\theta)$ is a convex function.

(iii) If $\widehat{\theta}$ is the MLE, show that

$$\mathrm{P}(\widehat{\theta} \leq z) = \mathrm{P}(\overline{Y} \leq b'(z)),$$

where $\bar{y}$ is the sample mean.

(iv) Show that $\mathrm{P}(\widehat{\theta} \leq z)$ converges to 0 or 1 depending on whether $z < \theta^*$, where θ^* is the true parameter value.

(v) What does the result in part (iv) imply for $\widehat{\theta}$ as $n \to \infty$?

Q3.1.3 – Likelihood-ratio test for the exponential density

Introduction. The likelihood-ratio test (LRT) is a classical approach to hypothesis testing. The objective in a LRT is to assess the fit of two competing models by comparing their likelihood functions evaluated at the estimated parameters. Typically, one model is obtained via maximum likelihood estimation, and the other model considers a constraint on the parameter space; this constraint is usually referred to as the null hypothesis. This problem considers a LRT for exponential distributions and explores the decision-making criteria under different alternative hypotheses.

Question. Consider the exponential model

$$f(y \mid \theta) = \theta\, e^{-y\theta}, \quad y > 0,$$

where $\theta > 0$ is an unknown parameter. The likelihood ratio statistic for testing $H_0 : \theta = \theta_0$ versus $H_1 : \theta \neq \theta_0$ is given by

$$\mathcal{L}(y_1, \ldots, y_n) = l(\widehat{\theta})/l(\theta_0),$$

where $\widehat{\theta}$ is the maximum likelihood estimator and $l(\theta)$ is the likelihood function.

(i) Show that $\mathcal{L}(\boldsymbol{y}) = (\theta_0\overline{y})^{-n} \exp\{n(\theta_0\overline{y} - 1)\}$, where $\boldsymbol{y} = \{y_i\}$ is the sample of size n.

(ii) If θ_0 is the correct value, show that $2\log\mathcal{L}(\boldsymbol{y}) \to \chi_1^2$, a chi-squared random variable with one degree of freedom.

(iii) Find the critical region for testing H_0 at the α level of significance.

(iv) If θ^* is the true value and $\theta^* \neq \theta_0$, what happens to $\log \mathcal{L}(y)$?

(v) What would the asymptotic version of the test be?

Q3.1.4 – Likelihood-ratio test for the exponential family

Introduction. From Question 3.1.3, it is known that, if the null hypothesis is true, twice the log-likelihood ratio converges to a chi-squared random variable; this fact motivates the derivation of a likelihood-ratio test for a chosen parametric value for the Poisson distribution. If the Poisson is replaced by a normal distribution, then the log-likelihood ratio for a similar test on the mean is exactly a chi-squared random variable under the null hypothesis.

Question. Consider the family of density functions belonging to the exponential class

$$f(y \mid \theta) = c(y)\exp\{\theta y - b(\theta)\}$$

with suitable functions $b(\cdot)$ and $c(\cdot)$, where $b(\cdot)$ is a convex function. The aim is to test the hypotheses $H_0 : \theta = \theta_0$ vs $H_1 : \theta \neq \theta_0$ using a likelihood-ratio test from data $y = (y_1, \ldots, y_n)'$.

(i) Find $c(y)$ and $b(\theta)$ for the Poisson model with mean $\phi = e^{\theta}$.

(ii) Let $f(y \mid H_1) = \sup_\theta f(y \mid \theta)$ and $f(y \mid H_0) = f(y \mid \theta_0)$. Show that the log-likelihood ratio, $\log \mathcal{L} = \log\{f(y \mid H_1)/f(y \mid H_0)\}$, is given by

$$\log \mathcal{L} = n\bar{y}\log\bar{y} - n\bar{y} - n\bar{y}\log\phi_0 + n\phi_0,$$

where $\bar{y}$ is the sample mean and $\phi_0 = e^{\theta_0}$.

(iii) If H_0 is correct, find the approximate normal distribution for $\bar{y}$.

(iv) Show that, under the null hypothesis, $2\log\mathcal{L}$ converges to a χ^2_1-distributed random variable as $n \to \infty$.

(v) Find the critical region for the approximate test with significance level α.

Q3.1.5 – Attaining the Cramér–Rao lower bound

Introduction. The Cramér–Rao lower bound is a well-known result that links the variance of an unbiased estimator with the Fisher information. The inequality that makes up the lower bound becomes an equality for the exponential family. This question establishes how the Cramér–Rao lower bound is attained.

Question. The density function for a random variable Y has exponential family form

$$f(y \mid \theta) = c(y)\exp\{y\theta - b(\theta)\}.$$

(i) Prove that $b(\theta)$ is a convex function.

(ii) Find the maximum likelihood estimator of θ (i.e., $\widehat{\theta}$) based on a sample of size n.

(iii) Find $\mathrm{E}\{b'(\widehat{\theta})\}$ and $\mathrm{Var}\{b'(\widehat{\theta})\}$, where b' is the derivative of b with respect to θ.

(iv) Suppose $T(y_1, \ldots, y_n)$ is an estimator of θ for which

$$\mathrm{E}\{T(Y_1, \ldots, Y_n)\} = b'(\theta).$$

Show that

$$\mathrm{E}\{T(Y_1, \ldots, Y_n)\,\overline{Y}\} = b''(\theta)/n + \{b'(\theta)\}^2.$$

(v) Using part (iv), show that

$$\mathrm{Var}\{T(Y_1, \ldots, Y_n)\} \geq b''(\theta)/n.$$

Explain the significance of this result.

Q3.1.6 – Parameter estimation using a score-based divergence

Introduction. The exponential family is a well-studied class of models, yet there are some models in the exponential family for which maximum likelihood estimation is infeasible due to an intractable normalizing constant. An alternative estimation strategy uses the score function, which does not depend on the normalizing constant. The idea is to minimize the Fisher information distance between the correct model and the exponential family.

Question. Consider the exponential family model

$$f_t(y \mid \theta) = \frac{\exp\{\theta\, t(y)\}}{Z(\theta)},$$

where the normalizing constant $Z(\theta)$ is not available. Define $h(\theta, t, y) = f_t'(y \mid \theta)/f_t(y \mid \theta)$, where differentiation is with respect to y. Similarly, define $h(\theta, s, y) = f_s'(y \mid \theta)/f_s(y \mid \theta)$, where $f_s(y \mid \theta) \propto \exp\{\theta\, s(y)\}$ for some function s.

(i) If interest is in estimating θ, explain why it is appropriate to minimize

$$D(\theta) = \int f_t(y \mid \theta^*)\, \{h(\theta, t, y) - h(\theta^*, t, y)\}\, \{h(\theta, s, y) - h(\theta^*, s, y)\}\, dy,$$

where θ^* is the true parametric value and $\int f_t(y \mid \theta^*)\, s'(y)\, t'(y)\, dy > 0$.

(ii) To estimate θ, provide a Monte Carlo estimator for $D(\theta)$ based on an independent and identically distributed sample $\{y_i\}$ taken from $f_t(y \mid \theta^*)$. List any regularity conditions required.

(iii) Show that an estimator for θ could be

$$\widehat{\theta}_s = \frac{-\sum_{i=1}^n s''(y_i)}{\sum_{i=1}^n s'(y_i)\, t'(y_i)}.$$

(iv) When $s = t$, what is $D(\theta)$ known as?

(v) If $t(y) = -\frac{1}{2}y^4$, consider several choices of s. Is it safe to assume that $s = t$ is the "best" choice?

Q3.1.7 – Bayesian predictives from the exponential family

Introduction. The previous exercises explored the exponential family from the classical perspective. Now, properties of the family are investigated for the Bayesian framework. Such properties usually arise for the predictive density when a conjugate prior is chosen.

Question. Consider the exponential family model

$$f(y \mid \theta) = c(y)\exp\{\theta y - b(\theta)\}$$

with prior

$$\pi(\theta) \propto \exp\{\theta y_0 - n_0 b(\theta)\}$$

for some choice of (n_0, y_0).

(i) Show that the prior expectation for $b'(\theta)$ is y_0/n_0. What regularity conditions are required?

(ii) Compute the prior expectation of y:

$$\mathrm{E}(Y) = \int\int y f(y \mid \theta)\,\pi(\theta)\,d\theta\,dy.$$

(iii) What is the posterior for θ based on the sample $\boldsymbol{y} = (y_1, \ldots, y_n)'$?

(iv) Write the predictive density $p(y \mid \boldsymbol{y})$ in terms of φ, where

$$\varphi(n_0, y_0) = \int \exp\{y_0\theta - n_0 b(\theta)\}\,d\theta.$$

(v) Show that

$$\mathrm{E}(Y \mid \boldsymbol{y}) = \frac{y_0 + n\bar{y}}{n_0 + n}.$$

S3.1 Solutions – Exponential Family

S3.1.1 – MLE for the exponential family

(i) Consider $b'(\theta)$, which suggests using the derivative of $f_Y(y \mid \theta)$ with respect to θ. The key starting point is that $\int f_Y(y \mid \theta)dy = 1$, so it follows that

$$\frac{\partial}{\partial\theta}\int f_Y(y \mid \theta)dy = 0.$$

Allowing the switch between differentiation and integration (Apostol, 1991),

$$\begin{aligned}\int c(y)\frac{\partial}{\partial\theta}\exp\{y\theta - b(\theta)\}dy &= \int c(y)\exp\{y\theta - b(\theta)\}\{y - b'(\theta)\}dy\\ &= \int y\cdot c(y)\exp\{y\theta - b(\theta)\}dy - b'(\theta)\\ &= \int y\, f_Y(y \mid \theta)dy - b'(\theta)\\ &= \mathrm{E}(Y) - b'(\theta) = 0,\end{aligned}$$

which implies that $\mathrm{E}(Y) = b'(\theta)$.

(ii) Differentiate $b'(\theta) = E(Y)$ once more:

$$\begin{aligned}b''(\theta) &= \frac{\partial}{\partial\theta}\int y\,c(y)\exp\{y\theta - b(\theta)\}dy\\ &= \int y\,c(y)\exp\{y\theta - b(\theta)\}\{y - b'(\theta)\}dy\\ &= \mathrm{E}(Y^2) - b'(\theta)E(Y)\\ &= \mathrm{E}(Y^2) - \{\mathrm{E}(Y)\}^2\\ &= \mathrm{Var}(Y),\end{aligned}$$

which is non-negative. Thus, $b(\theta)$ is a convex function.

(iii) This was proven directly in part (ii).

(iv) To obtain θ^*, differentiate the given integral with respect to θ:

$$\begin{aligned}\frac{\partial}{\partial\theta}\int f^*(y)\log f(y \mid \theta)dy &= \int \frac{\partial}{\partial\theta}f^*(y)\log f(y \mid \theta)dy\\ &= \int f^*(y)\frac{\partial}{\partial\theta}\left\{\log c(y) + y\theta - b(\theta)\right\}dy\\ &= \int f^*(y)\left\{y - b'(\theta)\right\}dy \overset{\text{set}}{=} 0.\end{aligned}$$

Thus, θ^* satisfies $\mathrm{E}\,(Y) = b'(\theta^*)$. By the law of large numbers, $b'(\widehat{\theta}) = \overline{y}$ converges to $\mathrm{E}(Y)$. Thus,

$$b'(\widehat{\theta}) \to \mathrm{E}(Y) = b'(\theta^*).$$

Suitable continuity conditions on b imply that $\widehat{\theta} \to \theta^*$.

(v) The interpretation is that the MLE converges to the parameter that minimizes the Kullback–Leibler divergence between $f^*(\cdot)$ and the exponential family model of density functions $f(\cdot \mid \theta)$.

Further Reading

Historical Background. Work on the exponential family began in the mid-1930s (Darmois, 1935; Koopman, 1936; Pitman, 1936). See Casella and Berger (2021) for a brief historical perspective on the family.

Points of Interest.

1. The main takeaway from this exercise is the robustness of the MLE. If the model is misspecified, not all hope is lost. If the specified distribution is part of the exponential family, the MLE will still converge to the best possible parameter that minimizes the Kullback–Leibler divergence from the true density. That is, maximizing

$$\int f^*(y) \log f(y \mid \theta)\, dy$$

corresponds to minimizing

$$\int f^*(y) \log \frac{f^*(y)}{f(y \mid \theta)}\, dy.$$

This property holds even when the dimension of θ is greater than 1.

2. A convenient property of the exponential family is closure under convolutions. Many well-known processes (including Brownian motion, the gamma process, and the Poisson process) rely on the convolution closure property for the existence of the process; this effectively ensures that a law of total probability holds between three points in time. To be convolution closed, the generating function needs to satisfy

$$\mathrm{E}(e^{\phi Y}) = \exp\{\theta\, h(\phi) + g(\phi)\}$$

for some functions h and g. In this case, the variance must be a quadratic function of the mean. With such a generating function, it is also easy to see the connected property of infinite divisibility: to get $y = \sum_{i=1}^n y_i$, the common generating function for y_i is $(\theta/n)\, h(\phi) + (g/n)(\phi)$.

3. Another important feature of the exponential family is the ease in identifying sufficient statistics based on an independent and identically distributed sample $\{y_i\}$. Denote the sufficient statistic as $S = \sum_{i=1}^n t(y_i)$. Note that

$$f(\boldsymbol{y} \mid S = s) \propto 1\left(\sum_{i=1}^{n} t(y_i) = s\right) \prod_{i=1}^{n} c(y_i)$$

does not depend on θ, which satisfies the formal definition of a sufficient statistic (Casella and Berger, 2021).

Demonstration. The aim of this demonstration is to show that the MLE converges to θ^*. First, take a sample of size 10 from the exponential distribution with rate parameter $\theta^* = 2$.

Then, compute the MLE for the rate parameter $\widehat{\theta} = 1/\overline{y}$, where $\overline{y}$ is the sample mean. Repeat this process 1000 times, where each new sample has 10 more observations than the previous sample. As depicted in Fig. 3.1, $\widehat{\theta}$ converges to θ^* as the sample size increases.

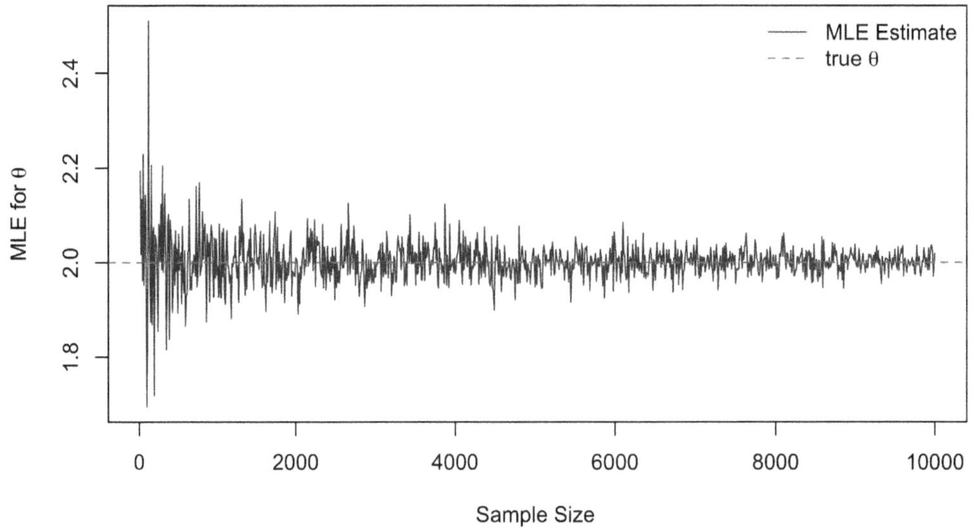

FIGURE 3.1
The MLE for the rate parameter of the exponential distribution converges to the true value as the sample size increases.

S3.1.2 – Properties of the MLE for the exponential family

(i) First, find the log-likelihood: $L(\theta, y) = \log c(y) + y\theta - b(\theta)$. Then, obtain the first and second derivatives:

$$\frac{\partial L}{\partial \theta} = y - b'(\theta) \quad \text{and} \quad \frac{\partial^2 L}{\partial \theta^2} = -b''(\theta).$$

Thus, $I(\theta) = b''(\theta)$ and $\mathrm{E}\left[\{\partial L(\theta, Y)/\partial\theta\}^2\right] = \mathrm{E}\{Y - b'(\theta)\}^2 = \mathrm{Var}(Y)$ because $\mathrm{E}(Y) = b'(\theta)$. From Question 3.1.1, it is known that $\mathrm{Var}(Y) = b''(\theta)$, so $I(\theta) = \mathrm{E}\left[\{\partial L(\theta, Y)/\partial\theta\}^2\right]$.

(ii) Because $b''(\theta) = \mathrm{Var}(Y) \geq 0$, $b(\theta)$ is a convex function. See Question 3.1.1 for a proof that $b''(\theta) = \mathrm{Var}(Y) \geq 0$.

(iii) The log-likelihood $L(\theta)$ is a strictly concave function:

$$L'(\theta) = n\overline{y} - nb'(\theta) \quad \text{and} \quad L''(\theta) = -nb''(\theta) < 0.$$

Thus, if $\widehat{\theta} \leq z$ it must be that $L'(z) \leq 0$ because $\widehat{\theta}$ maximizes the log-likelihood. Finally, $L'(z) = n\overline{y} - nb'(z) \leq 0$ implies $\overline{y} \leq b'(z)$.

(iv) First, find $\mathrm{E}(\overline{Y})$ and $\mathrm{Var}(\overline{Y})$ to get an asymptotic normal distribution for $\overline{y}$ based on the central limit theorem. The former is $b'(\theta^*)$ and the latter is $b''(\theta^*)/n$. Therefore,

$$\mathrm{P}(\overline{Y} \leq b'(z)) - \Phi\left(\frac{b'(z) - b'(\theta^*)}{b''(\theta^*)/n}\right) \to 0,$$

where $\Phi(\cdot)$ denotes the standard normal cumulative distribution function. Note that $b''(\theta^*)$ is positive. As $n \to \infty$, this probability will converge to 0 if $b'(z) < b'(\theta^*)$ or will converge to 1 if $b'(z) > b'(\theta^*)$. Because $b'(\cdot)$ is an increasing function, the probability will converge to 0 if $z < \theta^*$ or will converge to 1 if $z > \theta^*$.

(v) The result in part (iv) implies that the distribution of $\widehat{\theta}$ converges to a distribution function with a jump of size 1 at θ^*. This implies that $\widehat{\theta}$ converges to θ^* in probability.

Further Reading

Historical Background. Fisher information measures the amount of information that observable random variables carry about the parameters. In the case of the exponential family with a canonical parameter, the Fisher information is the variance. More generally, the Fisher information is defined as the variance of the score function (Gelman et al., 1995). The Fisher information features in the Cramér–Rao lower bound for the variance of an unbiased estimator: $\mathrm{Var}\left(\widehat{\theta}\right) \geq \{n\,I(\theta)\}^{-1}$. See Cramer (1946) and Rao (1945) for further details on this lower bound and its connection with Fisher information.

Points of Interest.

1. There is a reparameterization of the canonical exponential family for which the variance of the estimator attains the Cramér–Rao lower bound. To this end, define $\phi = b'(\theta)$. Then,
$$f(y \mid \phi) = c(y)\,\exp\{ya(\phi) - b(a(\phi))\},$$
where $a(\phi) = (b')^{-1}(\theta)$. Then, the Fisher information is
$$I(\phi) = \{a'(\phi)\}^2\, b''(a(\phi)) = 1/b''(\theta) = 1/\mathrm{Var}(Y)$$
because $b'(a(\phi)) = \phi$. Thus, $a'(\phi)\, b''(a(\phi)) = 1$. Finally, because $\widehat{\phi} = \overline{Y}$ and $\mathrm{Var}(Y) = b''(\theta)/n = \{n\,I(\phi)\}^{-1}$, the estimator attains the Cramér–Rao lower bound.

2. To see how the alternative parameterization works, consider the Bernoulli model
$$f(y \mid \phi) = \phi^y(1-\phi)^{1-y}, \quad 0 < \phi < 1, \quad y \in \{0,1\}.$$
In terms of the canonical parameter, this is written as
$$f(y \mid \phi) = \exp\{y\log\{\phi/(1-\phi)\} + \log(1-\phi),$$

so $\theta = \log\{\phi/(1-\phi)\}$ and $b(\theta) = \log(1+e^\theta)$. Therefore, $E(Y) = \phi = b'(\theta)$ and

$$\mathrm{Var}(Y) = \phi(1-\phi) = \frac{e^\theta}{(1+e^\theta)^2} = b''(\theta).$$

Further, $I(\phi) = \{\phi(1-\phi)\}^{-1}$, which is equivalent to $1/b''(\theta)$. One must decide whether to estimate ϕ or θ. It is easy to estimate $\widehat{\phi}$ because $\widehat{\phi} = \overline{y}$ for which the properties are well-known.

3. To see an example in which the original parameterization is more difficult to work with, consider the exponential density with canonical parameter

$$f(y \mid \theta) = \exp(-y\theta + \log\theta).$$

Then, $\widehat{\theta} = 1/\overline{y}$, which has a distribution with moments that are not easy to obtain.

S3.1.3 – Likelihood-ratio test for the exponential density

(i) First, find the MLE by maximizing

$$l(\theta) = \prod_{i=1}^{n} \theta e^{-y_i\theta} = \theta^n e^{-\theta\sum_{i=1}^n y_i}.$$

Because log is a monotone function, maximizing the likelihood is equivalent to maximizing the log-likelihood, $\log l(\theta) = n\log\theta - \theta\sum_{i=1}^n y_i$. Thus, the MLE is $\widehat{\theta} = 1/\overline{y}$ and the likelihood-ratio statistic is

$$\mathcal{L}(y) = l(\widehat{\theta})/l(\theta_0) = (\theta_0\overline{y})^{-n}\exp\{n(\theta_0\overline{y}-1)\}.$$

(ii) Using the central limit theorem, $\overline{y}$ is approximately normal with mean $1/\theta_0$ and variance $1/(n\theta_0^2)$ as $n \to \infty$. Thus, $\theta_0\overline{y} \sim \mathrm{N}(1, 1/n)$; this distribution can also be constructed with $1 + z/\sqrt{n}$, where z is a standard normal random variable. Therefore,

$$\begin{aligned}\log\mathcal{L}(y) &= -n\log(1+z/\sqrt{n}) + n(1+z/\sqrt{n}) - n \\ &= -n\left(z/\sqrt{n} - z^2/(2n) + o(z^2/n)\right) + n(1+z/\sqrt{n}) - n \\ &= \tfrac{1}{2}z^2 - n\,o(z^2/n),\end{aligned}$$

which converges to $\frac{1}{2}z^2$ as $n \to \infty$. The second equality is obtained using the expansion $\log(1+c) = c - \frac{1}{2}c^2 + o(c^2)$. Finally, $2\log\mathcal{L}(y) \to z^2$, where z^2 is a χ_1^2-random variable.

(iii) The function $\mathcal{L}(y)$ is convex in $\overline{y}$ and minimized at $1/\theta_0$. Thus, H_0 cannot be rejected for the two-sided alternative $\mathcal{L}(y) \geq k$ unless there is symmetry about $1/\theta_0$. The idea now is that H_0 would be rejected if $\overline{y}$ were too small or too large compared with $1/\theta_0$. The critical region would be of the form $(0, c_1) \cup (c_2, \infty)$, where

$$\frac{1}{\Gamma(n\theta_0)}\int_{nc_1}^{nc_2} z^{n\theta_0 - 1} e^{-z}\,dz = 1-\alpha.$$

The choice of (c_1, c_2) is limitless, but the pair that is typically chosen solves the above equation such that $c_2 - c_1$ is minimized. Using Lagrange multipliers, the appropriate (c_1, c_2) are those that satisfy $f(nc_1) = f(nc_2)$, where f is the density function of $n\overline{y}$.

(iv) Repeat the asymptotic analysis done in part (ii), though now with $\bar{y} = 1/\theta^* + z/(\theta^*\sqrt{n})$. The new dominant term is

$$n\left\{\theta_0/\theta^* - 1 - \log(\theta_0/\theta^*)\right\}.$$

The inequality $\log y < y - 1$ for all $y \neq 1$ implies that $\log \mathcal{L}(y) \to \infty$ with speed n as $n \to \infty$.

(v) The asymptotic version of the test would use the fact that, under the null hypothesis, $-2\log \mathcal{L}(y)$ is a χ_1^2-distributed random variable.

Further Reading

Historical Background. The likelihood-ratio test is commonly used to test hypothesized parametric values associated with some model using the observed data. Much of the original work was done by Neyman and Pearson (1933). In fact, the Neyman–Pearson lemma shows the existence and uniqueness of the likelihood ratio as a uniformly most powerful test. However, sometimes determining critical values for relevant hypothesis tests can prove difficult.

Points of Interest.

1. The likelihood-ratio test is just one classical hypothesis test; alternatives include the score test and the Wald test. The score test uses the statistic

$$S = L'(\theta_0)^2/I(\theta_0),$$

where $L(\theta)$ is the log-likelihood function and $I(\theta)$ is the usual Fisher information (Rao, 2005). The Wald test uses the statistic

$$W = (\widehat{\theta} - \theta_0)^2/\mathrm{Var}(\widehat{\theta})$$

(Lafontaine and White, 1986). These tests are asymptotically equivalent to the LRT. However for smaller samples, it has been found that the LRT is the preferred choice.

2. A well-known asymptotic result for the LRT is as follows: when testing $H_0 : \theta \in \Theta_0$ vs $H_1 : \theta \in \Theta_1 = \Theta - \Theta_0$, the statistic $\Lambda = -2\{l(\widehat{\theta}_0) - l(\widehat{\theta}_1)\}$ is asymptotically a χ_d^2-distributed random variable under the null hypothesis, where d is the difference in dimension between Θ_1 and Θ_0; this result is known as the Wilks' theorem (Wilks, 1938).

Demonstration. To illustrate the Wilks' theorem, 10,000 experiments are taken with $n = 10$ samples from the standard exponential distribution. Then, $2\log \mathcal{L}$ is computed with the $\mathcal{L}$ in part (i). A histogram of the samples are presented in Fig. 3.2 with the χ_1^2 density function overlaid. In this case, the Wilks' theorem approximation is quite good.

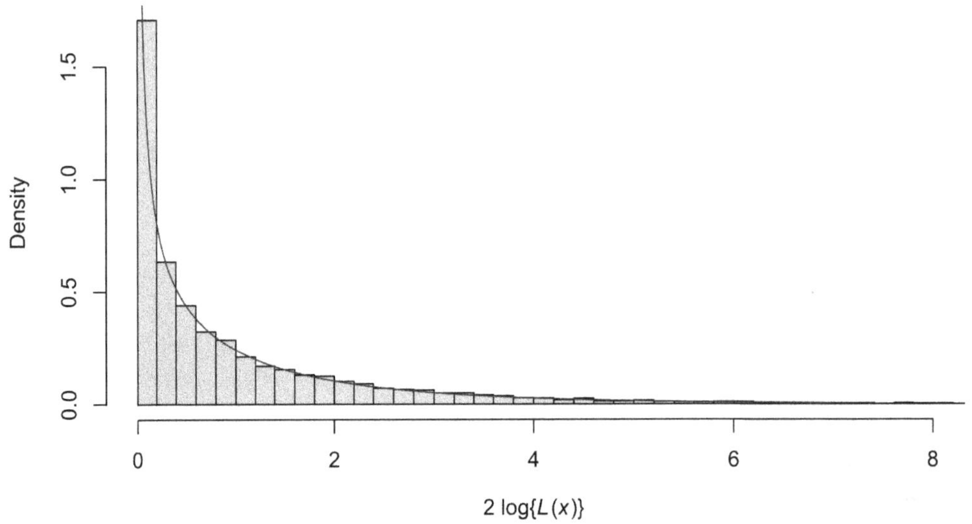

FIGURE 3.2
A histogram of 10,000 samples of $2\log\mathcal{L}$ with sample size 10 from a standard exponential distribution. The χ^2_1 density is overlaid.

S3.1.4 – Likelihood-ratio test for the exponential family

(i) By the definition of the Poisson distribution,

$$f(y_i \mid \phi) = \phi^{y_i}\exp\{-\phi\}/y_i! \\ = \frac{1}{y_i!}\exp\left\{\theta y_i - e^{\theta}\right\},$$

where $c(y_i) = 1/y_i!$ and $b(\theta) = \exp\{\theta\}$.

(ii) Denote the estimator under H_1 as $\widehat{\theta}$ and take the logarithm of

$$\frac{f(y \mid H_1)}{f(y \mid H_0)} = \frac{\exp(n\overline{y}\widehat{\theta} - ne^{\widehat{\theta}})}{\exp(n\overline{y}\theta_0 - ne^{\theta_0})}$$

to obtain

$$\log\mathcal{L} = n\overline{y}\log(\widehat{\phi}) - n\widehat{\phi} - n\overline{y}\log(\phi_0) + n\phi_0,$$

where $\widehat{\theta} = \log(\widehat{\phi})$ and $\theta_0 = \log(\phi_0)$. Finally, solve for $\widehat{\phi}$ by maximizing the log-likelihood; the solution is $\widehat{\phi} = \overline{y}$.

(iii) Because $\overline{y}$ is the sample mean, the central limit theorem applies. If the null hypothesis is true, $\overline{y}$ will be approximately normal with mean ϕ_0 and variance ϕ_0/n.

(iv) Consider $\overline{y} = \phi_0 + z\sqrt{\phi_0/n} + o(1/n)$, where z is a standard normal random variable; the first two terms have the same distribution as that in part (iii). Without loss of generality, neglect the $o(1/n)$ term as it will disappear as $n \to \infty$. Then,

$$\log \mathcal{L} = n\,(\phi_0 + z\sqrt{\phi_0/n})\log(1 + z/\sqrt{\phi_0\, n}) - n(\phi_0 + z\sqrt{\phi_0/n}) + n\phi_0.$$

Use the log equality $\log(1+\epsilon) = \epsilon - \frac{1}{2}\epsilon^2 + o(\epsilon^2)$ to get

$$\log \mathcal{L} = \tfrac{1}{2}z^2 - \tfrac{1}{2}z^3/\sqrt{\phi_0\, n}.$$

The result follows because z^2 is a χ^2_1-distributed random variable and the second term goes to 0 as $n \to \infty$.

(v) The null hypothesis is rejected at level α approximately if $2\log\mathcal{L} > c$, where c is such that $\mathrm{P}(\chi^2_1 > c) = \alpha$.

Further Reading

Historical Background. Wilks (1938) first proved that twice the log-likelihood ratio has a χ^2 distribution under the correctness of the hypothesis for a nested test. This is an important result and forms the basis of many asymptotic-based tests. For example, Fan et al. (2001) and Arnak and Olivier (2012) develop generalizations to this result.

Points of Interest.

1. A rigorous proof of Wilks' result is beyond the scope of this text. However, the underlying concepts are now presented. Assume $\widehat{\theta}$ is asymptotically normal and write $\widehat{\theta} = \theta_0 + z\, I^{-1}(\theta_0)/\sqrt{n}$, where z is a standard normal random variable and $I(\theta_0)$ is the usual Fisher information

$$I(\theta_0) = -\int \frac{\partial^2}{\partial\theta_0^2} \log f(y \mid \theta_0)\, f(y \mid \theta_0)\, dy.$$

Note that $I(\theta_0)$ is the asymptotic variance of the MLE. Using a standard Taylor expansion, the log-likelihood ratio becomes

$$\sum_{i=1}^{n} \log \frac{f(y_i \mid \theta_0)}{f(y_i \mid \widehat{\theta})} = \tfrac{1}{2}\,(\theta_0 - \widehat{\theta})^2 \sum_{i=1}^{n} \frac{\partial^2}{\partial\theta_2} \log f(y_i \mid \widehat{\theta}) + o(1/n),$$

because $\sum_{i=1}^{n} \partial/\partial\theta \log f(y_i \mid \widehat{\theta}) = 0$. Under usual assumptions,

$$n^{-1} \sum_{i=1}^{n} \frac{\partial^2}{\partial\theta_2} \log f(y_i \mid \widehat{\theta}) \to -I(\theta_0),$$

so the likelihood ratio converges to $-\frac{1}{2}z^2$.

2. The classic likelihood-ratio test has the form $H_0 : \theta = \theta_0$ vs $H_1 : \theta = \theta_1$ with likelihood-ratio statistic

$$\mathcal{L} = \prod_{i=1}^{n} \frac{f(y_i \mid \theta_1)}{f(y_i \mid \theta_0)}.$$

This test does not lead to $-2\log\mathcal{L}$ being asymptotically χ^2-distributed under the correctness of θ_0 because it would depend on θ_1.

S3.1.5 – Attaining the Cramér–Rao lower bound

(i) It is known that $\mathrm{Var}(Y) = b''(\theta)$ because the density function belongs to the exponential family; see Question 3.1.1 for this proof. Thus, $b''(\theta) \geq 0$ and b is a convex function.

(ii) The log-likelihood is given by

$$L(\theta) = \sum_{i=1}^{n} \log c(y_i) + \theta \sum_{i=1}^{n} y_i - n\, b(\theta).$$

Differentiate with respect to θ and set to 0 to obtain $\widehat{\theta} = (b')^{-1}(\overline{y})$. Thus, $b'(\widehat{\theta}) = \overline{y}$.

(iii) Apply the definitions of expectation and variance:

$$\mathrm{E}\{b'(\widehat{\theta})\} = \mathrm{E}(\overline{Y}) = b'(\theta), \quad \mathrm{Var}\{b'(\widehat{\theta})\} = \mathrm{Var}(\overline{Y}) = b''(\theta)/n.$$

See Question 3.1.1 for these equalities.

(iv) Differentiate

$$b'(\theta) = \int T(y_1, \ldots, y_n) \prod_{i=1}^{n} f(y_i \mid \theta)\, d\boldsymbol{y}$$

with respect to θ to get

$$b''(\theta) = \int T(y_1, \ldots, y_n) \{n\overline{y} - n\, b'(\theta)\} \prod_{i=1}^{n} f(y_i \mid \theta)\, d\boldsymbol{y}.$$

Therefore,

$$b''(\theta) = n\, \mathrm{E}\{\overline{Y}\, T(Y_1, \ldots, Y_n)\} - n\, \{b'(\theta)\}^2,$$

and the required result follows. Note that these steps were made to obtain the quantities in the equation of interest, namely to get $\overline{y}$ and $b''(\theta)$.

(v) To ease the notation, write $T_n \equiv T(y_1, \ldots, y_n)$. Note that the result from part (iv) can be written as $b''(\theta)/n = \mathrm{E}(T_n\overline{Y}) - b'(\theta)^2 = \mathrm{Cov}(T_n, \overline{Y})$ because $\mathrm{Cov}(T_n, \overline{Y}) = \mathrm{E}(T_n\overline{Y}) - \mathrm{E}(T_n)\mathrm{E}(\overline{Y})$. The correlation coefficient is bounded by 1, so

$$\mathrm{Cov}(T_n, \overline{Y}) \leq \sqrt{\mathrm{Var}(T_n)\mathrm{Var}(\overline{Y})}.$$

Rearrange the terms to obtain

$$\mathrm{Var}(T_n) \geq \{\mathrm{Cov}(T_n, \overline{Y})\}^2 / \mathrm{Var}(\overline{Y}).$$

The covariance is given by $b''(\theta)/n$, and the variance of $\overline{Y}$ is also $b''(\theta)/n$. Thus, $\mathrm{Var}(T_n) \geq b''(\theta)/n$ and the best unbiased estimator is $T_n = \overline{y}$, which has the smallest possible variance: $b''(\theta)/n$.

Further Reading

Historical Background. The inequality found in part (iv) is the Cramér–Rao lower bound for unbiased estimators (Rao, 1945; Cramer, 1946). Here, the focus was on the exponential family, but the steps for the general proof are similar. This question shows that, even if the theorem for the Cramér–Rao lower bound looks daunting, the proof is based upon one argument: the covariance between two random variables is bounded by the square root of the product of their variances.

Points of Interest.

1. In the general case, it is the Cauchy–Schwarz inequality that becomes useful for proving the Cramér–Rao lower bound. For some variable B and estimator T with $\mathrm{E}(T) = \theta$,

$$\mathrm{E}[\{T - \mathrm{E}(T)\}\, B] \leq \sqrt{\mathrm{Var}(T)\, \mathrm{E}(B^2)},$$

which leads to

$$\mathrm{Var}(T) \geq \{\mathrm{Cov}((T, B)\}^2 / \mathrm{E}(B^2).$$

Now, take

$$B = \sum_{i=1}^{n} \partial \log f(y_i \mid \theta) / \partial\theta$$

with $\mathrm{E}(B^2) = n\, I(\theta)$. It is left as a mathematical exercise to show that $\mathrm{Cov}(T, B) = 1$; use the starting point

$$\int \cdots \int T(y_1, \ldots, y_n) \prod_{i=1}^{n} f(y_i \mid \theta)\, d\boldsymbol{y} = \theta$$

and proceed by differentiating with respect to θ.

S3.1.6 – Parameter estimation using a score-based divergence

(i) Note that $h(\theta, t, y) = \theta\, t'(y)$ and $h(\theta, s, y) = \theta s'(y)$. Therefore,

$$D(\theta) = (\theta - \theta^*)^2 \int f_t(y \mid \theta^*)\, s'(y)\, t'(y)\, dy.$$

Minimizing $D(\theta)$ with respect to θ is equivalent to minimizing $(\theta - \theta^*)^2$ because $\int f_t(y \mid \theta^*)\, s'(y)\, t'(y)\, dy > 0$. A common approach to estimate θ^* with θ is to minimize $(\theta - \theta^*)^2$.

(ii) Two of the terms in $D(\theta)$ that involve θ are

$$\int f_t(y \mid \theta^*)\, \theta\, t'(y) \theta^*\, s'(y)\, dy \quad \text{and} \quad \int f_t(y \mid \theta^*)\, \theta^*\, t'(y)\, \theta\, s'(y)\, dy,$$

which are identical. Use integration by parts with $u = s'(y)$ and $v' = f_t(y \mid \theta^*)\theta^* t'(y)$ to show that both integrals are equal to $-\theta \int f_t(y \mid \theta^*)\, s''(y)\, dy$; this assumes that

$s'(y)f_t(y \mid \theta^*)$ vanishes at the end points. Note that $v = f_t(y \mid \theta^*)$ and the Monte Carlo estimator for $D(\theta)$ is

$$\theta^2 n^{-1} \sum_{i=1}^{n} t'(y_i)\, s'(y_i) + 2\theta n^{-1} \sum_{i=1}^{n} s''(y_i).$$

(iii) This follows by differentiating the expression in part (ii), then setting it to 0.

(iv) If $s = t$, then

$$D(\theta) = \int f_t(y \mid \theta^*) \left\{ \frac{f_t'(y \mid \theta)}{f_t(y \mid \theta)} - \frac{f_t'(y \mid \theta^*)}{f_t(y \mid \theta^*)} \right\}^2 dy$$

is known as the Fisher information distance.

(v) If $t(y) = -\frac{1}{2}y^4$, one possible $s \neq t$ is given by $s(y) = -\frac{1}{2}y^2$. Then,

$$\widehat{\theta} = \frac{n/2}{\sum_{i=1}^{n} y_i^4}$$

is the MLE. Therefore, it is not safe to assume that $s = t$ is the "best" choice.

Further Reading

Historical Background. The exponential family is widely used in statistical problems (Efron, 2022). This question reveals a large class of estimators for the exponential family that may be used when the normalizing constant is intractable. The class of estimators extends beyond those offered by the Fisher distance (Hyvarinen, 2005). In fact, the class is large enough that the optimal choice in s is not always $s = t$.

Points of Interest.

1. The score function is a very useful tool when working with intractable likelihoods because intractable normalizing constants disappear. Note that maximizing the score function is connected to minimizing Kullback–Leibler divergence. An alternative to $D(\theta)$ in part (i) is based on the Kullback–Leibler divergence:

$$D_{KL}(\theta) = -\theta \int f_t(y \mid \theta^*)\, dy + \log Z(\theta),$$

which is difficult to maximize directly due to the presence of $Z(\theta)$.

2. A more general Fisher estimator is based on minimizing

$$\int s(y)\, f(y \mid \theta^*) \left\{ \frac{f'(y \mid \theta^*)}{f(y \mid \theta^*)} - \frac{f'(y \mid \theta)}{f(y \mid \theta)} \right\}^2 dy$$

for some function $s(y) \geq 0$. The objective function to be minimized becomes

$$\int f(y \mid \theta^*) \left[\left\{ s(y) \frac{f'(y \mid \theta)}{f(y \mid \theta)} \right\}^2 + 2 \left\{ s(y) \frac{f'(y \mid \theta)}{f(y \mid \theta)} \right\}' \right] dy.$$

The Fisher estimator minimizes

$$\sum_{i=1}^{n}\left[\left\{s(y_i)\frac{f'(y_i \mid \theta)}{f(y_i \mid \theta)}\right\}^2 + 2\left\{s(y_i)\frac{f'(y_i \mid \theta)}{f(y_i \mid \theta)}\right\}'\right].$$

Note that, if $f'(y \mid \theta)/f(y \mid \theta) = \theta t'(y)$, as in this question, then

$$\hat{\theta} = -\frac{\sum_{i=1}^{n}(s(y_i)t'(y_i))'}{\sum_{i=1}^{n}(t'(y_i))^2 s(y_i)}.$$

S3.1.7 – Bayesian predictives from the exponential family

(i) Write the prior as $\pi(\theta) = c\,\exp\{y_0\theta - n_0 b(\theta)\}$, so

$$\mathrm{E}\{b'(\theta)\} = c\int b'(\theta)\exp\{y_0\theta - n_0 b(\theta)\}\,d\theta.$$

Using integration by parts with $u = \exp(y_0\theta)$ and $v' = b'(\theta)\exp\{-n_0 b(\theta)\}$, the integral becomes

$$\frac{y_0}{n_0}c\int \exp\{y_0\theta - n_0 b(\theta)\}\,d\theta = \frac{y_0}{n_0}$$

with the regularity condition that $\pi(\theta)$ disappears at the boundaries (i.e., $\pm\infty$).

(ii) Note that $\mathrm{E}(Y \mid \theta) = b'(\theta)$; to see this, review S3.1.1(i). Therefore, from part (i), $\mathrm{E}(Y) = \int\int y f(y \mid \theta)\,\pi(\theta)\,d\theta\,dy = \mathrm{E}\{b'(\theta)\} = y_0/n_0$.

(iii) The posterior is given by

$$\begin{aligned}\pi(\theta \mid \boldsymbol{y}) &\propto \prod_{i=1}^{n} f(y_i \mid \theta)\pi(\theta)\\ &= \frac{\exp\{(y_0 + n\bar{y})\theta - (n_0 + n)b(\theta)\}}{\int \exp\{(y_0 + n\bar{y})\theta - (n_0 + n)b(\theta)\}d\theta},\end{aligned}$$

where the denominator is written as $\varphi(n_0 + n, y_0 + n\bar{y})$.

(iv) The predictive density is

$$\begin{aligned}p(y \mid \boldsymbol{y}) &= \int f(y \mid \theta)\pi(\theta \mid \boldsymbol{y})d\theta\\ &= c(y)\frac{\varphi(n_0 + n + 1, y_0 + n\bar{y} + y)}{\varphi(n_0 + n, y_0 + n\bar{y})}.\end{aligned}$$

(v) From part (ii) and the conjugacy of prior to posterior, the mean of the predictive is

$$\mathrm{E}(Y \mid \boldsymbol{y}) = \frac{y_0 + n\bar{y}}{n_0 + n}.$$

Further Reading

Historical Background. As seen in part (v), the specification of the prior has a natural interpretation: n_0 represents a prior sample size that yielded a sample sum of y_0. There is an interesting characterization of the conjugate prior that was established by Diaconis and Ylvisaker (1979). If $\mathrm{E}(Y \mid y_1) = a + by_1$, then the prior is the conjugate prior with a particular (n_0, y_0) depending on (a, b).

Points of Interest.

1. In the Bayesian framework, the natural conjugate prior for the canonical exponential family is
$$\pi(\theta) \propto \exp\{\theta\, t_0 - n_0\, b(\theta)\}$$
for some t_0 and $n_0 > 0$. Then, the conjugate posterior is given by
$$\pi(\theta \mid \boldsymbol{y}) \propto \exp\left\{\theta\left(t_0 + \sum_{i=1}^{n} t(y_i)\right) - (n_0 + n)\, b(\theta)\right\},$$
and the predictive mean is
$$\mathrm{E}(Y \mid \boldsymbol{y}) = \int y\, p(y \mid \boldsymbol{y})\, dy = \int b'(\theta)\, \pi(\theta \mid \boldsymbol{y})\, d\theta,$$
which simplifies to
$$\mathrm{E}(Y \mid \boldsymbol{y}) = \frac{t_0 + \sum_{i=1}^{n} t(y_i)}{n_0 + n}.$$
Note that $\mathrm{E}(Y \mid \boldsymbol{y})$ is a weighted convex combination of the mean of the prior predictive distribution, t_0/n_0, and the data mean, $\sum_{i=1}^{n} t(y_i)/n$. The posterior representation is so elegant that a number of characterizations exists for the exponential family and the conjugate prior that are based on features of the posterior and predictive distributions (Diaconis and Ylvisaker, 1979; MacEachern, 1993).

2. When the model is misspecified, θ^* remains the target parameter, which is relevant when performing Bayesian inference. It is thought that specifying a prior for the parameter θ when the model is misspecified can be problematic. Indeed, if the prior is represented by $\pi(\theta)$, then the interpretation for $\Pi(A) = \int_A \pi(\theta)\, d\theta$ is not clear. What object is believed to lie in the set A with probability $\Pi(A)$ according to the prior? The answer is more clear when thinking about the object of interest as θ^*, a well-defined and targetable parameter. For more on misspecified Bayesian models, see Fudenberg et al. (2017).

4

Hypothesis Testing & Decision Theory

It can be argued that all statistics problems should result in an informed decision. For example, researchers in finance, medicine, and public policy use data to decide which actions to take. In fact, to many practitioners, decision making is the end goal – not statistical inference. Hypothesis testing and decision theory are the statistical fields that concern the use of data to make informed decisions. This chapter partitions the study of these fields into two sections: classical and Bayesian. The reason for this partition is that the decision making process under these two paradigms is conceptually quite different.

Classical decision making is concerned with determining whether the observed outcome is compatible with a hypothesis. To illustrate, consider the hypothesis that a forthcoming random outcome is standard normal. The hypothesis may seem incorrect if the outcome lays in either tail of the hypothesized density; this idea motivates the construction of a rejection region, which is the support of the outcome that leads to the rejection of the hypothesis. To determine the rejection region, one must specify the level of significance (α), which is the probability of incorrectly rejecting the null hypothesis (i.e., the probability of Type I error). In general, a test statistic is computed based on the observed sample, which is then compared to the critical value(s) defined by the rejection region to determine whether to reject a hypothesis. Thus, in classical hypothesis testing, decision making is concerned with the likelihood of the observations given the hypothesis and level of significance.

For many classical tests, normal approximations are required for the distribution of the test statistic under the null hypothesis. The normal approximation can be used if (i) the test statistic can be represented as a sum of independent and identically distributed random variables, (ii) the central limit theorem applies, and (iii) the sample size is sufficiently large. Question 4.1.1 uses the normal approximation to test the median of an arbitrary distribution function.

There are many classical hypothesis tests that can aid in decision making. Question 4.1.2 considers the t- and F-tests for the linear model and establishes a comparison between the tests. Alternatives to the t- and F-tests often incorporate likelihood ratios, which quantify the difference between likelihoods resulting from competing hypotheses. Question 4.1.3 relates the square root of the likelihood ratio to a well-known distance and investigates the asymptotic properties of the ratio. Question 4.1.4 shows that the LRT for competing simple hypotheses is uniformly most powerful for a given level of significance; Question 4.1.5 expands on this by determining when uniformly most powerful tests exist under the presence of a composite alternative hypothesis. Question 4.1.6 develops a test for the variance of the normal model; the approach in this exercise can also be used for testing hypotheses for a suite of parameters in non-Gaussian models.

In contrast to classical hypothesis testing, the Bayesian approach asserts a posterior probability on the hypothesis given the observations. The key object in Bayesian hypothesis testing is the Bayes factor, which is a ratio of the marginal likelihoods resulting from

DOI: 10.1201/9781003493471-4

the competing hypotheses. If the Bayes factor is too large, the null hypothesis is questionable. However, there is no formal procedure to determine how large the Bayes factor must be to reject the hypothesis; this lack of formality fits into the overall notion that Bayesian procedures work along the lines of subjectivity. Less charitably, the Bayes factor could be considered a test statistic that depends on the data and a prior distribution. If the Bayes factor is a monotonic function of a classical test statistic, then the Bayesian test via the Bayes factor and the classical test are identical; this fact is explored in Question 4.2.1.

During the early development of Bayesian statistics, there was substantial debate over whether the prior distributions for competing models could be improper (i.e., priors that do not impose strong beliefs). Today, the consensus is that the priors must be proper and hence carry a degree of subjective knowledge. Question 4.2.2 explores the impact that priors for a normal linear model have on decision making. The following two questions continue with Bayesian testing for normal models. Question 4.2.3 investigates the properties of the Bayes factor for normal models, and Question 4.2.4 concerns Bayesian variable selection and the invariance property for the normal linear model.

Q4.1 Questions – Classical Framework

Q4.1.1 – Hypothesis testing for a distribution median

Introduction. In hypothesis testing, there is often interest in the asymptotic distribution of a test statistic. This question tests the median of an arbitrary distribution function using a normal approximation to the test statistic. Typically, the central limit theorem (CLT) is used to obtain the asymptotic distribution, yet the reasoning behind its use is often poorly understood. This question motivates the use of the CLT with a Taylor expansion of the Laplace transform of the test statistic.

Question. Suppose $x_1, \ldots, x_n$ are independent and identically distributed from some continuous distribution function $F(x)$, and the aim is to test

$$H_0 : F(0) = \tfrac{1}{2} \quad \text{vs} \quad H_1 : F(0) \neq \tfrac{1}{2}.$$

That is, the aim is to test whether the median of F is 0.

(i) If H_0 is true, what is the distribution of

$$T = \sum_{i=1}^{n} 1(x_i \leq 0)?$$

Here, $1(x_i \leq 0)$ is an indicator function that is 1 if $x_i \leq 0$ and 0 otherwise.

(ii) Describe how to test H_0 using a normal approximation to T/n.

(iii) Find the Laplace transform $\mathrm{E}\left(e^{\theta X}\right)$ of a normal density function with mean μ and variance σ^2.

(iv) Let $y_1, \dots, y_n$ be independent variables with common mean 0 and common variance 1. Using a Laplace transform, show that

$$\frac{1}{\sqrt{n}}(y_1 + \cdots + y_n)$$

converges in distribution to a standard normal random variable.

(v) Using the result in part (iv), show that

$$\frac{T/n - F(0)}{\sqrt{F(0)\{1 - F(0)\}/n}}$$

converges in distribution to a standard normal random variable.

Q4.1.2 – t-test and F-test for the linear model

Introduction. Statistical tests are used to determine whether the observed data support a hypothesis. Some of the most common statistical hypotheses concern the mean of a population, the difference in means between two populations, or the value of a parameter in a regression model. This question explores the latter hypothesis through t- and F-tests. In particular, the distribution of the t and F statistics under the null hypothesis are compared, and conditions for their equivalence are sought.

Question. Consider the linear model $y_i = x_i \beta + \sigma \epsilon_i$, where the $\{\epsilon_i\}$ are independent standard normal random variables and the predictor variables $\{x_i\}$ satisfy $\sum_{i=1}^n x_i^2 = 1$ for $n > 1$. Interest is in testing the hypothesis $H_0 : \beta = 0$.

(i) Show that the least squares estimator of β is $\widehat{\beta} = \eta$, where $\eta = \sum_{i=1}^n x_i y_i$.

(ii) What is the distribution of η under the null hypothesis?

(iii) Let $\widehat{\sigma} = \sqrt{\text{RSS}/(n-1)}$ be an estimator of σ, where

$$\text{RSS} = \sum_{i=1}^{n} (y_i - x_i \widehat{\beta})^2$$

is the residual sum of squares. Note that RSS/σ^2 is a chi-squared random variable with $n-1$ degrees of freedom, and RSS/σ^2 is independent of η. It is well-known that a Student-t random variable with ν degrees of freedom, written T_ν, can be constructed as $T_\nu = Z/\sqrt{\chi_\nu^2/\nu}$, where Z is a standard normal random variable and χ_ν^2 is a chi-squared random variable with ν degrees of freedom that is independent of Z. What is the relevant statistic for the t-test of H_0?

(iv) The F-test of H_0 uses the test statistic

$$F = \frac{\text{RSS}_{(\beta=0)} - \text{RSS}}{\text{RSS}/(n-1)},$$

where $\text{RSS}_{(\beta=0)} = \sum_{i=1}^n y_i^2$. Write down the form of the F statistic.

(v) Show that the t-test and the F-test of H_0 are the same test.

Q4.1.3 – Likelihood-ratio asymptotics with relation to distances

Introduction. Likelihood ratios quantify the goodness of fit of two competing statistical models. They are often used in hypothesis testing and model selection to determine which model better fits a set of observations. The likelihood-ratio test specifically evaluates the ratio of the maximum likelihood of the data under one model to the maximum likelihood under another, where the models represent the null and alternative hypotheses. In general, likelihood ratios provide a method to compare models.

This question explores the properties of a likelihood ratio $\mathcal{L}(\theta)$. In particular, the expectation of $\mathcal{L}(\theta)$ and $\sqrt{\mathcal{L}(\theta)}$ are computed, the latter of which can be represented as a well-known distance. Finally, there is an investigation of the asymptotic properties of $\mathcal{L}(\theta)$.

Question. Suppose $\{x_i\}_{i=1}^n$ are independent and identically distributed from the density function $f(x \mid \theta^*)$ and consider the likelihood ratio

$$\mathcal{L}(\theta) = \prod_{i=1}^{n} \frac{f(x_i \mid \theta)}{f(x_i \mid \theta^*)}.$$

(i) What is the expected value of $\mathcal{L}(\theta)$?

(ii) What is the expected value of $\sqrt{\mathcal{L}(\theta)}$?

(iii) Show that the answer to part (ii) can be represented as $1 - d(\theta^*, \theta)$ for some pseudo-distance d that satisfies the conditions of a distance measure except for the triangle inequality.

(iv) If $\theta \neq \theta^*$ implies $d(\theta^*, \theta) > 0$, prove that $\mathcal{L}(\theta) \to 0$ with probability 1 as $n \to \infty$ (i.e., almost sure convergence).

(v) How else could it be shown that $\mathcal{L}(\theta)$ converges to 0?

Q4.1.4 – Most powerful likelihood-ratio test

Introduction. Likelihood-ratio tests are commonly used in hypothesis testing because they are intuitive, compatible with simple and composite hypotheses, and the optimal (most powerful) test under certain conditions. Additionally, they do not require the data to follow a specific distribution and may be used for comparing nested models. This question considers a likelihood-ratio test for two competing simple hypotheses. Throughout the question, it is shown that the likelihood-ratio test is the most powerful test for a given significance level. Finally, there is a discussion on the general compatibility of the likelihood-ratio test for composite hypotheses.

Question. A likelihood-ratio test for the hypotheses

$$H_0 : \theta = \theta_0 \quad \text{vs} \quad H_1 : \theta = \theta_1$$

is based on the value of

$$\mathcal{L}(t) = \frac{f_T(t \mid \theta_1)}{f_T(t \mid \theta_0)}$$

for some sufficient statistic T with density $f_T(t \mid \theta)$. The test rejects H_0 if $\mathcal{L}(t) > k$ using the decision rule $\delta(t) = 1(\mathcal{L}(t) > k)$ for some $k > 0$.

(i) What condition needs to be met for the test to be of size α?

(ii) What is the power of the test?

(iii) Let $\widetilde{\delta}(t)$ be another decision rule that is 1 if H_0 is rejected and 0 otherwise. Show that

$$\{\delta(t) - \widetilde{\delta}(t)\}\,\{f_T(t \mid \theta_1) - k\, f_T(t \mid \theta_0)\} \geq 0.$$

(iv) Show that the test using the decision rule $\delta(t)$ is most powerful.

(v) Why is it not possible to conclude that the decision rule $\delta(t)$ is uniformly most powerful for the test

$$H_0 : \theta = \theta_0 \quad \text{vs} \quad H_1 : \theta > \theta_0\,?$$

Q4.1.5 – Foundations for uniformly most powerful tests

Introduction. In Question 4.1.4, the likelihood-ratio test for competing simple hypotheses was shown to be uniformly most powerful (UMP). This question considers a likelihood-ratio test involving a composite alternative hypothesis, and the aim is to identify when this new test is uniformly most powerful.

Question. Consider the likelihood-ratio test for $H_0 : \theta = \theta_0$ vs $H_1 : \theta > \theta_0$, where the test statistic T has density function $f_T(t \mid \theta)$.

(i) Show that the test rejecting H_0 when $f_T(t \mid \theta_1)/f_T(t \mid \theta_0) > k$ is most powerful for the point alternative $H_1 : \theta = \theta_1$, where $\theta_1 > \theta_0$ and $k > 0$.

(ii) For the composite alternative $H_1 : \theta > \theta_0$, consider a decision rule of the type $\delta(t) = 1(t \in C)$, where $C = (a, b)$ is the rejection region. What is the condition for the size of the test to be α?

(iii) Now, assume that the likelihood ratio $f_T(t \mid \theta)/f_T(t \mid \theta_0)$ is increasing in t for all $\theta > \theta_0$. To find a uniformly most powerful test, write down the power function and show that the aim would be to maximize $p_\theta(b) = F_\theta(b) - F_\theta(a(b))$ with respect to b, where $a(b) = F_{\theta_0}^{-1}(F_{\theta_0}(b) - \alpha)$.

(iv) Show that $p_\theta(b)$ is increasing in b.

(v) Show that the uniformly most powerful test rejects H_0 if $T > a$ for some $a > 0$.

Q4.1.6* – Test for a variance

Introduction. This question investigates the foundational aspects of a one-sided test for the variance of a normal distribution with mean zero. Students often know the test for a normal mean by memory, yet struggle to find the right statistic if the test is on the variance. Often when the mean is known, the sample variance is wrongly chosen as the test statistic. The appropriate test statistic is obtained by looking at the likelihood function and finding in which tail the test statistic would fall for the null hypothesis to be rejected. Then, one must determine the critical value and how the Type II error behaves as the sample size increases.

Question. Consider independent and identically distributed observations $y_1, \ldots, y_n$ from a normal distribution with mean zero and unknown variance σ^2. The objective is to test the hypotheses

$$H_0 : \sigma = \sigma_0 \quad \text{vs} \quad H_1 : \sigma > \sigma_0$$

for some specified σ_0.

(i) What would the appropriate test statistic be for testing the hypothesis?

(ii) If H_0 is true, what is the distribution of the test statistic?

(iii) Determine the rejection region for the test with a Type I error of α.

(iv) If the true variance is $\sigma_1^2 > \sigma_0^2$, derive an expression for the Type II error.

(v) What happens to the error found in part (iv) as $n \to \infty$?

S4.1 Solutions – Classical Framework

S4.1.1 – Hypothesis testing for a distribution median

(i) It is that $T \sim \text{Bin}(n, \frac{1}{2})$ because T is the sum of n independent Bernoulli random variables with probability $\text{P}(X_i \leq 0) = \frac{1}{2}$ under the null hypothesis.

(ii) Note that $T/n = \sum_{i=1}^{n} 1(x_i \leq 0)/n$ is a sample mean of independent Bernoulli random variables. By the central limit theorem,

$$T/n \approx \text{N}\left(\frac{1}{2}, \frac{1}{4n}\right)$$

under H_0, because $\text{E}\{1(X_i \leq 0)\} = 1/2$ and $\text{Var}\{1(X_i \leq 0)\} = 1/4$. Now, consider the test statistic

$$S_n = 2\sqrt{n}\left(T/n - \tfrac{1}{2}\right).$$

Thus, if H_0 is true, S_n is approximately standard normal. If the observed statistic falls too far in the tails of a standard normal, there is reason to reject H_0.

(iii) Note that

$$\text{E}\left(e^{\theta X}\right) = \frac{1}{\sigma\sqrt{2\pi}} \int \exp\left\{\theta x - \frac{1}{2\sigma^2}(x-\mu)^2\right\} dx.$$

Then, complete the square for x in the exponent and integrate to get

$$\text{E}\left(e^{\theta X}\right) = \exp\{\mu\theta + (\sigma\theta)^2/2\}.$$

(iv) The $\{y_i\}$ are independent and identically distributed, so the Laplace transform is

$$\text{E}\left\{e^{\theta(Y_1+\cdots+Y_n)/\sqrt{n}}\right\} = \left\{\text{E}\left(e^{\theta Y_1/\sqrt{n}}\right)\right\}^n.$$

Use a Taylor expansion to obtain

$$\begin{aligned}\left\{\text{E}\left(e^{\theta Y_1/\sqrt{n}}\right)\right\}^n &= \left[\text{E}\left\{1 + \theta Y_1/\sqrt{n} + \theta^2 Y_1^2/(2n) + o(1/n)\right\}\right]^n \\ &= \{1 + \theta^2/(2n) + o(1/n)\}^n,\end{aligned}$$

which converges to $\exp(\theta^2/2)$ because $\{1 + \xi/n + o(1/n)\}^n \to e^{\xi}$. Therefore, the Laplace transform converges to $\exp(\theta^2/2)$, which is the Laplace transform of a standard normal random variable; see the solution to part (iii).

(v) First, note that the $\{y_i\}$ from part (iv) can be constructed as

$$y_i = \frac{1(x_i \leq 0) - F(0)}{\sqrt{F(0)\{1 - F(0)\}}},$$

which are independent. Further, note that $\text{E}\{1(X_i \leq 0)\} = \text{P}(X_i \leq 0) = F(0)$ and $\text{Var}\{1(X_i \leq 0)\} = F(0)\{1 - F(0)\}$, so $\text{E}(Y_i) = 0$ and $\text{Var}(Y_i) = 1$. From part (iv),

$$(y_1 + \cdots + y_n)/\sqrt{n} = \{T/n - F(0)\}/\sqrt{F(0)\{1 - F(0)\}/n}$$

converges in distribution to a standard normal random variable.

Further Reading

Historical Background. Exact hypothesis tests require knowledge about the distribution of the test statistic under the null hypothesis. Although many desirable test statistics do not have tractable distributions under the null, they may be represented as sums of independent random variables. Thus, when the central limit theorem is applicable, an approximate normal distribution can be used.

In 1738, French mathematician Abraham de Moivre published the second edition of *The Doctrine of Chances*, which introduced the concept of normal distributions approximating binomial distributions; his book is widely considered to be the first textbook on probability theory (de Moivre, 1738). Then in 1812, French polymath Pierre–Simon Laplace expanded upon de Moivre's work and established the de Moivre–Laplace theorem, which is considered to be the earliest version of the CLT (Laplace, 1812). The de Moivre–Laplace theorem is more restrictive than the modern version of the CLT, which was defined and proved in general terms by Russian mathematician Aleksandr Lyapunov in 1901.

Points of Interest.

1. A popular example of classical hypothesis testing uses the result from de Moivre (1738). Consider the test of a Bernoulli probability, which has a test statistic with a binomial distribution. The distribution of the test statistic can be approximated by a normal distribution, as done in part (v).

2. The CLT is used to approximate the distribution of test statistics in many hypothesis tests. For example, several tests operate on rank data, where comparative outcomes are presented rather than absolute values; one such test is the Wilcoxon rank-sum test (Lehmann, 1998). Normal approximations are often used for test statistics comprising rank data. Additionally, normal approximations are useful in several nonparametric tests. Consider the crossmatch test, which is a nonparametric test for assessing whether two samples arise from the same distribution; the test statistic for this test is based on pairwise distances between data points and is typically approximated with normal distributions (Rosenbaum, 2005).

3. A simple method to prove the central limit theorem is to use Laplace transforms, which are effectively generating and characteristic functions. For more on the Laplace transform and the central limit theorem, see Williams (1973) and Fischer (2011), respectively. Finally, Casella and Berger (2021) illustrated several applications of the central limit theorem.

Demonstration How good is a normal approximation for a test statistic? As an example, consider the standard normal approximation to

$$S = 2\sqrt{n}\left(T/n - \tfrac{1}{2}\right),$$

where T is the number of negative-valued samples and the n samples are taken from a distribution with $F(0) = \frac{1}{2}$. A sample of 10,000 statistics S are presented in Fig. 4.1 based on a sample of size $n = 100$. The normal density is overlaid. A correction factor can be used to reduce the observed bias.

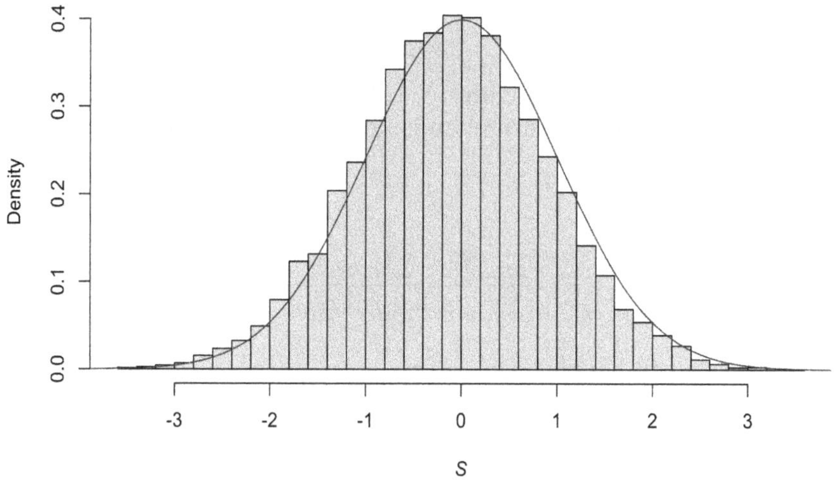

FIGURE 4.1
Histogram of S values for $n = 100$ with a standard normal density overlaid.

S4.1.2 – t-test and F-test for the linear model

(i) The least squares estimator is given by

$$\widehat{\beta} = \arg\min_{\beta} \left\{ \sum_{i=1}^{n} (y_i - x_i\beta)^2 \right\},$$

which can be found by differentiating with respect to β:

$$\frac{\partial}{\partial\beta} \left\{ \sum_{i=1}^{n} \left(y_i^2 - 2x_iy_i\beta + x_i^2\beta^2 \right) \right\} = -2\sum_{i=1}^{n} x_iy_i + 2\beta\sum_{i=1}^{n} x_i^2.$$

Set the above expression to 0 to obtain the desired $\widehat{\beta}$.

(ii) The η is a weighted sum of independent normal random variables and, consequently, is normal itself. Hence, the mean and variance of η are all that is required. If $\beta = 0$, then $\mathrm{E}(Y_i) = 0$ for all i and $\mathrm{E}(\eta) = 0$. The variance of η is

$$\mathrm{Var}(\eta) = \mathrm{Var}\left(\sum_{i=1}^{n} x_iY_i \right) = \sum_{i=1}^{n} x_i^2 \,\mathrm{Var}(Y_i) = \sigma^2,$$

because $\sum_{i=1}^{n} x_i^2 = 1$. Therefore, $\eta \sim \mathrm{N}(0, \sigma^2)$.

(iii) If σ is known, then the distribution of η is known under the null hypothesis. Further, it is observable, and the test statistic is η. If σ is unknown, it is estimated as

$$\widehat{\sigma}^2 = \frac{1}{n-1} \sum_{i=1}^{n} (y_i - x_i\widehat{\beta})^2.$$

It is well-known that $\mathrm{RSS}/\sigma^2 = (n-1)\widehat{\sigma}^2/\sigma^2$ is a χ^2_{n-1} random variable that is independent of η. Hence,

$$t_{n-1} = \frac{\eta/\sqrt{\sigma^2}}{\sqrt{\widehat{\sigma}^2/\sigma^2}} = \frac{\eta}{\widehat{\sigma}}$$

is a Student-t random variable with $n-1$ degrees of freedom and can be used as the test statistic because its distribution is known under the null hypothesis and is observable.

(iv) The F statistic is

$$F = \frac{\sum_{i=1}^n y_i^2 - \sum_{i=1}^n (y_i - x_i\widehat{\beta})^2}{\sum_{i=1}^n (y_i - x_i\widehat{\beta})^2/(n-1)}.$$

(v) Simplify the F statistic to obtain

$$\begin{aligned}F &= \frac{\sum_{i=1}^n \left(y_i^2 - y_i^2 + 2y_i x_i\widehat{\beta} - x_i^2\widehat{\beta}^2\right)}{\widehat{\sigma}^2} \\ &= \frac{\sum_{i=1}^n \left(2y_i x_i\widehat{\beta} - x_i^2\widehat{\beta}^2\right)}{\widehat{\sigma}^2} \\ &= \frac{2\widehat{\beta}\sum_{i=1}^n y_i x_i - \widehat{\beta}^2 \sum_{i=1}^n x_i^2}{\widehat{\sigma}^2} \\ &= \frac{2\widehat{\beta}^2 - \widehat{\beta}^2}{\widehat{\sigma}^2} = \frac{\eta^2}{\widehat{\sigma}^2} = t_{n-1}^2.\end{aligned}$$

Thus, the F statistic is equivalent to the squared t_{n-1} statistic under the null hypothesis. Therefore, the t-test and F-test for testing $H_0 : \beta = 0$ are equivalent.

Further Reading

Historical Background. According to Stanton (2001), the first data analysis using a linear regression model was done by Galton and Pearson (Galton, 1894; Pearson, 1896). Over a century later, the linear model is still one of the most widely used models in statistics. However, despite being relatively simple, it can be difficult to understand every test associated with the linear model, namely the normal test, t-test, chi-squared test, and F-test. The aim of this question was to draw connections between a couple of these tests.

Points of Interest.

1. The last step of this question requires knowledge of the relationship between the Student's t and Fisher–Snedecor's F distributions. A well-known result is that a t_ν^2 random variable, where t_ν is a Student-t with ν degrees of freedom, is the same as a $F_{1,\nu}$ random variable, where $F_{1,\nu}$ is an F variable with 1 and ν degrees of freedom. In general, a random variable is F-distributed if it can be written as the ratio of two independent χ^2 random variables divided by their respective degrees of freedom. A t-distributed random variable can be represented as a standard normal random variable divided by an independent χ_1^2-distributed variable. It is also possible to relate the standard normal distribution with the χ^2 distribution, the χ^2 distribution with the t-distribution, and the F-distribution with two independent χ^2 distributions or the t-distribution.

2. An interesting exercise is to show that the RSS is independent of η (see part (iii)). As a starting point, show that $e_i = y_i - x_i\eta$ and η are independent. Because their joint distribution is normal, this exercise simplifies to showing that $\mathrm{E}(e_i\,\eta) = 0$ for all i.

3. To elaborate on the previous point, consider $\{x_i\}$ to be independent standard normal random variables. To show that the sample mean and sample variance are independent, it would be enough to show that $x_1 - \overline{x}$ is independent of $\overline{x}$. The best way to show this is to demonstrate that

$$\mathrm{E}\left[\exp\left\{\theta\overline{X} + \phi(X_1 - \overline{X})\right\}\right]$$

separates into a function of θ and ϕ. This is equivalent to

$$\mathrm{E}\left[\exp\left\{X_1(\theta/n + \phi - \phi/n)\right\}\right]\ \mathrm{E}\left[\exp\left\{\sum_{i\neq 1} X_i(\theta/n - \phi/n)\right\}\right].$$

Using the Laplace transform of a standard normal random variable, this becomes

$$\exp\{-\phi^2(1-1/n)\} \times \exp(\theta^2/n).$$

Hence, the independence follows.

S4.1.3 – Likelihood-ratio asymptotics with relation to distances

(i) Let $\mathcal{X}$ denote the support for the $\{x_i\}$. Take the expectation of $\mathcal{L}(\theta)$ with respect to $\mathbf{x}$ to get

$$\begin{aligned}
\mathrm{E}\{\mathcal{L}(\theta)\} &= \int_{\mathcal{X}} \cdots \int_{\mathcal{X}} \prod_{i=1}^{n} \frac{f(x_i \mid \theta)}{f(x_i \mid \theta^*)} \prod_{i=1}^{n} f(x_i \mid \theta^*) d\mathbf{x} \\
&= \int_{\mathcal{X}} \cdots \int_{\mathcal{X}} \prod_{i=1}^{n} f(x_i \mid \theta) d\mathbf{x} \\
&= \int_{\mathcal{X}} f(x_1 \mid \theta) dx_1 \cdots \int_{\mathcal{X}} f(x_n \mid \theta) dx_n \\
&= 1,
\end{aligned}$$

where the penultimate equality follows because the $f(x_i \mid \theta)$ are independent. The last equality follows because the integral of each probability density over the entire support $\mathcal{X}$ equals 1.

(ii) Let $\mathcal{X}^n$ denote the support for $\mathbf{x} = (x_1, \ldots, x_n)'$. As in part (i), apply the definition of expectation for continuous random variables:

$$\begin{aligned}
\mathrm{E}\left\{\sqrt{\mathcal{L}(\theta)}\right\} &= \int_{\mathcal{X}^n} \sqrt{\prod_{i=1}^{n} \frac{f(x_i \mid \theta)}{f(x_i \mid \theta^*)}} \prod_{i=1}^{n} f(x_i \mid \theta^*) d\mathbf{x} \\
&= \int_{\mathcal{X}^n} \prod_{i=1}^{n} \left\{ \sqrt{\frac{f(x_i \mid \theta)}{f(x_i \mid \theta^*)}} f(x_i \mid \theta^*) \right\} d\mathbf{x} \\
&= \int_{\mathcal{X}^n} \prod_{i=1}^{n} \sqrt{f(x_i \mid \theta) f(x_i \mid \theta^*)} d\mathbf{x} \\
&= \int_{\mathcal{X}^n} \sqrt{f(\mathbf{x} \mid \theta) f(\mathbf{x} \mid \theta^*)} d\mathbf{x},
\end{aligned}$$

where $f(\mathbf{x} \mid \theta) = \prod_{i=1}^{n} f(x_i \mid \theta)$.

(iii) First, solve for $d(\theta^*, \theta)$:

$$d(\theta^*, \theta) = 1 - \int_{\mathcal{X}^n} \sqrt{f(\mathbf{x} \mid \theta) f(\mathbf{x} \mid \theta^*)} d\mathbf{x}.$$

For $d(\theta^*, \theta)$ to be a pseudo-distance measure, it must be shown that:

(a) It is non-negative for any $\theta, \theta^* \in \Theta$.
Note that

$$\begin{aligned}\int_{\mathcal{X}^n} \sqrt{f(\mathbf{x} \mid \theta) f(\mathbf{x} \mid \theta^*)} d\mathbf{x} &= \int_{\mathcal{X}^n} f(\mathbf{x} \mid \theta^*) \sqrt{\frac{f(\mathbf{x} \mid \theta)}{f(\mathbf{x} \mid \theta^*)}} d\mathbf{x} \\ &\leq \left(\int_{\mathcal{X}^n} f(\mathbf{x} \mid \theta^*) \frac{f(\mathbf{x} \mid \theta)}{f(\mathbf{x} \mid \theta^*)} d\mathbf{x} \right)^{1/2} = 1,\end{aligned}$$

where the inequality follows by Jensen's inequality. Additionally, it must be that $0 \leq f(\mathbf{x} \mid \theta)$ for all $\theta \in \Theta$. Therefore, $0 \leq \sqrt{f(\mathbf{x} \mid \theta) f(\mathbf{x} \mid \theta^*)} \leq 1$ and, consequently, $d(\theta, \theta^*) \geq 0$.

(b) $d(\theta, \theta^*) = 0$ if and only if $\theta = \theta^*$.
If $\theta = \theta^*$, then

$$d(\theta^*, \theta) = 1 - \int_{\mathcal{X}^n} f(\mathbf{x} \mid \theta) d\mathbf{x} = 1 - 1 = 0.$$

Additionally, $d(\theta^*, \theta) = 0$ implies that $\int_{\mathcal{X}^n} \sqrt{f(\mathbf{x} \mid \theta) f(\mathbf{x} \mid \theta^*)} d\mathbf{x} = 1$; this must be true because the two distributions belong to the same family and must be proportional to each other. Thus, Jensen's inequality becomes an equality.

(c) It is symmetric (i.e., $d(\theta^*, \theta) = d(\theta, \theta^*)$), and this is trivial to see.

Therefore, $d(\theta^*, \theta)$ is a pseudo-distance. In particular, it is half the squared Hellinger distance.

(iv) A common method to prove almost sure convergence is to use the Borel–Cantelli lemma, which requires the specification of events $\{E_n\}$ such that $\sum_{n=1}^{\infty} \mathrm{P}(E_n) < \infty$. If such an event is specified, then the converse of the event (assuming it exists) occurs for all large n almost surely. With some foresight, define E_n to be $\mathcal{L}(\theta) \geq e^{-nc}$, so the converse event is $\mathcal{L}(\theta) < e^{-nc}$. If $\sum_{n=1}^{\infty} \mathrm{P}(E_n) < \infty$, then $\mathcal{L}(\theta) < e^{-nc}$ occurs almost surely for all large n, which implies that $\mathcal{L}(\theta) \overset{\text{a.s}}{\to} 0$ as $n \to \infty$. Thus, it must be shown that $\sum_{n=1}^{\infty} \mathrm{P}(E_n) < \infty$. This can be shown using Markov's inequality, which requires knowledge of $\mathrm{E}\{\sqrt{\mathcal{L}(\theta)}\}$.

From part (iii), $\mathrm{E}\{\sqrt{\mathcal{L}(\theta)}\}$ is a function of the squared Hellinger distance. Let $c = \frac{1}{2}d_H^2$ denote the squared Hellinger distance for a single observation and compute

$$\begin{aligned}\mathrm{E}\left\{\sqrt{\mathcal{L}(\theta)}\right\} &= \mathrm{E}\left\{\prod_{i=1}^{n} \sqrt{\frac{f(x_i \mid \theta)}{f(x_i \mid \theta^*)}}\right\} \\ &= \left(1 - \tfrac{1}{2}d_H^2\right)^n \\ &\leq e^{-nc},\end{aligned}$$

where the inequality follows from $1 - x \leq e^{-x}$. Now, use Markov's inequality to get

$$\begin{aligned} P(E_n) &= P(\mathcal{L}(\theta) \geq e^{-nc}) \\ &= P\left(\sqrt{\mathcal{L}(\theta)} \geq e^{-nc/2}\right) \\ &\leq E\left\{\sqrt{\mathcal{L}(\theta)}\right\} e^{nc/2} \\ &\leq e^{-nc/2}, \end{aligned}$$

where the first inequality is a result of Markov's inequality and the second inequality uses $E\{\sqrt{\mathcal{L}(\theta)}\} \leq e^{-nc}$. Consequently, $\sum_{n=1}^{\infty} P(E_n) < \infty$ and, by the Borel–Cantelli lemma, $\mathcal{L}(\theta) \overset{a.s.}{\to} 0$ as $n \to \infty$.

(v) If the Kullback–Leibler divergence between the density functions exists, then

$$n^{-1} \log \mathcal{L}(\theta) \to -\int f(x \mid \theta^*) \log\{f(x \mid \theta^*)/f(x \mid \theta)\}\, dx$$

by the law of large numbers. This divergence is non-zero and $\log \mathcal{L}(\theta) \to -\infty$ almost surely, implying $\mathcal{L}(\theta) \to 0$ almost surely.

The use of the Hellinger distance is preferable because it does not require an assumption of finite Kullback–Leibler divergence.

Further Reading

Historical Background. Early work on the likelihood ratio dates back to Neyman and Pearson (1933) and Wilks (1938). They identified that likelihood ratios could be used for hypothesis testing. Subsequent study on the asymptotics of likelihood ratios further supported their use in statistical hypothesis testing.

The likelihood ratio $\mathcal{L}(\theta)$ is related to various measures of distance and dissimilarity between probability distributions. For example, it is shown in part (ii) that the expectation of $\sqrt{\mathcal{L}(\theta)}$ is

$$\int_{\mathcal{X}^n} \sqrt{f(\mathbf{x} \mid \theta) f(\mathbf{x} \mid \theta^*)} d\mathbf{x},$$

which is the Bhattacharyya coefficient (Bhattacharyya, 1943). This coefficient approximately quantifies the closeness of two random statistical samples and is essential to several distance measures, including the Bhattacharyya distance, Hellinger distance, and Mahalanobis distance. In part (iv), $d(\theta^*, \theta)$ is revealed to be a function of the Bhattacharyya coefficient. In this case, $d(\theta^*, \theta)$ is the squared Hellinger distance, another measure used to quantify the similarity between probability distributions. Hellinger distances are used commonly in sequential and asymptotic statistics. Squared Mahalanobis distance, a special case of Bhattacharyya distance up to a multiplicative factor, finds extensive use in cluster analysis.

Points of Interest.

1. In part (v), the log-likelihood ratio may be used, which takes the form of negative Kullback–Leibler (KL) divergence; this alternative solution utilizes the strong law of large numbers, which, along with the Borel–Cantelli lemma, constitute the two primary

methods of proving almost sure convergence. The strong law of large numbers states that, for a sequence of random variables, the sample average converges almost surely to the expected value. Note that $\mathcal{L}(\theta)$ can be written in the form of a sample average by considering

$$W_n = \frac{1}{n} \log \mathcal{L}(\theta) = \frac{1}{n} \sum_{i=1}^{n} \log \left\{ \frac{f(x_i \mid \theta)}{f(x_i \mid \theta^*)} \right\},$$

which assumes the form of Kullback–Leibler divergence between $f(x_i \mid \theta)$ and $f(x_i \mid \theta^*)$. From here, use the strong law of large numbers to conclude that W_n converges almost surely to $\mathrm{E}[\log\{f(X_i \mid \theta)/f(X_i \mid \theta^*)\}]$ as $n \to \infty$, where

$$\begin{aligned} \mathrm{E}\left[\log\left\{\frac{f(X_i \mid \theta)}{f(X_i \mid \theta^*)}\right\}\right] &= \int_{\mathcal{X}} \log\left\{\frac{f(x_i \mid \theta)}{f(x_i \mid \theta^*)}\right\} f(x_i \mid \theta^*) dx_i \\ &< \int_{\mathcal{X}} \left\{\frac{f(x_i \mid \theta)}{f(x_i \mid \theta^*)} - 1\right\} f(x_i \mid \theta^*) dx_i \\ &= \int_{\mathcal{X}} f(x_i \mid \theta) dx_i - \int_{\mathcal{X}} f(x_i \mid \theta^*) dx_i \\ &= 0. \end{aligned}$$

Note that the inequality follows because $\log(a) < a - 1$ for $a > 0$; this result is used to remove the logarithm in the integral. Thus, by the strong law of large numbers,

$$\lim_{n\to\infty} W_n \overset{\text{a.s.}}{\to} \mathrm{E}\left[\log\left\{\frac{f(X_i \mid \theta)}{f(X_i \mid \theta^*)}\right\}\right] < 0$$

for $\theta \neq \theta^*$. Multiply both sides by n, which tends to infinity, to obtain

$$\lim_{n\to\infty} nW_n = \lim_{n\to\infty} \log \mathcal{L}(\theta) \overset{\text{a.s.}}{\to} -\infty$$

because W_n converges almost surely to a negative quantity. Finally, exponentiate both sides to get

$$\lim_{n\to\infty} \mathcal{L}(\theta) \overset{\text{a.s.}}{\to} 0.$$

Both the KL divergence and a function of the Bhattacharyya coefficient are special cases of Rényi divergence (Rényi, 1961).

2. Consider the convergence of $\mathcal{L}(\theta)$ as $n \to \infty$. The condition that $\theta \neq \theta^*$ implies $d(\theta^*, \theta) > 0$ underpins a heuristic argument that reflects the proposed solution: as more observations are made, there is more evidence against θ being the correct model parameter and the KL divergence approaches 0. Then, the strong law of large numbers can be applied to W_n as in the previous point. If a single θ value is considered, it is easy to evaluate the asymptotics of the likelihood ratio. However, as the parameter space Θ increases, the asymptotic analysis of the likelihood ratio (i.e., $\sup_{\theta\in\Theta} \mathcal{L}(\theta)$) becomes more tricky because point-wise convergence for all θ does not imply uniform convergence, which would be needed to show that the MLE converged to the true parameter value.

Demonstration. In what follows, the point-wise convergence of $\mathcal{L}(\theta)$ is demonstrated. For concreteness, assume that the limit is 0 and the functions are deterministic. If point-wise

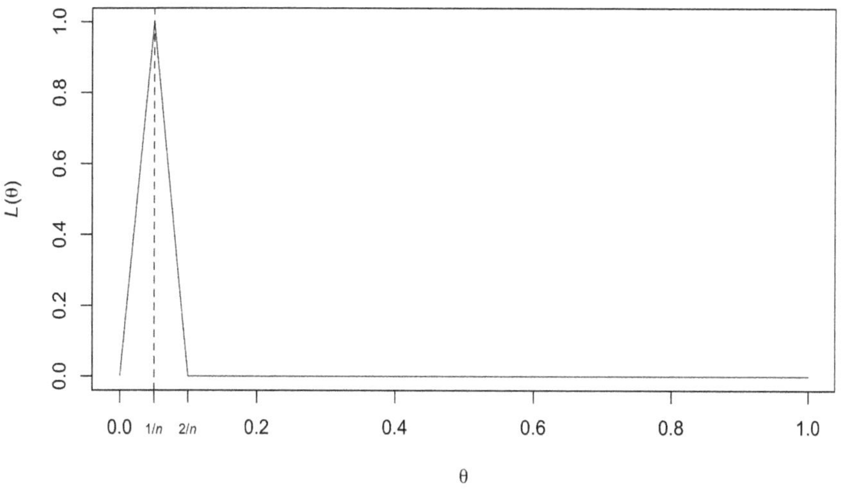

FIGURE 4.2
The function $\mathcal{L}(\theta)$ that converges to 0 pointwise but not uniformly. The spike is centered at $\theta = 1/n$ and $\mathcal{L}(\theta) = 0$ for $\theta > 2/n$.

convergence holds, showing uniform convergence (i.e., $\sup_{\theta \in \Theta} |\mathcal{L}(\theta)| \to 0$) can be awkward, despite the perception that it is typically true. Notably, the existence of oscillations and spikes prove to be the challenge. For example, consider the function

$$\mathcal{L}(\theta) = \begin{cases} n\theta, & 0 \leq \theta \leq 1/n, \\ 2 - n\theta, & 1/n < \theta \leq 2/n, \\ 0, & \theta > 2/n. \end{cases}$$

Note that $\mathcal{L}(\theta) \to 0$ for all θ, whereas $\sup_{\theta \in [0,1]} |\mathcal{L}(\theta)| = 1$. This function is plotted in Fig. 4.2, which shows that the existence of the spike is problematic for proving uniform convergence.

S4.1.4 – Most powerful likelihood-ratio test

(i) The size of a test is the probability of falsely rejecting H_0; this probability is called the Type I error rate. For a fixed α, choose k to satisfy

$$\alpha = \mathrm{P}(\text{reject } H_0 \mid H_0 \text{ is true}) = \mathrm{P}(\mathcal{L}(t) > k \mid \theta = \theta_0).$$

(ii) The power of the test with decision rule $\delta(t)$ is the probability of rejecting H_0 given that H_1 is true:

$$\beta_{\delta(t)}(\theta_1) = \mathrm{P}(\text{reject } H_0 \mid H_1 \text{ is true}) = \mathrm{P}(\mathcal{L}(t) > k \mid \theta = \theta_1).$$

(iii) Both decision rules are binary, so there are four possible combinations of decisions. If the two decisions are identical ($\delta(t) = 0$, $\tilde{\delta}(t) = 0$ or $\delta(t) = 1$, $\tilde{\delta}(t) = 1$), there is equality. If $\delta(t) = 0$ and $\tilde{\delta}(t) = 1$, then

$$-\{f_T(t \mid \theta_1) - kf_T(t \mid \theta_0)\} = kf_T(t|\theta_0) - f_T(t|\theta_1) \geq 0,$$

where the inequality follows because $\delta(t) = 0$ implies that $\mathcal{L}(t) \leq k$ and, thus, $f_T(t|\theta_1) \leq kf_T(t \mid \theta_0)$. If $\delta(t) = 1$ and $\tilde{\delta}(t) = 0$, then

$$f_T(t \mid \theta_1) - kf_T(t \mid \theta_0) > 0$$

because $\delta(t) = 1$ implies $\mathcal{L}(t) > k$ and, thus, $f_T(t \mid \theta_1) > kf_T(t \mid \theta_0)$.

(iv) The decision rule $\delta(t)$ is most powerful if $\beta_{\delta(t)}(\theta_1) \geq \beta_{\tilde{\delta}(t)}(\theta_1)$ for any other test $\tilde{\delta}(t)$ of the same size or smaller. By the definition of power and size,

$$\begin{aligned}
\beta_{\delta(t)}(\theta_1) = \mathrm{E}(\delta(t) \mid \theta = \theta_1) &= \int \delta(t) f_T(t \mid \theta = \theta_1) dt, \\
\alpha = \mathrm{E}(\delta(t) \mid \theta = \theta_0) &= \int \delta(t) f_T(t \mid \theta = \theta_0) dt, \\
\tilde{\alpha} = \mathrm{E}(\tilde{\delta}(t) \mid \theta = \theta_0) &= \int \tilde{\delta}(t) f_T(t \mid \theta = \theta_0) dt,
\end{aligned}$$

where $\tilde{\alpha}$ is the size of the test using decision rule $\tilde{\delta}(t)$. Note that the inequality in part (iii) contains all of these integral expressions, so integrate that inequality with respect to t:

$$\int \{\delta(t) - \tilde{\delta}(t)\} \{f_T(t \mid \theta_1) - k\, f_T(t \mid \theta_0)\}\, dt \geq 0.$$

Rearrange terms to get

$$\int \{\delta(t) - \tilde{\delta}(t)\}\, f_T(t \mid \theta_1) dt \geq k \left(\int \{\delta(t) - \tilde{\delta}(t)\} f_T(t \mid \theta_0) dt \right),$$

which is equivalent to

$$\beta_{\delta(t)}(\theta_1) - \beta_{\tilde{\delta}(t)}(\theta_1) \geq k(\alpha - \tilde{\alpha}).$$

The right side of the last inequality is the scaled difference in size between the tests. When discussing the most powerful test, compare the test associated with $\delta(t)$ with tests of equal or smaller size (i.e., $\tilde{\alpha} \leq \alpha$). Therefore, $\beta_{\delta(t)}(\theta_1) - \beta_{\tilde{\delta}(t)}(\theta_1) \geq k(\alpha - \tilde{\alpha}) \geq 0$ implies $\beta_{\delta(t)}(\theta_1) \geq \beta_{\tilde{\delta}(t)}(\theta_1)$, so the test using $\delta(t)$ is most powerful.

(v) For a test with a composite alternative hypothesis to be uniformly most powerful, the critical region (k, ∞) for the likelihood ratio for each test $H_0 : \theta = \theta_0$ vs $H_1 : \theta = \theta_1$ should not depend on θ_1. This generally will not be the case. For more details, see Question 4.1.5.

Further Reading

Historical Background. The origins of this question date back to the Neyman–Pearson lemma, which identifies the uniformly most powerful (UMP) test for simple hypotheses

(Neyman and Pearson, 1933). Specifically, the lemma is used to identify the test that minimizes Type II error among all α-level tests. The Neyman–Pearson lemma also guarantees that the UMP test is unique. Outside of decision theory and hypothesis testing, this lemma is used in consumer theory, engineering, physics, and signal processing.

The Karlin–Rubin theorem extends the Neyman–Pearson lemma for composite hypotheses and requires a monotone non-decreasing likelihood ratio (Karlin and Rubin, 1956). This theorem can be used for competing composite hypotheses or a simple null and composite alternative hypothesis. Although the Karlin–Rubin theorem is restricted to scalar parameters and measurements, it is applicable for the one-dimensional exponential family

$$f(x \mid \theta) = g(\theta)h(x)\exp(\eta(\theta)T(x))$$

for sufficient statistic $T(x)$ and non-decreasing $\eta(\theta)$.

Points of Interest.

1. Neyman–Pearson hypothesis testing is a common tool in decision theory, which is the field concerning the process of making decisions under uncertainty. In decision theory, measures of loss, risk, and utility are computed for competing decisions, and the decisions are often selected based on a predefined criteria. The Neyman–Pearson lemma and Karlin–Rubin theorem (Karlin and Rubin, 1956) can be cast as decision theoretic procedures, where hypotheses are tested through the analysis of decisions.

2. An alternative to the Neyman–Pearson framework is Bayesian hypothesis testing, which incorporates prior information into the testing procedure. Bayesian hypothesis testing calculates posterior probabilities of hypotheses given the data. Typically, the hypothesis with the highest posterior probability is selected. The Bayesian analogue to the likelihood ratio for hypothesis testing is the Bayes factor, which is a ratio of competing statistical models represented by their evidence. Bayes factors quantify the support of one hypothesis over another using the observed data and prior information. While the Neyman–Pearson lemma uses likelihood ratios as a test statistic to make decisions regarding hypotheses, the Bayes factor uses the ratio of posterior odds to update beliefs about the hypotheses. Additionally, the Neyman–Pearson lemma provides a binary decision outcome based on a predetermined level of significance, whereas Bayes factors provide a continuous scale of evidence that lends itself to a more nuanced interpretation of the validity of each hypothesis.

3. A more direct look at the most powerful test would be to consider the idea of maximizing

$$\int \delta(t)\, f(t \mid \theta_1)\, dt - \lambda \left(\int \delta(t)\, f(t \mid \theta_0)\, dt - \alpha \right)$$

over $\delta(t)$; that is, to maximize the power subject to the constraint of the Type I error. Re-writing as

$$\int \delta(t)\ \{f(t \mid \theta_1) - \lambda f(t \mid \theta_0)\}\ dt + \lambda\alpha,$$

it is possible to see that the best $\delta(t)$ would be 1 for $f(t \mid \theta_1) > \lambda f(t \mid \theta_0)$ and 0 otherwise. Of course, the Lagrange multiplier λ would be determined by the constraint involving α.

Demonstration. Consider testing the mean of a Gaussian with a sample of 30 observations. The null hypothesis H_0 is that $\mu = 0$ and the alternative hypothesis is that $\mu > 0$. The

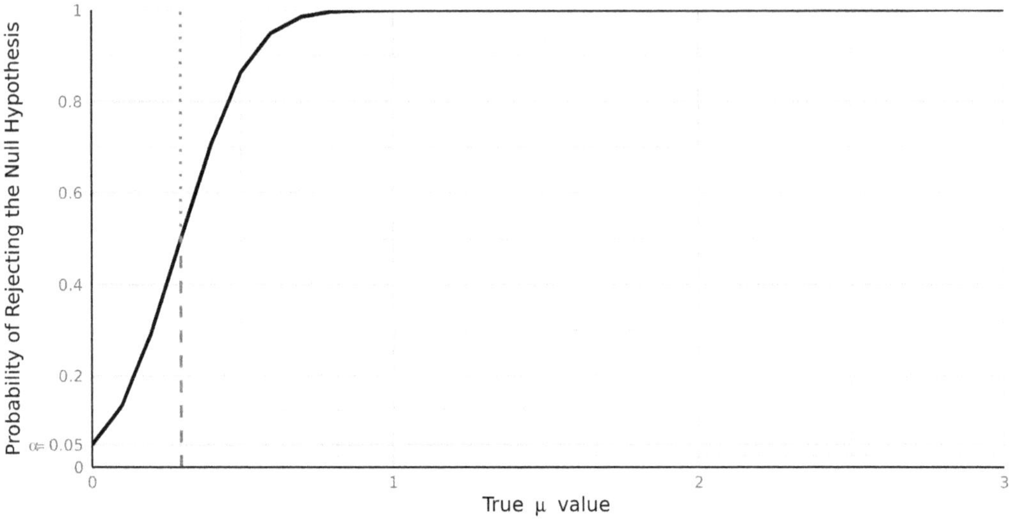

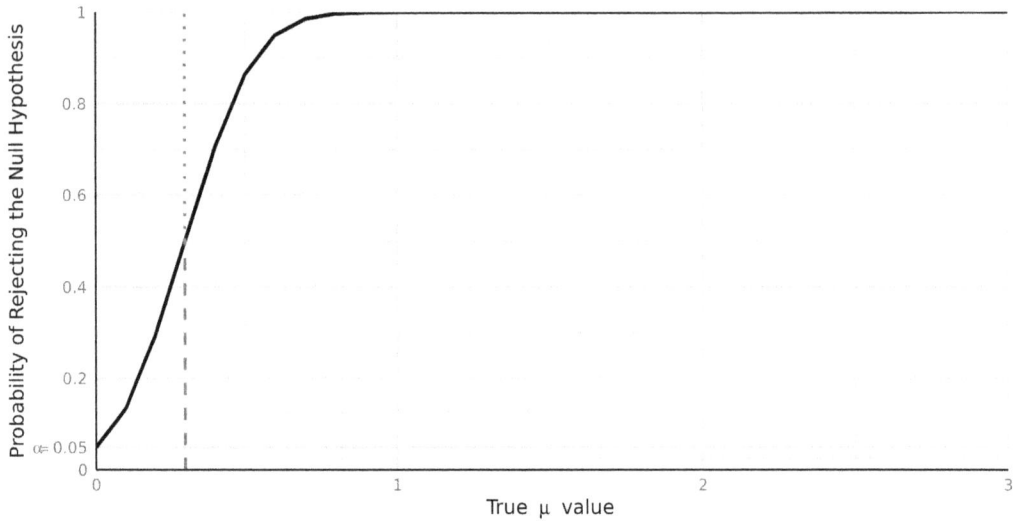

FIGURE 4.3
Visualization of the power function for testing the mean of a Gaussian.

power function for this test is plotted in Fig. 4.3. The power function takes the value of the size of the test (α) when $\mu = 0$. As the true value of μ increases, the probability of rejecting H_0 increases. The height of the dashed gray line represents the power of the test when the true value of μ is 0.3. The length of the dotted gray line represents the probability of failing to reject H_0 despite a true value of $\mu = 0.3$ (i.e., Type II error).

S4.1.5 – Foundations for uniformly most powerful tests

(i) Denote the decision rule for the test as $\delta(t) = 1\{f_T(t \mid \theta_1)/f_T(t \mid \theta_0) > k\}$, and let C be the corresponding critical region. A test is most powerful when it has the highest power among tests of the same size or smaller. Hence, the claim is that the test having decision rule $\delta(t)$ maximizes

$$\beta_{\tilde{\delta}(t)}(\theta_1) = \int_{\tilde{C}} f_T(t \mid \theta_1)dt \quad \text{subject to} \quad \int_{\tilde{C}} f_T(t \mid \theta_0)\, dt = \tilde{\alpha},$$

where $\beta_{\tilde{\delta}(t)}(\theta_1)$ denotes the power of an arbitrary test of size $\tilde{\alpha} \leq \alpha$ with decision rule $\tilde{\delta}(t)$ and critical region $\tilde{C}$. To show $\beta_{\delta(t)}(\theta_1) \geq \beta_{\tilde{\delta}(t)}(\theta_1)$, integrate

$$\{\delta(t) - \tilde{\delta}(t)\}\{f_T(t \mid \theta_1) - k\, f_T(t \mid \theta_0)\} \geq 0$$

with respect to t. More explicit steps can be found in the solution to part (iv) of Question 4.1.4.

(ii) The test with decision rule $\delta(t) = 1(t \in C)$ has size α if

$$\text{P(reject } H_0 \mid H_0 \text{ is true)} = \int_C f_T(t \mid \theta_0)dt = \int_a^b f_T(t \mid \theta_0)dt = F_{\theta_0}(b) - F_{\theta_0}(a) = \alpha,$$

where $a = F_{\theta_0}^{-1}\left(F_{\theta_0}(b) - \alpha\right) < b$. The notation $F_{\theta_0}(b)$ is used in favor of $F_T(b)$ for clarity in later steps.

(iii) Consider the power function $\beta_{\delta(t)}(\theta) = \int_C f_T(t \mid \theta)dt = F_\theta(b) - F_\theta(a)$ subject to $F_{\theta_0}(b) - F_{\theta_0}(a) = \alpha$. The uniformly most powerful test maximizes $F_\theta(b) - F_\theta(a)$ with respect to a and b such that $F_{\theta_0}(b) - F_{\theta_0}(a) = \alpha$; this maximization problem can be written solely in terms of b because a can be expressed as a function of b: $a = F_{\theta_0}^{-1}\left(F_{\theta_0}(b) - \alpha\right)$. Hence, the goal becomes maximizing

$$p_\theta(b) = F_\theta(b) - F_\theta\left(F_{\theta_0}^{-1}(F_{\theta_0}(b) - \alpha)\right)$$

with respect to b.

(iv) Differentiate $p_\theta(b)$ with respect to b:

$$\begin{aligned} p'_\theta(b) &= F'_\theta(b) - F'_\theta(a(b)) \cdot \frac{F'_{\theta_0}(b)}{F'_{\theta_0}(F_{\theta_0}^{-1}(F_{\theta_0}(b) - \alpha))} \\ &= F'_\theta(b) - F'_\theta(a(b)) \cdot \frac{F'_{\theta_0}(b)}{F'_{\theta_0}(a(b))} \\ &= f_T(b \mid \theta_0)\left\{\frac{f_T(b \mid \theta)}{f_T(b \mid \theta_0)} - \frac{f_T(a(b) \mid \theta)}{f_T(a(b) \mid \theta_0)}\right\}. \end{aligned}$$

Because $a(b) < b$ and the likelihood ratio $f_T(t \mid \theta)/f_T(t \mid \theta_0)$ is increasing in t, it is that $p'_\theta(b) \geq 0$ and $p_\theta(b)$ is increasing in b. Note that the maximum value is $b = \infty$, so the value of a is fully determined by α: $a = F_{\theta_0}^{-1}(1 - \alpha)$.

(v) Because the likelihood ratio $f_T(t \mid \theta)/f_T(t \mid \theta_0)$ is increasing in t for all $\theta > \theta_0$, there exists some $a > 0$ such that $f_T(t \mid \theta)/f_T(t \mid \theta_0) > k$ if and only if $t > a$. Hence, from part (i), the most powerful test for $H_0 : \theta = \theta_0$ vs $H_1 : \theta = \theta_1$ for all $\theta_1 > \theta_0$ has critical region $t > a$. Thus, this test is uniformly most powerful for $H_0 : \theta = \theta_0$ vs $H_1 : \theta > \theta_0$.

Further Reading

Historical Background. The notion of a uniformly most powerful test was introduced by Neyman and Pearson (1933) and later extended by Karlin and Rubin (1956). A test of size α is uniformly most powerful for $H_0 : \theta = \theta_0$ vs $H_1 : \theta \in \Theta$ if $\mathrm{E}_{\theta^*}\{\delta(T)\} \geq \mathrm{E}_{\theta^*}\{\tilde{\delta}(T)\}$ for all $\theta^* \in \Theta$, where $\tilde{\delta}(T)$ denotes all other tests of size α evaluated using the test statistic T.

Points of Interest.

1. A more direct approach to part (v) could argue that $f_T(t \mid \theta_1)/f_T(t \mid \theta_0) > k$ if and only if $t > a$ and a is the same for all θ_1 because it is determined by the Type I error. However, this question tackles the problem from a more fundamental perspective with the notion that $p_\theta(b)$ is maximized at $b = \infty$ for all $\theta > \theta_0$. Otherwise, b would depend on θ, so the test would not be the same for each $\theta > \theta_0$.

2. It is not possible for a two-sided test to be uniformly most powerful. As in this problem, if $f(t \mid \theta)/f(t \mid \theta_0)$ is increasing for $\theta > \theta_0$, then H_0 is rejected if $t > a$ (i.e., if t is too big). Consider the alternative hypothesis $\theta < \theta_0$. If $f(t \mid \theta)/f(t \mid \theta_0)$ is decreasing for

$\theta < \theta_0$, then H_0 is rejected if t is too small. Now, consider the log-likelihood ratio for the exponential family in canonical form:

$$\log\{f(t \mid \theta)/f(t \mid \theta_0)\} = \eta(t)\{\theta - \theta_0\} + b(\theta_0) - b(\theta).$$

This log-likelihood ratio is increasing for all $\theta > \theta_0$ if and only if η is an increasing function. Therefore, $f(t \mid \theta)/f(t \mid \theta_0)$ must be decreasing for $\theta < \theta_0$. Then, $f(t \mid \theta_0)/f(t \mid \theta^*)$ is increasing and $f(t \mid \theta^*)/f(t \mid \theta_0)$ is decreasing. Therefore, the test $H_0 : \theta = \theta_0$ vs $H_1 : \theta = \theta^* < \theta_0$ can not be the same test as $H_0 : \theta = \theta_0$ vs $H_1 : \theta = \theta^* > \theta_0$.

3. If the likelihood ratio is not increasing in t, then reject the null hypothesis in the test $H_0 : \theta = \theta_0$ vs $H_1 : \theta \neq \theta_0$ if

$$\mathcal{L}(\widehat{\theta}) = \frac{f(t \mid \widehat{\theta})}{f(t \mid \theta_0)} > k,$$

where $\widehat{\theta}$ is the MLE. The Type I error for this test can be set using Wilks' theorem, which gives

$$-2\log\mathcal{L}(\widehat{\theta}) \approx \chi^2_d,$$

where d is the dimension of the model.

S4.1.6* – Test for a variance

(i) The likelihood function is given by

$$l(\sigma) = \sigma^{-n}\exp\left(-\tfrac{1}{2}\sum_{i=1}^{n} y_i^2/\sigma^2\right).$$

The relevant test statistic is $T = \sum_{i=1}^{n} y_i^2$, which is how the information in the data carries through to σ^2.

(ii) If each y_i is normal with mean 0 and variance σ_0^2, then each y_i/σ_0 is an independent standard normal random variable. Therefore, each y_i^2/σ_0^2 is an independent χ_1^2 random variable, and their sum is a χ_n^2 random variable. The mean and variance of a χ_1^2 random variable is 1 and 2, respectively. Thus, the mean and variance of a χ_n^2 random variable is n and $2n$, respectively.

(iii) The null hypothesis would be rejected when T is too large, so the aim is to find c for which $\mathrm{P}(\chi_n^2 > c) = \alpha$. Write c as $\chi^2_{n,\alpha}$.

(iv) The Type II error is one minus the power function evaluated at σ_1^2. If σ_1^2 is the true variance, then the power value is $\mathrm{P}(T/\sigma_0^2 > c)$, and T/σ_1^2 is χ_n^2. Thus, the power function at σ_1^2 is $\mathrm{P}(\chi_n^2 > qc)$, where $q = \sigma_0^2/\sigma_1^2 < 1$.

(v) One could use the central limit theorem because a χ_n^2 random variable is the sum of n independent χ_1^2 random variables. Therefore,

$$\frac{\chi_n^2 - n}{\sqrt{2n}} \to z,$$

where z is a standard normal random variable. Thus, $\chi^2_{n,\alpha} = n + z_\alpha \sqrt{2n}$ and

$$P(\chi^2_n > q\chi^2_{n,\alpha}) \to P\left(Z > n(q-1)/\sqrt{2n} + qz_\alpha\right) \to 1$$

because $n(q-1)/\sqrt{2n} \to -\infty$.

Further Reading

Historical Background. This exercise examines one of the simplest tests that was developed by Fisher, Neyman and Pearson for the parameters of a normal model. The test for the variance uses the chi-squared distribution because this is the distribution of the estimator and sufficient statistic.

Points of Interest.

1. The crucial point about the power function converging to 1 is that $\chi^2_{n,\alpha}$ always sits in the right tail of the χ^2_n density function. However, for any $q < 1$, it is that $q\chi^2_{n,\alpha}$ disappears down the left tail of the density function as $n \to \infty$.

2. The χ^2_1 density can be derived from the normal distribution function. Let Z be standard normal. For $x > 0$,

$$P(Z^2 \le x) = P(-\sqrt{x} < Z < \sqrt{x}) = 2\Phi(\sqrt{x}) - 1.$$

   The corresponding density function is

$$f_{\chi^2_1}(x) = \frac{1}{\sqrt{2\pi}} \exp(-\tfrac{1}{2}x)\, x^{-\frac{1}{2}},$$

   which is a $\text{Ga}(\frac{1}{2}, \frac{1}{2})$ random variable where $\frac{1}{2}^{\frac{1}{2}}/\Gamma(\frac{1}{2}) = 1/\sqrt{2\pi}$.

3. The fact that a χ^2_n random variable can be represented as the sum of n independent χ^2_1 random variables follows from the summable property of the gamma distribution with the same scale parameter.

4. The connection with the well-known Pearson χ^2 test is as follows: If x is multinomial with n objects and m categories with corresponding probabilities $\{p_j\}_{j=1}^m$, then

$$\sum_{j=1}^{m} \frac{(x_j - np_j)^2}{np_j} \approx \chi^2_{m-1}.$$

   This approximation follows because $E(x_j) = np_j$ and $\text{Var}(x_j) = np_j(1-p_j)$, so each term in the sum is approximately a χ^2_1 random variable. Further, the sum is a χ^2_{m-1} random variable rather than a χ^2_m random variable due to the constraint $\sum_{j=1}^m x_j = n$. The dropping of the degree of freedom explains why

$$S = \sum_{i=1}^{m} (z_i - \bar{z})^2$$

   is χ^2_{m-1} distributed when the $\{z_i\}$ are independent standard normal. This is actually an interesting result to prove, along with the additional result that S and $\bar{z}$ are independent.

Q4.2 Questions – Bayesian Framework

Q4.2.1 – Bayes factors and hypothesis testing

Introduction. This question compares classical and Bayesian hypothesis testing for simple hypotheses on the mean of a normal distribution. First, the critical value and power function for the classical test are computed. Then, the classical test is compared to the corresponding Bayesian test, which uses a Bayes factor. Finally, conditions for their equivalence are sought.

Question. Suppose observations $\{y_i\}$ for $i = 1, \ldots, n$ have a normal distribution with unknown mean θ and known variance σ^2. The aim is to test

$$H_0 : \theta = 0 \quad \text{vs} \quad H_1 : \theta \neq 0$$

with test statistic $T = \bar{y} = \sum_{i=1}^{n} y_i / n$.

(i) Describe the classical test in full. That is, find the critical value with a corresponding Type I error of α, and derive the power function for the test.

(ii) A Bayesian assigns the prior $\pi(\theta) \equiv \mathrm{N}(\theta \mid 0, \tau^2)$. Find the Bayes factor for the test given by

$$B = \frac{m(\text{data} \mid H_1)}{m(\text{data} \mid H_0)},$$

where $m(\text{data} \mid H)$ is the marginal likelihood of the data given hypothesis H.

(iii) The Bayesian rejects H_0 if $B > \lambda$ for some $\lambda > 0$. How would the Bayesian set λ such that the Type I error of the test is α?

(iv) Using the choice of λ from part (iii), find the corresponding power function for the Bayesian test.

(v) Comment on the findings.

Q4.2.2 – Bayes factor for a linear regression model

Introduction. In Bayesian hypothesis testing, prior probabilities are specified for competing hypotheses. These priors represent initial beliefs about the hypotheses' correctness. Decisions concerning the hypotheses are often made using Bayes factors, which are ratios of the evidence of the competing models. This question uses Bayes factors to assess hypotheses for a linear regression model.

Question. Consider the linear model $\mathbf{y} \sim \mathrm{N}_n(\mathbf{X}\boldsymbol{\beta}, \sigma^2 \mathbf{I}_n)$, where $\mathbf{y} = (y_1, \ldots, y_n)'$ is a $n \times 1$ vector of observations, $\mathbf{X}$ is a $n \times p$ matrix of explanatory variables, $\boldsymbol{\beta}$ is a $p \times 1$ vector of unknown regression parameters, and σ^2 is a known variance term. Further, let $\mathbf{I}_n$ denote the $n \times n$ identity matrix and N_n denote the n–multivariate normal distribution. The prior for $\boldsymbol{\beta}$ is given by the p–variate normal distribution with mean $\mathbf{0}$ and variance–covariance matrix given by $g^{-1}\sigma^2(\mathbf{X}'\mathbf{X})^{-1}$ for some $g > 0$.

(i) Find the posterior distribution for $\boldsymbol{\beta}$.

(ii) For $p = 1$, find the Bayes factor for comparing the fully specified model and the model with $\beta = 0$:

$$\mathrm{B} = \frac{p(\mathbf{y})}{p(\mathbf{y} \mid \beta = 0)},$$

where $p(\mathbf{y}) = \int \mathrm{N}_n(\mathbf{y} \mid \mathbf{x}\beta, \sigma^2\,\mathbf{I}_n)\,\pi(\beta)\,d\beta$ and $\mathbf{x} = (x_1, \ldots, x_n)'$. Further, show that

$$\log(B) = \frac{1}{2}\frac{z^2}{(1+g)\,\sigma^2\,\sum_{i=1}^n x_i^2} + \tfrac{1}{2}\log\left(\frac{g}{1+g}\right),$$

where $z = \sum_{i=1}^n x_i\,y_i$.

(iii) What is the expected value of $\log(B)$ under the model with $\beta = 0$?

(iv) Given the prior for $\boldsymbol{\beta}$, show that it is possible to write $\log(B) = \alpha\chi_1^2 + \gamma$, where χ_1^2 is a chi-squared random variable with 1 degree of freedom.

(v) How would one decide whether to include the single covariate using the Bayes factor?

Q4.2.3 – Properties of a Bayes factor for a normal mean

Introduction. This question looks at properties of the Bayes factor for testing a normal mean to be 0. Different representations of the Bayes factor are used depending on the validity of the null hypothesis. Additionally, a symmetric mean on the prior is considered.

Question. Suppose $\{y_1, \ldots, y_n\}$ are independent and identically distributed from the normal distribution with unknown mean θ and known variance σ^2. The Bayes factor for comparing the model with prior $\pi(\theta)$ and the model with $\theta = 0$ is given by

$$B = \frac{\int \prod_{i=1}^n \mathrm{N}(y_i \mid \theta, \sigma^2)\,\pi(\theta)\,d\theta}{\prod_{i=1}^n \mathrm{N}(y_i \mid 0, \sigma^2)}.$$

Let $m(\overline{y}) = \int \mathrm{N}(\overline{y} \mid \theta, \sigma^2/n)\,\pi(\theta)\,d\theta$, where $\overline{y} = \sum_{i=1}^n y_i/n$.

(i) Show that it is possible to write

$$\mathrm{E}(\theta \mid \overline{y}) = \overline{y} + \frac{\sigma^2 m'(\overline{y})}{n\,m(\overline{y})}.$$

(ii) Show that the Bayes factor can be written as a function of $\overline{y}$:

$$B(\overline{y}) = \frac{m(\overline{y})}{\mathrm{N}(\overline{y} \mid 0, \sigma^2/n)}.$$

(iii) Show that $\mathrm{E}(\theta \mid \overline{y}) = (\sigma^2/n)\ B'(\overline{y})/B(\overline{y})$.

(iv) Show that the expected value of the Bayes factor is 1 if $\theta = 0$.

(v) Let θ^* denote the true value of θ. Show that the expected value of the Bayes factor is

$$\mathrm{E}\{B(\overline{Y})\} = \int \exp\{n\,\theta\,\theta^*/\sigma^2\}\,\pi(\theta)\,d\theta.$$

Derive the special case if $\pi(\theta)$ is a symmetric density about 0.

Q4.2.4* – Invariance properties for Bayesian variable selection

Introduction. Variable selection is a widely studied problem within statistics, yet an often neglected question in variable selection is whether there should be any rules to follow. For example, if the variance is unknown, should rescaling the data have an effect on the decision as to which covariates are kept? This exercise considers a linear transformation to the data and argues that the choice to keep a subset of covariates should not change. Indeed, the choice does not change with a classical test based on the likelihood evaluated at the MLE. However, the Bayesian test based on the Bayes factor can lead to different decisions.

Question. Consider the linear model

$$y_i = x_{i1}\beta_1 + x_{i2}\beta_2 + \sigma\epsilon_i,$$

where the $\boldsymbol{\beta} = (\beta_1, \beta_2)'$ are unknown, σ is known, and the $\{\epsilon_i\}$ are independent standard normal random variables for $i = 1, \ldots, n$. Write this model as $M(\beta_1, \beta_2)$. The model with $\beta_2 = 0$ is written as $M(\beta_1, 0)$.

(i) Derive the least squares estimator for $\boldsymbol{\beta}$, written as $\widehat{\boldsymbol{\beta}}$.

(ii) The data are transformed to $\widetilde{y}_i = y_i + c\,x_{i1}$ for some known $c > 0$. Show that the least squares estimator for $\boldsymbol{\beta}$ with data $\widetilde{\mathbf{y}} = (\widetilde{y}_1, \ldots, \widetilde{y}_n)'$ is given by

$$\widetilde{\boldsymbol{\beta}} = \widehat{\boldsymbol{\beta}} + c\begin{pmatrix} 1 \\ 0 \end{pmatrix}.$$

(iii) Show that the residual sum of squares with data $\widetilde{\mathbf{y}}$ and estimator $\widetilde{\boldsymbol{\beta}}$ does not depend on c.

(iv) With the transformed data $\widetilde{\mathbf{y}}$, explain why the usual classical test for hypotheses $H_0 : \beta_2 = 0$ vs $H_1 : \beta_2 \neq 0$ does not depend on c.

(v) Show that the Bayes factor for comparing the model $M(\beta_1, \beta_2)$ with the model $M(\beta_1, 0)$ does depend on c. Here, assume independent normal priors for β_1 and β_2. Is there a choice for the prior of β_1 that removes the dependence of the Bayes factor on c?

S4.2 Solutions – Bayesian Framework

S4.2.1 – Bayes factors and hypothesis testing

(i) The test statistic $T = \bar{y}$ has distribution $\mathrm{N}\left(\theta, \sigma^2/n\right)$. Under $H_0, \bar{y} \sim \mathrm{N}\left(0, \sigma^2/n\right)$, so define the standard normal random variable $z_0 = \sqrt{n}\,\bar{y}/\sigma$. The critical value k_α for a test with Type I error α satisfies $\mathrm{P}(|\bar{y}| > k_\alpha) = \alpha$. Multiply both terms in the inequality by $\sqrt{n}/\sigma$ to obtain $\mathrm{P}(|z_0| > z_{\alpha/2})$, where $z_{\alpha/2} = \sqrt{n}\,k_\alpha/\sigma$. Then, the critical value can be computed using a z-table.

Under H_1, consider the standard normal random variable $z_1 = \sqrt{n}(\bar{y} - \theta)/\sigma$. The corresponding power function is

$$\begin{aligned}\beta(\theta) &= \mathrm{P}(|\overline{Y}| > k_\alpha \mid H_1)\\ &= \mathrm{P}\left(\frac{\sigma}{\sqrt{n}} z_1 + \theta > k_\alpha \ \text{ or } \ \frac{\sigma}{\sqrt{n}} z_1 + \theta < -k_\alpha\right)\\ &= 1 - \Phi\left(z_{\alpha/2} - \frac{\sqrt{n}\theta}{\sigma}\right) + \Phi\left(-z_{\alpha/2} - \frac{\sqrt{n}\theta}{\sigma}\right),\end{aligned}$$

where $\Phi(\cdot)$ denotes the standard normal cumulative distribution function.

(ii) Let $\mathbf{y} = (y_1, \ldots, y_n)'$. The Bayes factor can be written as

$$\begin{aligned}B &= \frac{m(\mathbf{y} \mid H_1)}{m(\mathbf{y} \mid H_0)} = \frac{\int p(\mathbf{y} \mid \theta \neq 0)\pi(\theta \mid H_1)d\theta}{\int p(\mathbf{y} \mid \theta = 0)\pi(\theta \mid H_0)d\theta}\\ &= \frac{\int \mathrm{N}(\bar{y} \mid \theta, \sigma^2/n)\mathrm{N}(\theta \mid 0, \tau^2)d\theta}{\mathrm{N}(\bar{y} \mid 0, \sigma^2/n)}\\ &= \frac{\mathrm{N}(\bar{y} \mid 0,\, \sigma^2/n + \tau^2)}{\mathrm{N}(\bar{y} \mid 0,\, \sigma^2/n)}\\ &= \left(1 + \tau^2 n/\sigma^2\right)^{-\frac{1}{2}} \exp\left\{\frac{n\tau^2}{2\sigma^2(\sigma^2/n + \tau^2)}\, \bar{y}^2\right\}.\end{aligned}$$

(iii) The Bayes factor B is a monotone increasing function of the classical test statistic $\bar{y}$. Thus, rejecting the null hypothesis when $B > \lambda_\alpha$ is equivalent to rejecting H_0 when $|\bar{y}| > k_\alpha$, where

$$\lambda_\alpha = \left(1 + \tau^2 n/\sigma^2\right)^{-\frac{1}{2}} \exp\left\{\frac{n\tau^2}{2\sigma^2(\sigma^2/n + \tau^2)}\, k_\alpha^2\right\}.$$

Note that λ_α has a subscript to explicitly show its dependence on α. Finally, replace the test statistic $\bar{y}$ with the frequentist critical value k_α to obtain the Bayesian critical value λ_α.

(iv) The power function under this choice of λ_α is

$$\begin{aligned}\beta(\theta) &= \mathrm{P}(B > \lambda_\alpha \mid H_1) = \mathrm{P}(|\overline{Y}| > k_\alpha \mid H_1)\\ &= 1 - \Phi\left(z_{\alpha/2} - \sqrt{n}\theta/\sigma\right) + \Phi\left(-z_{\alpha/2} - \sqrt{n}\theta/\sigma\right),\end{aligned}$$

which is the same as in part (i) for the classical hypothesis test.

(v) In this exercise, the Bayes factor is a monotone function of the classical test statistic. Therefore, the Bayesian and classical tests are equivalent for the same Type I error α. However, the difference between the two frameworks is that the frequentist will select the critical region based on a specified α, whereas the Bayesian will select the critical region (based on rules-of-thumb or prior beliefs) and be left with a corresponding α which may be quite large.

Further Reading

Historical Background. The Bayesian test with a composite alternative hypothesis H_1 has always been problematic because the specification of the prior $\pi(\theta)$ introduces subjectivity. In fact, the prior for composite hypotheses can be overly influential when used in combination with Bayes factors (Young and Pettit, 1996). The impact that priors have on decision making is often discussed in the context of the Jeffreys–Lindley paradox, which draws on the conflict between the Bayes factor and the p-value associated with the classical test (Jeffreys, 1935; Lindley, 1957; Bartlett, 1957). Additionally, a variety of commentaries have been written that contrast Bayesian and classical testing approaches (Held and Ott, 2018).

Points of Interest.

1. Traditionally, the classical approach to testing is summarized by the p-value. Despite this, the interpretation of a p-value has been "made extraordinarily difficult because it is not part of any formal system of statistical inference" (Goodman, 2008). The American Statistical Association (ASA) released a statement on p-values, focusing on aspects and issues that are too often "misunderstood and misused" (Wasserstein and Lazar, 2016). The Bayesian counterpart makes direct probability statements about a hypothesis by using the observed sample to update prior beliefs on the hypothesis.

2. The historical definition of the p-value does not present it as a probability at all. Indeed, it is formally defined as the smallest α for which the test statistic T lies in the critical region C_α. It is then possible from this definition to show that the distribution of the p-value under the null hypothesis is a standard uniform random variable.

3. The Bayesian test for a point null hypothesis is often problematic because the prior on the alternative must necessarily include the null. For example, suppose the null hypothesis for a normal mean is $H_0 : \theta = 0$. The prior for the alternative is usually a density on $\mathbb{R}$, which includes the null point. Thus, it is difficult to assess the evidence for the null hypothesis with such a prior on the alternative. Continuing with the normal example, let the variance be 1 and the alternative hypothesis have a normal prior with mean 0 and variance $1/\lambda$. Then, the Bayes factor is given by

$$B = \sqrt{\frac{\lambda}{\lambda + n}} \exp\left\{\frac{1}{2}\frac{n^2\overline{x}^2}{n+\lambda}\right\},$$

where the null is rejected if B is "too large"; how large is a key concern. Naturally, one might look at the Type I error because errors can be made regardless of how the decision is implemented. However, the irony is that if a Type I error is to be set for size α, the decision reverts to the classical test: reject H_0 if $n\overline{x}^2 > \chi^2_{1,1-\alpha}$, regardless of the choice for λ. On the other hand, if the decision rule becomes $B > \tau$ for some $\tau > 0$, then H_0 is rejected if

$$n\overline{x}^2 > 2(1 + \lambda/n)\ \log\{\tau(1 + \lambda/n)\}.$$

Note that the right-hand side of the inequality becomes $2\log\tau$ for large n. Therefore, the probability of rejecting H_0 when it is true is $\mathrm{P}(\chi_1^2 > 2\log\tau)$. It then appears that setting τ in ignorance of this probability is not a good idea.

4. This question reveals the following main idea: if a Bayes factor is used as a test statistic that depends on information beyond the data (i.e., the prior), then the use of prior information to preserve a Type I error becomes redundant. On the other hand, if the prior information is pivotal, it is the Type I error that could become problematic.

5. Naturally, one might ask whether the Bayesian should be interested in Type I errors. Any decision, no matter how it is made, is always prone to being wrong. If the probability of rejecting the null given that it is true (i.e., α) can be computed, then one must decide whether the test should be determined by providing an upper bound for α – making any prior belief irrelevant – or whether the use of prior information should override a possibly large Type I error.

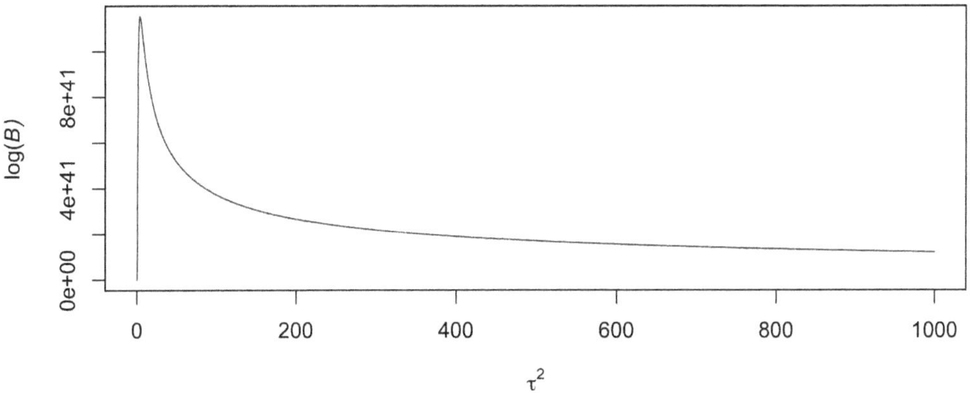

FIGURE 4.4
The log-Bayes factor as a function of τ^2 for the test of a normal mean with variance 1 and hypotheses $H_0 : \theta = 0$ vs $H_1 : \theta \neq 0$.

Demonstration. Consider the Bayes factor for the test of a normal mean with known variance 1 and hypotheses $H_0 : \theta = 0$ vs $H_1 : \theta \neq 0$. The prior for θ is normal with mean 0 and variance τ^2, the sample size is $n = 50$, and $\bar{y}$ is taken to be 1; this sample mean suggests that the alternative hypothesis is true. However, Fig. 4.4, which plots the log-Bayes factor as a function of τ^2, tells a different story. As $\tau^2 \to \infty$, representing a decreasingly informative prior, the Bayes factor decreases to 0. Notably, the log-Bayes factor

$$\log B \approx -\tfrac{1}{2}\log(1+\tau^2 n/\sigma^2) + \tfrac{1}{2}n\bar{y}^2/\sigma^2$$

converges to $-\infty$, which supports the null hypothesis.

S4.2.2 – Bayes factor for a linear regression model

(i) The kernel for the posterior distribution of $\boldsymbol{\beta}$ is

$$\pi(\boldsymbol{\beta} \mid \mathbf{y}) \propto \exp\left\{-\frac{1}{2\sigma^2}(\mathbf{y}-\mathbf{X}\boldsymbol{\beta})'(\mathbf{y}-\mathbf{X}\boldsymbol{\beta})\right\} \cdot \exp\left\{-\frac{g}{2\sigma^2}\boldsymbol{\beta}'\mathbf{X}'\mathbf{X}\boldsymbol{\beta}\right\}$$
$$\propto \exp\left\{-\frac{1}{2\sigma^2}\left[\boldsymbol{\beta}'\mathbf{X}'\mathbf{X}\boldsymbol{\beta} - 2\mathbf{y}'\mathbf{X}\boldsymbol{\beta}\right] - \frac{g}{2\sigma^2}\boldsymbol{\beta}'\mathbf{X}'\mathbf{X}\boldsymbol{\beta}\right\}$$
$$\propto \exp\left\{-\frac{1}{2}\left[-2\Bigg(\underbrace{\frac{1}{\sigma^2}\mathbf{y}'\mathbf{X}}_{\mathbf{b}'}\Bigg)\boldsymbol{\beta} + \boldsymbol{\beta}'\Bigg(\underbrace{\frac{1}{\sigma^2}\mathbf{X}'\mathbf{X}(1+g)}_{\mathbf{A}}\Bigg)\boldsymbol{\beta}\right]\right\}.$$

This kernel is uniquely identified as a multivariate normal distribution with mean $\mathbf{A}^{-1}\mathbf{b}$ and covariance matrix $\mathbf{A}^{-1}$.

(ii) First, compute the marginal likelihood $p(\mathbf{y}) = \int p(\mathbf{y} \mid \beta)p(\beta)d\beta$. Note that it is possible to write $\beta = (\sigma/\sqrt{g})(\mathbf{X}'\mathbf{X})^{-1/2}\eta$, where η is a standard normal random variable. Hence,

$$\mathbf{y} = \mathbf{X}(\sigma/\sqrt{g})\,(\mathbf{X}'\mathbf{X})^{-1/2}\eta + \sigma\varepsilon,$$

where $\mathrm{E}(\mathbf{Y}) = \mathbf{0}$ and ε is a vector of independent standard normal random variables. Additionally,

$$\mathrm{Cov}(\mathbf{Y}) = \sigma^2(\mathbf{X}(\mathbf{X}'\mathbf{X})^{-1}\mathbf{X}'/g + \mathbf{I}) \equiv \sigma^2\boldsymbol{\Sigma},$$

which gives

$$p(\mathbf{y}) = \frac{1}{|\boldsymbol{\Sigma}|^{1/2}\sigma^n(2\pi)^{n/2}}\exp\left(-\frac{1}{2\sigma^2}\mathbf{y}'\boldsymbol{\Sigma}^{-1}\mathbf{y}\right).$$

Note that $\boldsymbol{\Sigma} = \mathbf{I} + \mathbf{v}\mathbf{v}'/g$, where the ith component of $\mathbf{v}$ is

$$v_i = \frac{x_i}{\sum_{i=1}^n x_i^2}.$$

Therefore, the determinant and inverse of $\boldsymbol{\Sigma}$ are $|\boldsymbol{\Sigma}| = 1 + \mathbf{v}'\mathbf{v}/g = 1 + 1/g$ and $\boldsymbol{\Sigma}^{-1} = \mathbf{I} - \mathbf{v}\mathbf{v}'/(1+g)$, respectively. Now, if $\beta = 0$, then

$$p(\mathbf{y} \mid \beta = 0) = \frac{1}{\sigma^n(2\pi)^{n/2}}\exp\left(-\frac{1}{2\sigma^2}\mathbf{y}'\mathbf{y}\right).$$

Therefore,

$$B = \frac{p(\mathbf{y})}{p(\mathbf{y} \mid \beta = 0)} = \sqrt{\frac{g}{1+g}}\exp\left\{\frac{\mathbf{y}'\mathbf{v}\mathbf{v}'\mathbf{y}}{2\sigma^2(1+g)}\right\},$$

where

$$\mathbf{y}'\mathbf{v}\mathbf{v}'\mathbf{y} = \frac{z^2}{\sum_{i=1}^n x_i^2}.$$

Apply the logarithm to B to obtain the result.

(iii) If $\beta = 0$, then z is normal with mean 0 and variance $\sigma^2 \sum_{i=1}^{n} x_i^2$ because z is a linear combination of independent normal random variables with mean 0 and variance σ^2. Thus, $\mathrm{E}(z^2) = \mathrm{Var}(z)$ and

$$\mathrm{E}\left\{\log(B)\right\} = \frac{1}{2(1+g)} + \tfrac{1}{2}\log\left(\frac{g}{1+g}\right).$$

(iv) Under the null hypothesis, z is normal with mean 0 and variance $\sigma^2 \sum_{i=1}^{n} x_i^2$. Therefore,

$$\chi_1^2 = \frac{z^2}{\sigma^2 \sum_{i=1}^{n} x_i^2}$$

is a chi-squared random variable with 1 degree of freedom, so

$$\log(B) = \frac{\chi_1^2}{2(1+g)} + \frac{1}{2}\log\left(\frac{g}{1+g}\right),$$

where $\alpha = 1/\{2(1+g)\}$ and $\gamma = \log\{g/(1+g)\}/2$.

(v) The null hypothesis would be rejected if $\log(B)$ is too large. Often, the critical values for Bayes factors can be vaguely set or rely on rules of thumb. A practical approach to specifying critical values is to leverage the connection between $\log(B)$ and the χ_1^2 random variable to set the critical value to coincide with a Type I error of a predetermined size. Otherwise, the Type I error may be excessively large because such an error always exists.

Further Reading

Historical Background. Bayes factors are the most common tool for assessing model hypotheses in the Bayesian framework. However, it is well-known that the Bayes factor depends heavily on the choice of prior (Kass and Raftery, 1995). One issue that has received substantial attention is their use with improper priors; the Jeffreys–Lindley paradox discusses the problem with using an improper prior in Bayesian hypothesis testing (Jeffreys, 1935; Lindley, 1957; Bartlett, 1957). For example, if the normal model has unknown mean θ, then the improper prior for θ is normal with mean 0 and variance ∞: $\pi(\theta) \propto 1$. However, regardless of the true value of θ, the posterior for the null hypothesis tends to one; this motivates the use of proper priors for the linear model. For more discussion, see Question 4.2.1.

Points of Interest.

1. The prior used in this question is known as the g-prior and was introduced in Zellner (1986). The influence of the value of g is evident in the expression for $\log B$, as is demonstrated in Fig 4.5. There are various ways to choose g, including placing a prior on it (Liang et al., 2008).

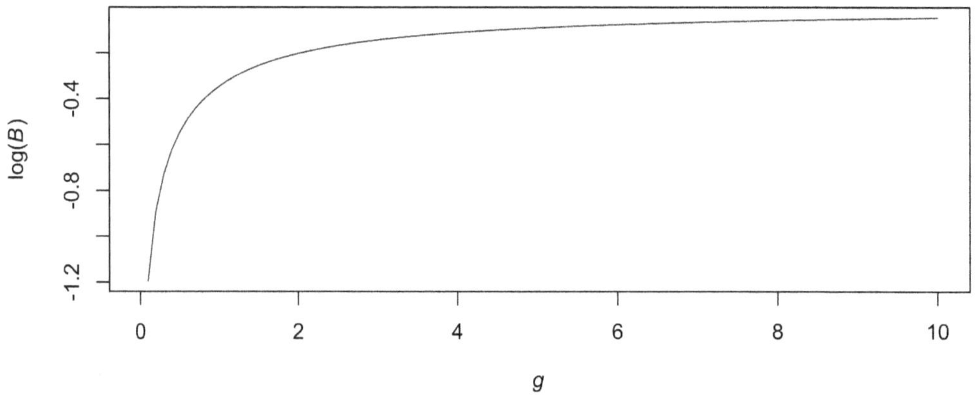

FIGURE 4.5
Log(B) as a function of g.

S4.2.3 – Properties of a Bayes factor for a normal mean

(i) Note that the desired expression contains $m'(\overline{y})$, so differentiate $m(\overline{y})$ with respect to $\overline{y}$. Assuming that the order of integration and differentiation can be swapped, $m'(\overline{y})$ is given by

$$m'(\overline{y}) = -\int \frac{n}{\sigma^2}(\overline{y}-\theta)\,\mathrm{N}(\overline{y} \mid \theta, \sigma^2/n)\,\pi(\theta)\,d\theta.$$

By definition, the posterior mean is

$$\mathrm{E}(\theta \mid \overline{y}) = \frac{\int \theta\,\mathrm{N}(\overline{y} \mid \theta, \sigma^2/n)\pi(\theta)d\theta}{\int \mathrm{N}(\overline{y} \mid \theta, \sigma^2/n)\pi(\theta)d\theta} = \frac{\int \theta\,\mathrm{N}(\overline{y} \mid \theta, \sigma^2/n)\pi(\theta)d\theta}{m(\overline{y})}.$$

Hence, $m'(\overline{y}) = -(n/\sigma^2)\,\{\overline{y}\,m(\overline{y}) - \mathrm{E}(\theta \mid \overline{y})\,m(\overline{y})\}$, which gives the required result.

(ii) Note that

$$\sum_{i=1}^{n}(y_i-\theta)^2 = \sum_{i=1}^{n} y_i^2 - 2\theta n\overline{y} + n\theta^2 = \sum_{i=1}^{n} y_i^2 - n\overline{y}^2 + n(\overline{y}-\theta)^2,$$

so

$$\prod_{i=1}^{n}\mathrm{N}(y_i \mid \theta, \sigma^2) = \underbrace{\frac{1}{(2\pi\sigma^2)^{n/2}}\exp\left\{-\frac{\sum_{i=1}^{n} y_i^2}{2\sigma^2}\right\}}_{a}\underbrace{\exp\left\{\frac{n\overline{y}^2}{2\sigma^2}\right\}}_{b}\underbrace{\exp\left\{-\frac{n(\overline{y}-\theta)^2}{2\sigma^2}\right\}}_{c}.$$

The a term is equivalent to $\prod_{i=1}^{n}\mathrm{N}(y_i \mid 0, \sigma^2)$. In the denominator, b is the kernel of $\mathrm{N}(\overline{y} \mid 0, \sigma^2/n)$ and, in the numerator, c is the kernel of $\mathrm{N}(\overline{y} \mid \theta, \sigma^2/n)$; note that the normalizing constants for these two kernels are equivalent. Thus, the result follows.

(iii) From part (ii), $m(\overline{y}) = B(\overline{y})\mathrm{N}(\overline{y} \mid 0, \sigma^2/n)$. Use the product rule to find the derivative of $m(\overline{y})$ with respect to $\overline{y}$:

$$m'(\overline{y}) = \left(-\frac{n\overline{y}}{\sigma^2}\right)B(\overline{y})\mathrm{N}(\overline{y} \mid 0, \sigma^2/n) + B'(\overline{y})\mathrm{N}(\overline{y} \mid 0, \sigma^2/n).$$

Plug these new expressions for $m(\overline{y})$ and $m'(\overline{y})$ in the answer to part (i) to obtain the solution.

(iv) Under the null hypothesis (i.e., $\theta = 0$), $\overline{y}$ is distributed as $\mathrm{N}(0, \sigma^2/n)$. Thus,

$$B(\overline{y}) = \frac{\int \mathrm{N}(\overline{y} \mid 0, \sigma^2/n)\, \pi(\theta)\, d\theta}{\mathrm{N}(\overline{y} \mid 0, \sigma^2/n)} = \int \pi(\theta) d\theta = 1.$$

(v) The form of $B(\overline{y})$ in part (ii) can be simplified to

$$B(\overline{y}) = \int \exp\{-n\theta^2/(2\sigma^2) + \theta\, \overline{y}\, n/\sigma^2\}\, \pi(\theta)\, d\theta.$$

If θ^* is the true value for θ, then $\overline{y}$ has distribution $\mathrm{N}(\theta^*, \sigma^2/n)$ with moment-generating function $M_{\overline{y}}(t) = \mathrm{E}(e^{t\overline{Y}}) = \exp\{\theta^* t + \sigma^2 t^2/(2n)\}$. Therefore,

$$\begin{aligned}
\mathrm{E}\{B(\overline{Y})\} &= \int \exp\{-n\theta^2/(2\sigma^2)\}\mathrm{E}\left(\exp\{\theta\, \overline{Y}\, n/\sigma^2\}\right) \pi(\theta) d\theta \\
&= \int \exp\{-n\theta^2/(2\sigma^2)\} \exp\left\{\frac{n\theta^*\theta}{\sigma^2} + \frac{n\theta^2}{2\sigma^2}\right\} \pi(\theta) d\theta \\
&= \int \exp\left\{\frac{n\theta^*\theta}{\sigma^2}\right\} \pi(\theta) d\theta,
\end{aligned}$$

where the moment-generating function evaluated at $t = \theta n/\sigma^2$ is used to obtain the second equality. Note that if $\pi(\theta)$ is symmetric about 0, then the expectation can be expressed as

$$\int_{\theta>0} \left\{\exp(-n\theta\theta^*/\sigma^2) + \exp(n\theta\theta^*/\sigma^2)\right\} \pi(\theta)\, d\theta,$$

which can be written as $2\int_{\theta>0} \cosh(n\theta\theta^*/\sigma^2)\, \pi(\theta)\, d\theta$.

Further Reading

Historical Background. The representations for the posterior mean in parts (i) and (iii) date back to Perrichi and Smith (1992); they were concerned with Bayesian robustness and studied what would happen to the posterior mean if the sample mean approached infinity. These representations explicitly reveal how the prior influences the posterior mean behavior with respect to the observed y. For further insights, see Mitchell (1994) and Choy and Smith (1997).

Points of Interest.

1. The key to understanding the influence that the prior has on the posterior mean is knowing how $m'(\overline{y})/m(\overline{y})$ behaves for large $\overline{y}$. Because the prior is influential, choices must be made to obtain desirable properties for the posterior mean as $\overline{y}$ grows (i.e., whether the posterior mean tracks $\overline{y}$ or at some point reverts to the prior mean). In part (iii), the Bayes factor is shown to be related to the posterior mean. For more on the properties of the Bayes factor, see Kass and Raftery (1995).

2. Another way to view the marginal likelihood for this exercise is

$$m(\overline{y}) = \int \mathrm{N}(\overline{y} \mid \theta, \sigma^2/n)\, \pi(\theta)\, d\theta = \mathrm{E}\{\pi(\theta)\},$$

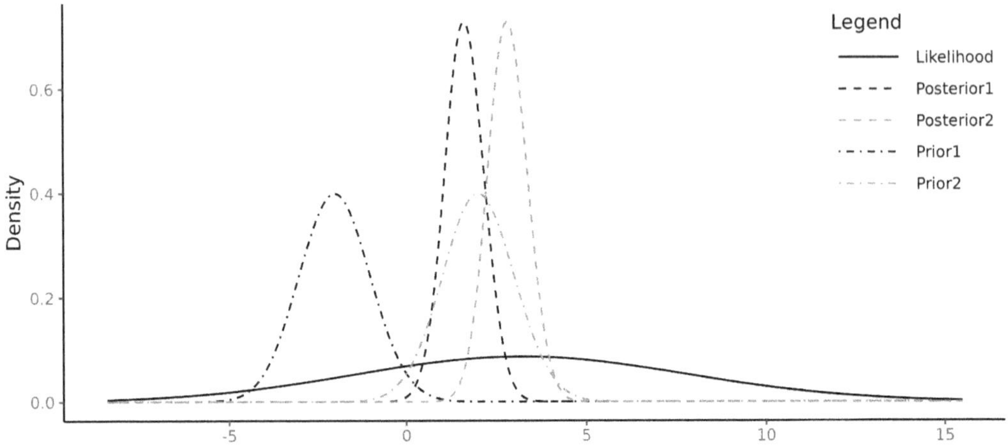

FIGURE 4.6
Gaussian likelihood and posteriors under different prior specifications.

where the expectation is with respect to $\mathrm{N}(\theta \mid \overline{y}, \sigma^2/n)$. Let $\theta = \overline{y} + z\,\sigma/\sqrt{n}$, where z is a standard normal random variable. Then,

$$m(\overline{y}) = \pi(\overline{y}) + \tfrac{1}{2}\,\sigma^2\pi''(\overline{y})/n + \text{higher order terms.}$$

This approach is also relevant for other normal likelihood problems, including the linear model.

Demonstration. Consider the case of 50 data points coming from a data generating function that corresponds to a Gaussian with $\mu = 3$ and $\sigma = 5$. The aim is to estimate μ. Fig. 4.6 shows the direct impact of prior choices on the posterior. For example, the black lines use a $\mathrm{N}(-2, 1)$ prior and the gray lines use a $\mathrm{N}(2, 1)$ prior. Because the black prior has a lower mean, the black posterior is centered on a value less than $\mu = 3$. In contrast, the mean of the gray prior is closer to the truth, so the gray posterior is centered closer to $\mu = 3$.

S4.2.4* – Invariance properties for Bayesian variable selection

(i) To find the least squares estimator for $\boldsymbol{\beta}$, set

$$\widehat{\boldsymbol{\beta}} = \arg\min_{\boldsymbol{\beta}} \sum_{i=1}^{n} \left(y_i - \mathbf{x}_i'\boldsymbol{\beta}\right)^2,$$

where $\mathbf{x}_i = (x_{i1}, x_{i2})'$. Minimizing this expression requires the derivative, i.e.

$$\frac{\partial}{\partial \boldsymbol{\beta}} \left\{ \sum_{i=1}^{n} \left(y_i - \mathbf{x}_i'\boldsymbol{\beta}\right)^2 \right\} = -2\sum_{i=1}^{n} \mathbf{x}_i y_i + 2\sum_{i=1}^{n} \mathbf{x}_i\mathbf{x}_i'\boldsymbol{\beta}.$$

Set the derivative to zero to obtain the least squares estimator:

$$\widehat{\boldsymbol{\beta}} = \left(\sum_{i=1}^{n} \mathbf{x}_i\mathbf{x}_i'\right)^{-1} \sum_{i=1}^{n} \mathbf{x}_i y_i.$$

(ii) If the data are transformed to $\widetilde{y}_i = y_i + c\, x_{i1}$, the least squares estimator for $\boldsymbol{\beta}$ will become

$$\widetilde{\boldsymbol{\beta}} = \left(\sum_{i=1}^{n} \mathbf{x}_i \mathbf{x}_i'\right)^{-1} \sum_{i=1}^{n} \mathbf{x}_i \widetilde{y}_i.$$

Rewrite the transformed data as $\widetilde{y}_i = y_i + c\mathbf{x}_i'\mathbf{u}$ with $\mathbf{u} = (1,0)'$. Then,

$$\begin{aligned}
\widetilde{\boldsymbol{\beta}} &= \left(\sum_{i=1}^{n} \mathbf{x}_i \mathbf{x}_i'\right)^{-1} \sum_{i=1}^{n} \mathbf{x}_i \left(y_i + c\mathbf{x}_i'\mathbf{u}\right) \\
&= \left(\sum_{i=1}^{n} \mathbf{x}_i \mathbf{x}_i'\right)^{-1} \sum_{i=1}^{n} \mathbf{x}_i y_i + c \left(\sum_{i=1}^{n} \mathbf{x}_i \mathbf{x}_i'\right)^{-1} \sum_{i=1}^{n} \mathbf{x}_i \mathbf{x}_i' \mathbf{u} \\
&= \widehat{\boldsymbol{\beta}} + c\mathbf{u}.
\end{aligned}$$

(iii) Let $\widehat{\boldsymbol{\beta}} = (\widehat{\beta}_1, \widehat{\beta}_2)'$, so $\widetilde{\boldsymbol{\beta}} = (\widehat{\beta}_1 + c, \widehat{\beta}_2)'$. To obtain the RSS, compute

$$\begin{aligned}
\text{RSS} &= \sum_{i=1}^{n} \left(\widetilde{y}_i - \mathbf{x}_i'\widetilde{\boldsymbol{\beta}}\right)^2 \\
&= \sum_{i=1}^{n} \left\{y_i + c x_{i1} - x_{i1}(\widehat{\beta}_1 + c) - x_{i2}\widehat{\beta}_2\right\}^2 \\
&= \sum_{i=1}^{n} \left(y_i - x_{i1}\widehat{\beta}_1 - x_{i2}\widehat{\beta}_2\right)^2 \\
&= \sum_{i=1}^{n} \left(y_i - \mathbf{x}_i'\widehat{\boldsymbol{\beta}}\right)^2,
\end{aligned}$$

which does not depend on c because the terms containing c in the transformed data $\widetilde{\mathbf{y}}$ and the corresponding least squares estimator $\widetilde{\boldsymbol{\beta}}$ cancel out.

(iv) The most common classical test for testing whether a parameter is equal to 0 is the t-test, which makes use of the following statistic:

$$F = \frac{\text{RSS}_{|\beta_2=0} - \text{RSS}}{\text{RSS}/(n-1)}.$$

In particular, the t-test makes use of the square root of F, which is a function of the RSS. As shown in part (iii), the RSS under the transformed data $\widetilde{\mathbf{y}}$ is the same as under the original data $\mathbf{y}$. So, the test statistic F and the classical test do not depend on c.

(v) The relevant Bayes factor uses independent priors $\text{N}(\beta_1 \mid 0, \tau_1^2)$ and $\text{N}(\beta_2 \mid 0, \tau_2^2)$ for the full model and $\text{N}(\beta_1 \mid 0, \tau_1^2)$ for the competing model, where $\tau_1, \tau_2 > 0$. Under the full model, the marginal model for $\mathbf{y}$ is multivariate normal with mean $\mathbf{0}$ and covariance matrix $\boldsymbol{\Sigma}$, where

$$\Sigma_{i,j} = x_{i1}x_{j1}\tau_1^2 + x_{i2}x_{j2}\tau_2^2 + \sigma^2\, 1\{i=j\}.$$

The marginal likelihood for $\mathbf{y}$ under the reduced model is multivariate normal with mean $\mathbf{0}$ and covariance matrix $\boldsymbol{\Omega}$, where

$$\Omega_{i,j} = x_{i1}x_{j1}\tau_1^2 + \sigma^2\, 1\{i=j\}.$$

The corresponding marginal distributions for $\tilde{\mathbf{y}}$ are as above but with mean $-c\mathbf{x}_1$ instead of $\mathbf{0}$, where $\mathbf{x}_1 = (x_{11}, \ldots, x_{n1})'$. The relevant term in the Bayes factor that contains c is

$$(\mathbf{y} + c\mathbf{x}_1)'(\boldsymbol{\Sigma}^{-1} - \boldsymbol{\Omega}^{-1})(\mathbf{y} + c\mathbf{x}_1).$$

For this term to not depend on c, it must be that $\boldsymbol{\Sigma}^{-1} - \boldsymbol{\Omega}^{-1} \to 0$. As $\tau_1 \to \infty$, the $x_{i1}x_{j1}\tau_1^2$ dominates in the definition of $\Sigma_{i,j}$ and $\Omega_{i,j}$, so $\boldsymbol{\Sigma}^{-1} - \boldsymbol{\Omega}^{-1} \to 0$. Thus, the prior for β_1 becomes improper.

Further Reading

Historical Background. Tests for linear models with unknown parameters often consider a number of principles. For example, if the scale of $\mathbf{y}$ changes, the outcome of variable selection procedures should not change. A change in location of $\mathbf{y}$ should not change the outcome of variable selection, as well. For more on these invariance principles for tests, see Lehmann and Romano (2005); for a Bayesian perspective, see Bayarri et al. (2012).

Points of Interest.

1. Adding the vector $c\mathbf{x}_1$ to the data does not change the model:

$$y_i = x_{i1}(\beta_1 - c) + x_{i2}\beta_2 + \sigma\varepsilon_i.$$

If β_1 is unknown, nothing has changed because $\beta_1 - c$ is also unknown. Consequently, any test of variable selection for $\mathbf{x}_2$ should not change. However for the Bayesian, using a normal prior for β_1 induces a dependence of the Bayes factor on c. Therefore, the Bayesian decision for β_2 does depend on c.

2. To better understand the Bayes factor, consider a decision problem that aims to maximize a utility function. If this is a 0-1 utility function, then the aim is to maximize $1\{a = a^*\}$, where a is the selected decision (or "action") and a^* is the correct decision. Let $\pi(a)$ denote the probability of making the correct decision. Then, the coherent choice of a is the decision that maximizes $\pi(a)$ (i.e., the mode). This is the idea behind the Bayes factor; select the set of covariates that maximizes $\mathrm{P}(S \mid \text{data})$, where S is a given subset of covariates. In this exercise, the prior became overly influential because $\mathrm{P}(S \mid \text{data})$ was a function of both RSS and c.

3. In general, one could use a smarter utility function. For example, consider the function

$$U(S) = \boldsymbol{\beta}'\mathbf{X}_S'\mathbf{X}_S\boldsymbol{\beta} - |S|\sigma^2,$$

where $\mathbf{X}_S$ are the covariates in subset S and the last term penalizes the dimension of S. If the posterior for $\boldsymbol{\beta}$ under the model with subset S is

$$\pi(\boldsymbol{\beta} \mid \text{data}) = \mathrm{N}\left(\boldsymbol{\beta} \mid \widehat{\boldsymbol{\beta}}_S, \sigma^2(\mathbf{X}_S'\mathbf{X}_S)^{-1}\right),$$

then

$$U(S) = \widehat{\boldsymbol{\beta}}_S'\mathbf{X}_S'\mathbf{X}_S\widehat{\boldsymbol{\beta}}_S = \mathbf{y}'\mathbf{H}_S\mathbf{y}.$$

Selecting the optimal model with the largest utility makes use of the RSS because RSS $= -(\mathbf{y}'\mathbf{H}_S\mathbf{y} - \mathbf{y}'\mathbf{y})$ and the $\mathbf{y}'\mathbf{y}$ term is common to all S.

5

Linear Models & Regression

The linear model is one of the most widely used models in statistical analysis due in large part to its mathematical simplicity and ease in interpretation. The main idea behind the linear model is to understand which of the recorded covariates has an influence on the response of interest. For example, a biostatistician may use a linear model to determine whether age, height, or weight (the covariates) can be used to predict a blood pressure reading (the response).

There are several fundamental tasks in linear modeling, all of which can be accomplished with linear regression: estimating the effect of each covariate; predicting an outcome at a chosen set of covariates; and selecting the covariates that are the most influential in the model. The most prevalent approach to linear regression is least squares estimation, which minimizes the sum of squared distances between the observed outcomes and the predicted values based on a set of regression parameters.

Least squares estimation is an unreliable tool on its own because its efficacy decreases as the number of covariates increases, regardless of whether the covariates are informative for the response of interest. To highlight this point, consider a design matrix of random noise that has nothing to do with the experiment, and let the rank be equivalent to the sample size. Least squares estimation with this design matrix will present a perfect fit that yields a zero residual sum of squares. In fact, the predicted values coincide exactly with the observed values because the only idempotent matrix (the hat matrix) of full rank is the identity matrix. Thus, in terms of goodness of fit, the model is perfect. However, adding an observation with new values for the covariates will result in a nonsensical prediction.

To avoid overfitting, penalties for the dimension of the covariates are often used. The history of penalization is quite extensive and there are many proposals for different penalty terms. Some of the most popular penalization approaches are the Akaike information criterion (AIC), the Bayesian information criterion (BIC), the least absolute shrinkage and selection operator (lasso), ridge regression, and elastic net regularization. Additionally, Mallow's C_p statistic is a criterion for variable selection that also improves predictive performance.

This chapter opens with Question 5.1.1, which considers a regularized linear model where estimation is subject to a constraint. In particular, this exercise considers lasso and ridge regression, two of the most popular approaches to regularization. The estimators in this exercise may also be viewed as Bayesian estimators. Ridge regression is also the subject of Question 5.1.2 and Question 5.1.3. If the signal-to-noise ratio is greater than one, then the ridge estimator can improve upon the least squares estimator. Ridge estimators were introduced to counter large variances for estimators under the presence of collinearity between covariates. This collinearity leads to a small eigenvalue in the absolute value for $\mathbf{X}'\mathbf{X}$, where $\mathbf{X}$ is the design matrix. As a result, there is at least one large diagonal element of $(\mathbf{X}'\mathbf{X})^{-1}$ that inflates the variance of the estimators. Question 5.1.2 considers ridge regression for only one covariate, and Question 5.1.3 extends the results in Question 5.1.2 for multiple covariates.

DOI: 10.1201/9781003493471-5

Question 5.1.4 discusses the challenges of variable selection in high-dimensional settings and outlines an algorithm for determining an optimal set of latent indicator variables that indicate which covariates should be included in the "best" model. This optimization algorithm is known as the Hopfield network and is popular within the field of neuroscience. Notably, the Hopfield network increases the objective function at each iteration.

Question 5.1.5 explores a derivation of the AIC variable selection procedure when the variance is assumed known. Although this exercise does not derive the AIC procedure exactly, the ideas are similar: find an unbiased estimator for a quantity that is useful to determine which covariates are important. Ideally, the quantity of interest increases if a helpful covariate is added but remains constant if a non-helpful covariate is added. This is also the theme of Question 5.1.6. Finally, Question 5.1.7 explores the lasso estimator, which plays the role of variable selection in linear models.

Q5.1 Questions – Linear Models & Regression

Q5.1.1 – Regularized linear regression

Introduction. An important topic in modern statistics is adapting ordinary least squares regression for settings with big data. If the number of variables surpasses the number of observations, the design matrix becomes rank-deficient; this causes several numerical problems, notably the lack of uniqueness for the ordinary least squares estimator. One possible remedy is to use restrictions that make the problem solvable. The most popular restrictions are based on the elastic-net models, including lasso and ridge regression. This question investigates both lasso and ridge regression for a regularized model and considers their Bayesian interpretation.

Question. Assume n independent observations $\{y_i\}$ arise from the model $y_i = \theta_i + \varepsilon_i$, where $\boldsymbol{\theta} = (\theta_1, \ldots, \theta_n)'$ are unknown parameters, the $\{\varepsilon_i\}$ are independent zero-mean normal random variables with $\mathrm{Var}(\varepsilon_i) = \sigma^2$, and $\sigma > 0$ is assumed known. Let the true value for $\boldsymbol{\theta}$ be denoted by $\boldsymbol{\theta}^*$.

(i) Find the least squares estimator of $\boldsymbol{\theta}$, denoted $\widehat{\boldsymbol{\theta}}$, subject to the constraint

$$\sum_{i=1}^{n} \theta_i = \tau$$

for some fixed τ.

(ii) Find $\mathrm{E}(\widehat{\boldsymbol{\theta}})$ and the covariance matrix $\mathrm{Cov}(\widehat{\boldsymbol{\theta}})$.

(iii) Now, consider the model

$$y_i = \mathbf{x}_i' \boldsymbol{\theta} + \varepsilon_i,$$

where the $\{\varepsilon_i\}$ are the same as before but $\boldsymbol{\theta}$ is an unknown parameter of dimension d and $\mathbf{x}_i = (x_{i1}, \ldots, x_{id})'$ is a vector of covariates. Find the least squares estimator of $\boldsymbol{\theta}$ subject to the constraint $\boldsymbol{\theta}'\boldsymbol{\theta} = \tau$ for some fixed $\tau > 0$.

(iv) Show that the estimator given in part (iii) has a Bayesian interpretation.

(v) Can all Bayesian estimators be seen as constrained maximum likelihood estimators?

Q5.1.2 – Ridge regression for linear models

Introduction. Linear regression models are fundamental within statistics and are applied in a variety of scientific fields. Because of its wide adoption, there are many methods for estimating linear regression parameters, one of the most popular being ridge regression. This question assesses the performance of ridge estimators by looking at the resulting mean squared error (MSE). Additionally, the bias-variance tradeoff is discussed in the context of ridge regression.

Question. Consider the linear regression model $\mathbf{y} = \mathbf{X}\boldsymbol{\beta} + \sigma\boldsymbol{\varepsilon}$, where $\mathbf{y} = (y_1, \dots, y_n)'$, $\mathbf{X} = (x_{ij})$ is a $n \times p$ design matrix, $\boldsymbol{\beta}$ is a $p \times 1$ vector of unknown coefficients, σ is a known standard deviation, and $\boldsymbol{\varepsilon}$ is a $n \times 1$ vector of independent standard normal random variables. A ridge estimator for $\boldsymbol{\beta}$ is of the form $\widehat{\boldsymbol{\beta}}_R = (\mathbf{X}'\mathbf{X} + \lambda\,\mathbf{I})^{-1}\,\mathbf{X}'\mathbf{y}$ for some $\lambda > 0$.

(i) Assume a normal prior distribution for $\boldsymbol{\beta}$. Show that the ridge estimator can be seen as a Bayes estimator.

(ii) Find the variance-covariance matrix of $\widehat{\boldsymbol{\beta}}_R$.

(iii) Now, let $p = 1$ and $\sum_{i=1}^{n} x_i^2 = 1$, where x_i is the single covariate for the ith observation. Find the MSE for the ridge estimator: $\mathrm{E}\{(\widehat{\beta}_R - \beta)^2\}$.

(iv) Under the same conditions as in part (iii), what is the variance of the maximum likelihood estimator $\widehat{\beta}$?

(v) Are there values of λ for which $\mathrm{E}\{(\widehat{\beta}_R - \beta)^2\} \leq \mathrm{E}\{(\widehat{\beta} - \beta)^2\}$? If so, what is the problem with using such a value for λ?

Q5.1.3 – Highly correlated covariates in linear regression

Introduction. This question explores the problem with using naïve least squares estimators for linear regression models with highly correlated covariates. When covariates are highly correlated, it is difficult to determine the individual effect of each covariate. Because of this collinearity, ordinary least squares estimators have inflated variances. A common remedy is to introduce a penalty term in estimation that penalizes the size of the coefficients. This question considers the ridge penalty term.

Question. Consider the linear regression model

$$y_i = x_{i1}\beta_1 + x_{i2}\beta_2 + \sigma\epsilon_i, \quad i = 1, \dots, n,$$

where σ is assumed known and the errors $\{\epsilon_i\}$ are independent standard normal random variables.

(i) Derive the least squares estimator $\widehat{\boldsymbol{\beta}} = (\mathbf{X}'\mathbf{X})^{-1}\mathbf{X}'\mathbf{y}$, where $\mathbf{y} = (y_1, \ldots, y_n)'$ is the observation vector and $\mathbf{X}$ is the $n \times 2$ matrix with columns (x_{i1}) and (x_{i2}).

(ii) If

$$\mathbf{X}'\mathbf{X} = \begin{pmatrix} 1 & \rho \\ \rho & 1 \end{pmatrix}$$

for some $0 < \rho < 1$, find Cov $(\widehat{\boldsymbol{\beta}})$.

(iii) Explain the potential problem with using $\widehat{\boldsymbol{\beta}}$ if ρ is too close to 1.

(iv) A solution to the problem in part (iii) is to use the ridge estimator

$$\boldsymbol{\beta}_R = (\mathbf{X}'\mathbf{X} + \lambda\,\mathbf{I})^{-1}\,\mathbf{X}'\mathbf{y}$$

for some $\lambda > 0$, where $\mathbf{I}$ is the 2×2 identity matrix. Show that it is possible to write $\boldsymbol{\beta}_R = u\,\widehat{\boldsymbol{\beta}} + v\,\mathbf{X}'\mathbf{y}$, where u and v are scalar functions of λ and ρ.

(v) If $u = 0$, what problem arises and can it be fixed?

Q5.1.4* – Variable selection with the Hopfield network

Introduction. Variable selection for linear models is a classic statistical problem. A challenging aspect of variable selection is the optimization and implementation of a selection criterion over a space that is potentially of high dimension; for instance, there are 2^p possible ways to select p variables within a model. This question constructs a fast algorithm for selecting variables based on the posterior mode of latent indicator variables. The algorithm is a sequence of conditional maximization steps that has found popularity in neuroscience: the Hopfield network.

Question. Consider the linear model

$$y_i = \sum_{j=1}^{p} x_{ij}\, z_j\, \beta_j + \sigma\epsilon_i, \quad i = 1, \ldots, n,$$

where σ is assumed known, the $\{\epsilon_i\}$ are independent standard normal random variables, and the $\{z_j\}$ are binary variables indicating whether the covariates $\{x_j\}$ are active.

(i) Show that the likelihood for the model can be written in matrix notation as

$$\exp\left\{-\tfrac{1}{2}\lambda(\mathbf{y} - \mathbf{X}\boldsymbol{\beta})'(\mathbf{y} - \mathbf{X}\boldsymbol{\beta})\right\},$$

where $\mathbf{y} = (y_1, \ldots, y_n)'$, $\boldsymbol{\beta} = (\beta_1, \ldots, \beta_p)'$, and $\lambda = 1/\sigma^2$. Find $\mathbf{X}$ in terms of $\mathbf{D}$ and $\mathbf{X}_0$, where $\mathbf{D}$ is the diagonal matrix with elements $\mathbf{z} = (z_1, \ldots, z_p)'$ and $\mathbf{X}_0$ is the $n \times p$ matrix with elements (x_{ij}).

(ii) The prior for $\boldsymbol{\beta}$ conditional on $\mathbf{z}$ is taken as $\mathrm{N}(\mathbf{0}, g\sigma^2(\mathbf{X}'\mathbf{X})^{-1})$ for some $g > 0$, where $\beta_j = 0$ if $z_j = 0$. Find the marginal density of $p(\mathbf{y} \mid \mathbf{z})$.

(iii) If the prior for $\mathbf{z}$ is uniform on the 2^p possible permutations, find $p(\mathbf{z} \mid \mathbf{y})$ up to the normalizing constant.

(iv) Find $p(z_j \mid \mathbf{z}_{-j}, \mathbf{y})$, where $\mathbf{z}_{-j} = (z_1, \ldots, z_{j-1}, z_{j+1}, \ldots, z_p)'$. Give a condition when the maximum value of $p(z_j \mid \mathbf{z}_{-j}, \mathbf{y})$ is 0 or 1.

(v) A maximization algorithm iterates through $j = 1, \ldots, p$, maximizing each $p(z_j \mid \mathbf{z}_{-j}, \mathbf{y})$ until convergence. For a single update, show that, if $(z_1, \ldots, z_p)$ is the current $\mathbf{z}$ and $\widehat{z}_j$ maximizes $p(z_j \mid \mathbf{z}_{-j}, \mathbf{y})$, then

$$p(z_1, \ldots, z_{j-1}, \widehat{z}_j, z_{j+1}, \ldots, z_p \mid \mathbf{y}) \geq p(z_1, \ldots, z_{j-1}, z_j, z_{j+1}, \ldots, z_p \mid \mathbf{y}).$$

Q5.1.5 – Variable selection with the value of the linear model

Introduction. Question 5.1.4 opened the discussion on variable selection for linear models. This question continues that discussion by exploring the value of the linear model, which is the quadratic form $\boldsymbol{\beta}\mathbf{X}'\mathbf{X}\boldsymbol{\beta}$; this scalar value is helpful for model diagnostics and variable selection because it indicates how the variability in $\boldsymbol{\beta}$ and the structure of $\mathbf{X}$ contribute to the overall variability of the predictions. As this question explores, the value of the linear model can help identify an optimal set of covariates to use for modeling.

Question. Consider the standard linear model

$$\mathbf{y} = \mathbf{X}\boldsymbol{\beta} + \sigma\boldsymbol{\epsilon},$$

where $\mathbf{y} = (y_1, \ldots, y_n)'$, $\boldsymbol{\beta} = (\beta_1, \ldots, \beta_d)'$, σ is assumed known, $\boldsymbol{\epsilon} = (\epsilon_1, \ldots, \epsilon_n)'$ are independent standard normal random variables, and $\mathbf{X}$ is the design matrix of dimension $n \times d$ with rank d. The true value of $\boldsymbol{\beta}$ is $\boldsymbol{\beta}^*$, and $\boldsymbol{\beta}^{*\prime}\mathbf{X}'\mathbf{X}\boldsymbol{\beta}^*$ is the value of the model with design matrix $\mathbf{X}$.

(i) Show that the least squares estimator for $\boldsymbol{\beta}$ is $\widehat{\boldsymbol{\beta}} = (\mathbf{X}'\mathbf{X})^{-1}\mathbf{X}'\mathbf{y}$.

(ii) The hat matrix is given by $\mathbf{H} = \mathbf{X}(\mathbf{X}'\mathbf{X})^{-1}\mathbf{X}'$. Show that the trace of $\mathbf{H}$ is d.

(iii) If $\mathbf{z}$ is a $n \times 1$ vector of independent standard normal random variables, show that $\mathrm{E}(\mathbf{z}'\mathbf{H}\mathbf{z}) = d$.

(iv) Find $\mathrm{E}(\widehat{\boldsymbol{\beta}}'\mathbf{X}'\mathbf{X}\widehat{\boldsymbol{\beta}})$.

(v) Find an unbiased estimator for the value of the model.

Q5.1.6 – Variable selection: A form of AIC

Introduction. Question 5.1.5 used the value of the linear model for variable selection when the variance was assumed known. This question extends that approach by assuming that the variance is unknown. First, a simulation study is conducted, which motivates the use of a penalty term that penalizes the addition of extraneous covariates to the model. Then, an unbiased estimator is proposed for a variable selection criterion that resembles the AIC.

Question. Consider the linear model

$$\mathbf{y} = \mathbf{X}\boldsymbol{\beta} + \sigma\boldsymbol{\epsilon},$$

where $\mathbf{y} = (y_1, \ldots, y_n)'$, $\boldsymbol{\beta} = (\beta_1, \ldots, \beta_p)'$, and $\boldsymbol{\epsilon}$ is a n-dimensional vector of independent standard normal random variables. Let σ^* be the true value of σ. Additionally, assume that the true covariates of the model are the first d_0 columns of $\mathbf{X}$; the true covariates form the true design matrix $\mathbf{X}_0$ with coefficients $\boldsymbol{\beta}^*$. For a model of size $d \leq p$, define

$$L(d) = \boldsymbol{\beta}^{*\,\prime}\mathbf{X}_0'\mathbf{H}_d\mathbf{X}_0\boldsymbol{\beta}^*,$$

where $\mathbf{H}_d$ is the hat matrix for the design matrix $\mathbf{X}_d$ that contains the first d covariates of $\mathbf{X}$.

(i) Let $n = 100$, $p = 5$, $d_0 = 3$, $\boldsymbol{\beta}^* = (1,1,1)'$, and the components of $\mathbf{X}$ be independent standard normal random draws. Simulate $L(d)$ for $d = 1, \ldots, 5$, assuming that the d_0 true covariates are the first three columns of $\mathbf{X}$.

(ii) Using the simulation from part (i), explain why a penalty term may be appropriate for variable selection.

(iii) Show that $\mathbf{y}'\mathbf{H}_d\mathbf{y}$ is an unbiased estimator of $\boldsymbol{\beta}^{*\,\prime}\mathbf{X}_0'\mathbf{H}_d\mathbf{X}_0\boldsymbol{\beta}^* + (\sigma^*)^2 d$.

(iv) Show that $\mathrm{E}(\mathbf{y}'\mathbf{y})/n > (\sigma^*)^2$.

(v) Consider a penalty of the form $\gamma\, d$ for some $\gamma > 0$. The aim is to maximize $L(d) - \gamma\, d$. Find a criterion for variable selection that has an expected value of $L(d) - \gamma\, d$.

Q5.1.7 – On the lasso estimator

Introduction. The lasso is an estimator for the regression coefficients of a linear model that, unlike the ridge estimator, can set some coefficients to 0. Hence, the lasso estimator plays the role of variable selection. This question considers the particular lasso estimator that arises when the observations have a symmetric density with mean θ^*.

Question. Consider the estimation problem of minimizing

$$l(\theta) = \tfrac{1}{2}(y - \theta)^2 + \lambda\,|\theta|,$$

where y is observed and $\lambda \geq 0$ is fixed.

(i) Show that the $\widehat{\theta}$ minimizing $l(\theta)$ is given by

$$\widehat{\theta} = y \max\{0, 1 - \lambda/|y|\}.$$

(ii) Let $\lambda > \theta^*$ and Y be a Laplace random variable with mean $\theta^* > 0$ and scale $b = 1$. Find the expected value of $\widehat{\theta}$.

(iii) Show that $\mathrm{E}(\widehat{\theta}) \leq \frac{1}{2}(1 - e^{-2\theta^*})$. For what value of θ^* is this an equality?

(iv) Now, find the $\widehat{\theta}$ that minimizes

$$\frac{1}{2}\sum_{i=1}^{n}(y_i - x_i\theta)^2 + n\lambda|\theta|$$

for a given $\{x_i, y_i\}$.

(v) What happens to the estimator in part (iv) as $n \to \infty$?

S5.1 Solutions – Linear Models & Regression

S5.1.1 – Regularized linear regression

(i) Using Lagrange multipliers, let

$$f(\theta_1,\ldots,\theta_n)=\sum_{i=1}^{n}(y_i-\theta_i)^2+\lambda\left(\sum_{i=1}^{n}\theta_i-\tau\right).$$

To find the maximizer $\widehat{\boldsymbol{\theta}}$, take the partial derivative of f with respect to θ_i for $i=1,\ldots,n$ and set to 0:

$$\frac{\partial f}{\partial\theta_i}=2(\theta_i-y_i)+\lambda\overset{\text{set}}{=}0,\qquad i=1,\ldots,n.$$

Note that λ can be obtained under the given constraint by solving

$$\sum_{i=1}^{n}\{2(\theta_i-y_i)+\lambda\}=0.$$

Therefore, $\lambda=2\overline{y}-2\tau/n$ and $\widehat{\theta}_i=y_i-\overline{y}+\tau/n$.

(ii) Because $\mathrm{E}(Y_i)=\theta_i^*$ and $\mathrm{Var}(Y_i)=\sigma^2$, it is that

$$\mathrm{E}(\widehat{\theta}_i)=\mathrm{E}\left(Y_i-\overline{Y}+\tau/n\right)=\theta_i^*-\overline{\theta}^*+\tau/n,$$

where $\overline{\theta}^*=\sum_{i=1}^{n}\theta_i^*/n$. Then, calculate each element in the covariance matrix:

$$\begin{aligned}\mathrm{Cov}(\widehat{\theta}_i,\widehat{\theta}_j)&=\mathrm{Cov}(Y_i-\overline{Y},Y_j-\overline{Y})\\&=\mathrm{E}\left\{(Y_i-\overline{Y})(Y_j-\overline{Y})\right\}-\mathrm{E}(Y_i-\overline{Y})\mathrm{E}(Y_j-\overline{Y})\\&=\mathrm{E}(Y_iY_j)-\mathrm{E}(Y_i\overline{Y})-\mathrm{E}(Y_j\overline{Y})+\mathrm{E}(\overline{Y}^2)-(\theta_i^*-\overline{\theta}^*)(\theta_j^*-\overline{\theta}^*),\end{aligned}$$

where

$$\begin{aligned}\mathrm{E}(\overline{Y}^2)&=\mathrm{Var}(\overline{Y})+\mathrm{E}(\overline{Y})^2\\&=\frac{1}{n^2}\sum_{i=1}^{n}\mathrm{Var}(Y_i)+(\overline{\theta}^*)^2\\&=\sigma^2/n+(\overline{\theta}^*)^2\end{aligned}$$

and

$$\begin{aligned}\mathrm{E}(Y_i\overline{Y})&=\frac{1}{n}\mathrm{E}\left(Y_i\sum_{j=1}^{n}Y_j\right)=\frac{1}{n}\left\{\mathrm{E}(Y_i^2)+\sum_{j\neq i}\mathrm{E}(Y_i)\mathrm{E}(Y_j)\right\}\\&=\frac{1}{n}\left\{\sigma^2+(\theta_i^*)^2+\sum_{j\neq i}\theta_i^*\theta_j^*\right\}\\&=\frac{\sigma^2}{n}+\frac{1}{n}\sum_{j=1}^{n}\theta_i^*\theta_j^*=\frac{\sigma^2}{n}+\theta_i^*\overline{\theta}^*.\end{aligned}$$

Thus, $\text{Cov}(\widehat{\theta}_i, \widehat{\theta}_j) = -\sigma^2/n$ for $i \neq j$. For $i = j$,

$$\text{Cov}(\widehat{\theta}_i, \widehat{\theta}_j) = \text{Var}(\widehat{\theta}_i) = \text{Var}\left(\frac{n-1}{n}Y_i\right) + \text{Var}\left(\frac{\sum_{j\neq i} Y_j}{n}\right) = \left(1 - \frac{1}{n}\right)\sigma^2.$$

(iii) Write $\mathbf{y} = (y_1, \ldots, y_n)'$ as a multivariate normal with mean $\mathbf{X}\boldsymbol{\theta}$ and covariance matrix $\sigma^2\mathbf{I}$, where $\mathbf{X}$ is the $n \times p$ design matrix and $\mathbf{I}$ is the $n \times n$ identity matrix. Using Lagrange multipliers, solve for

$$\widehat{\boldsymbol{\theta}} = \arg\min_{\boldsymbol{\theta}}\{(\mathbf{y} - \mathbf{X}\boldsymbol{\theta})'(\mathbf{y} - \mathbf{X}\boldsymbol{\theta}) + \lambda(\boldsymbol{\theta}'\boldsymbol{\theta} - \tau)\}.$$

Compute the derivative of the argument with respect to $\boldsymbol{\theta}$, and set it to 0:

$$2\mathbf{X}'\mathbf{X}\boldsymbol{\theta} - 2\mathbf{X}'\mathbf{y} + 2\lambda\mathbf{I}\boldsymbol{\theta} \overset{\text{set}}{=} 0.$$

Finally, $\widehat{\boldsymbol{\theta}} = (\mathbf{X}'\mathbf{X} + \lambda\mathbf{I})^{-1}\mathbf{X}'\mathbf{y}$.

(iv) To see the Bayesian interpretation, rewrite $\widehat{\boldsymbol{\theta}}$ as

$$\begin{aligned}
\widehat{\boldsymbol{\theta}} &= \arg\min_{\boldsymbol{\theta}}\{(\mathbf{y} - \mathbf{X}\boldsymbol{\theta})'(\mathbf{y} - \mathbf{X}\boldsymbol{\theta}) + \lambda(\boldsymbol{\theta}'\boldsymbol{\theta} - \tau)\} \\
&= \arg\max_{\boldsymbol{\theta}}\left\{-\frac{1}{2\sigma^2}(\mathbf{y} - \mathbf{X}\boldsymbol{\theta})'(\mathbf{y} - \mathbf{X}\boldsymbol{\theta}) - \frac{\lambda}{2\sigma^2}(\boldsymbol{\theta}'\boldsymbol{\theta} - \tau)\right\} \\
&= \arg\max_{\boldsymbol{\theta}}\left[\exp\left\{-\frac{1}{2\sigma^2}(\mathbf{y} - \mathbf{X}\boldsymbol{\theta})'(\mathbf{y} - \mathbf{X}\boldsymbol{\theta})\right\}\exp\left(-\frac{\lambda}{2\sigma^2}\boldsymbol{\theta}'\boldsymbol{\theta}\right)\right].
\end{aligned}$$

Note that the argument is the product of two multivariate normal kernels. Thus, $\widehat{\boldsymbol{\theta}}$ is equivalent to the posterior mode of the model

$$p(\mathbf{y} \mid \boldsymbol{\theta}) = \text{N}(y \mid \mathbf{X}\boldsymbol{\theta}, \sigma^2\mathbf{I}) \quad \text{with} \quad \boldsymbol{\theta} \sim \text{N}\left(\mathbf{0}, (\sigma^2/\lambda)\,\mathbf{I}\right).$$

(v) In general, the use of priors induces constraints in parameter estimation. For example, the posterior in part (iv) can be written as $\exp\{L(\theta) + \lambda\, g(\theta)\}$, where $L(\theta)$ is the log-likelihood and the prior is proportional to $\exp\{\lambda g(\theta)\}$ for some scalar λ. The posterior mode solves

$$L'(\theta) + \lambda g'(\theta) = 0, \tag{5.1}$$

which is equivalent to maximizing $L(\theta)$ subject to $g(\theta) = \tau$.

Further Reading

Historical Background. Constrained least squares, or regularization, is a regression technique that reduces the variance of regression coefficients and possibly reduces the mean squared error at the cost of introducing bias. In particular, regularization is appropriate when the dimension of the parameter space p is larger than the sample size n. Such situations arise, for instance, in genetics, where biologists might sample thousands of genes from a couple hundred people. However, if $p > n$, the coefficients become unstable and have high variance. Additionally, there would not exist a unique solution to the normal equations. By restricting the coefficients to all lie within a bounded set, such as the unit L_p ball, the problem becomes regularized; that is, the coefficients that are not relevant to the regression model are penalized. Such procedures are imperative because, if the design matrix has rank n, then the residual least squares are zero.

Points of Interest.

1. The most popular types of regularization include the lasso and ridge, which are the L_1 and L_2 penalties on the parameter, respectively. See Hoerl and Kennard (1970) and Tibshirani (1996) for ridge and lasso regularization, respectively. In the one-dimensional case, lasso regression involves minimizing $\sum_{i=1}^{n}(y_i - x_i\theta)^2 + \lambda|\theta|$ for some $\lambda > 0$. The lasso estimator is 0 when
$$\left|\sum_{i=1}^{n} y_i x_i\right| < \lambda \sum_{i=1}^{n} x_i^2.$$
The ridge estimator does not get to 0 but does shrink the estimator toward 0. A mix between lasso and ridge regularization may be achieved used the elastic net, which combines the L_1 and L_2 penalties (Zou and Hastie, 2005). Note that in this setting, the information criterion procedures (i.e., AIC, BIC) can be viewed as using a L_0 penalty, which amounts to penalizing the dimension of the parameter.

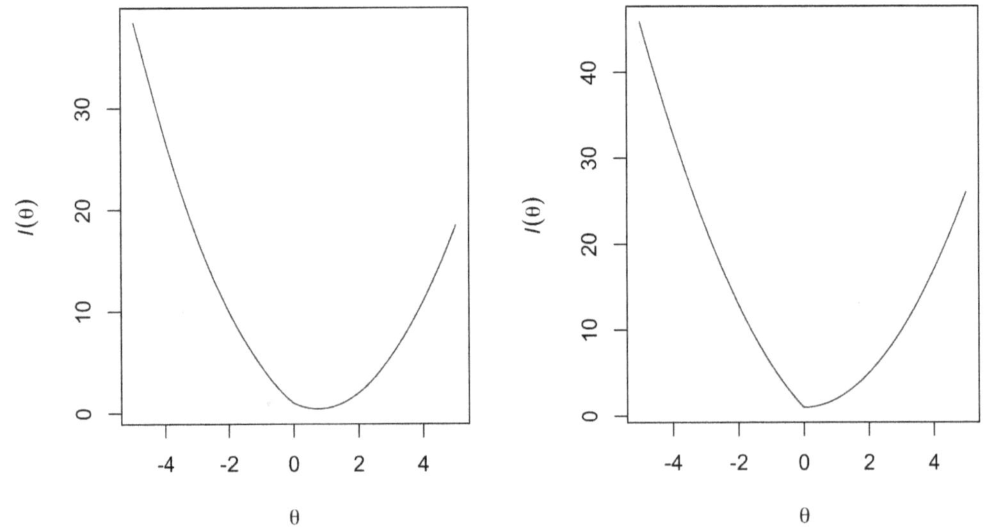

FIGURE 5.1
Lasso minimization with $\lambda = \frac{1}{2}$ (left) and $\lambda = 2$ (right).

Demonstration. Consider lasso minimization in its most basic form:
$$\arg\min_{\theta}\{(y-\theta)^2 + \lambda|\theta|\},$$
where $y = 1$ and θ is a univariate parameter. Fig. 5.1 depicts two scenarios: $\lambda = 1/2$ (left) and $\lambda = 2$ (right). Minimization occurs at $\theta = 3/4$ and $\theta = 0$, respectively.

S5.1.2 – Ridge regression for linear models

(i) Consider the prior for $\boldsymbol{\beta}$ which is $\mathrm{N}(\boldsymbol{\beta} \mid \mathbf{0}, \tau^2\mathbf{I})$, and which has the following posterior distribution:

$$\pi(\boldsymbol{\beta} \mid \mathbf{y}) \propto \exp\left[-\frac{1}{2}\left\{\boldsymbol{\beta}^T\left(\frac{\mathbf{X}'\mathbf{X}}{\sigma^2} + \frac{\mathbf{I}}{\tau^2}\right)\boldsymbol{\beta} - 2\left(\frac{\mathbf{y}'\mathbf{X}}{\sigma^2}\right)\boldsymbol{\beta}\right\}\right].$$

This is the kernel of a normal distribution with mean $(\mathbf{X}'\mathbf{X} + \lambda\mathbf{I})^{-1}\mathbf{X}'\mathbf{y}$ and covariance matrix $\sigma^2(\mathbf{X}'\mathbf{X} + \lambda\mathbf{I})^{-1}$, where $\lambda = \sigma^2/\tau^2$. Thus, the ridge estimator $\widehat{\boldsymbol{\beta}}_R$ can be seen as the posterior mean under a conjugate normal prior for $\boldsymbol{\beta}$.

(ii) Note that $\mathrm{Cov}(\mathbf{Az}) = \mathbf{A}\mathrm{Cov}(\mathbf{z})\mathbf{A}'$, where $\mathbf{A}$ is a constant matrix and $\mathbf{z}$ is a random column vector. Therefore,

$$\begin{aligned}\mathrm{Cov}(\widehat{\boldsymbol{\beta}}_R) &= \mathrm{Cov}\{(\mathbf{X}'\mathbf{X} + \lambda\mathbf{I})^{-1}\mathbf{X}'\mathbf{y}\} \\ &= (\mathbf{X}'\mathbf{X} + \lambda\mathbf{I})^{-1}\mathbf{X}'\,\mathrm{Cov}(\mathbf{y})\,\mathbf{X}(\mathbf{X}'\mathbf{X} + \lambda\mathbf{I})^{-1} \\ &= \sigma^2(\mathbf{X}'\mathbf{X} + \lambda\mathbf{I})^{-1}\mathbf{X}'\mathbf{X}(\mathbf{X}'\mathbf{X} + \lambda\mathbf{I})^{-1},\end{aligned}$$

because $\mathrm{Cov}(\mathbf{Y}) = \sigma^2\,\mathbf{I}$.

(iii) Because $p = 1$ and $\mathbf{x}'\mathbf{x} = 1$, it is true that $\widehat{\beta}_R = \sum_{i=1}^n x_i y_i/(1+\lambda)$. The expected value is $\beta/(1+\lambda)$, the variance is $\sigma^2/(1+\lambda)^2$, and the bias is $-\lambda\beta/(1+\lambda)$. The mean squared error of an estimator is the sum of its variance and squared bias, so

$$\mathrm{E}\left\{(\widehat{\beta}_R - \beta)^2\right\} = \frac{\sigma^2 + \lambda^2\beta^2}{(1+\lambda)^2}.$$

(iv) Note that the maximum likelihood estimator arises with $\lambda = 0$: $\widehat{\beta} = (\mathbf{x}'\mathbf{x})^{-1}\mathbf{x}'\mathbf{y}$. Because $\mathbf{x}'\mathbf{x} = 1$, the variance is σ^2.

(v) The MLE $\widehat{\beta}$ is unbiased, so its MSE is σ^2. The aim is to find the values of λ that satisfy

$$\frac{\sigma^2 + \lambda^2\beta^2}{(1+\lambda)^2} \le \sigma^2.$$

If $\beta^2/\sigma^2 \le 1$, then this holds true for all $\lambda \ge 0$. If $\beta^2/\sigma^2 > 1$, then it holds for all $\lambda < 2/(\beta^2/\sigma^2 - 1)$. Note that β^2/σ^2 is known as the signal-to-noise ratio.

Further Reading

Historical Background. Ridge regression was first introduced in Hoerl and Kennard (1970). The objective function to be minimized includes a penalty term in addition to the usual function for ordinary least squares estimation:

$$\widehat{\beta}_R = \arg\min_{\beta} \sum_{i=1}^n \left(y_i - \sum_{j=1}^p x_{ij}\beta_j\right)^2 + \lambda\sum_{j=1}^p \beta_j^2.$$

The original motivation for ridge regression was to avoid large variances for the estimators, which arise if at least one eigenvalue of $\mathbf{X}'\mathbf{X}$ is close to 0; this happens when two sets of covariates are highly collinear.

Points of Interest.

1. Without loss of generality, consider the case when

$$\mathbf{X}'\mathbf{X} = \begin{pmatrix} 1 & \rho \\ \rho & 1 \end{pmatrix}.$$

   The variance of the least squares estimator is $\sigma^2(\mathbf{X}'\mathbf{X})^{-1}$, which has a denominator of $1-\rho^2$. If ρ is close to -1 or 1, then the variance of the estimators will be very large; see Question 5.1.3 for more details on this particular case. Fortunately, bias can be introduced to bring down the variance, which is achieved by selecting a non-zero value for λ. The selection of λ has generated a substantial amount of interest and research; for example, cross-validation has been used to select λ (Golub et al., 1979).

2. Ridge estimation does not reduce bias so much that the estimator for some components are 0; this phenomenon is actually characteristic of the lasso estimator (Tibshirani, 1996). The difference between lasso and ridge regression is that the penalty for the former uses the magnitude of the coefficients and the penalty for the latter uses the square of the coefficients. Lasso regression can also be used for model selection, whereas ridge regression cannot.

3. Regularization is a major topic in statistics and is commonly used for linear and generalized linear models (Finch, 2022). Bayesian methods regularize automatically. The choice of a prior informs the penalty placed on the coefficients; consequently, much care must be taken to chose adequate priors when attempting to regularize.

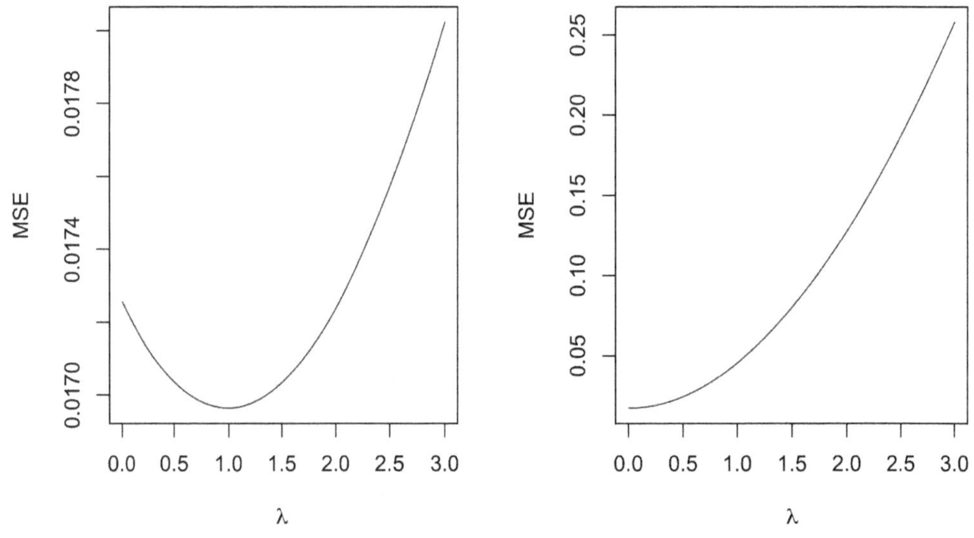

FIGURE 5.2
MSE as a function of λ for $\beta = 1$ (left) and $\beta = 10$ (right).

Demonstration. For the model $y_i = x_i\beta + \sigma\varepsilon_i$, the MSE is given by

$$\frac{\lambda^2\beta^2 + \sigma^2\sum_{i=1}^n x_i^2}{(\lambda + \sum_{i=1}^n x_i^2)^2}.$$

Let $\sigma = 1$. In Fig. 5.2, the MSE is plotted as a function of λ when $\beta = 1$ (left) and $\beta = 10$ (right). When $\beta = 1$ (i.e., the signal-to-noise ratio is 1), the MSE is minimized at $\lambda = 1$. When $\beta = 10$, a value of $\lambda = 0$ minimizes the MSE. Thus, for β close to 0, the MSE is minimized at larger values of λ.

S5.1.3 – Highly correlated covariates in linear regression

(i) The least squares estimator is given by

$$\widehat{\boldsymbol{\beta}} = \arg\min_{\boldsymbol{\beta}} (\mathbf{y} - \mathbf{X}\boldsymbol{\beta})'(\mathbf{y} - \mathbf{X}\boldsymbol{\beta}),$$

where $\boldsymbol{\beta} = (\beta_1, \beta_2)'$. Set the derivative to 0 to obtain $\widehat{\boldsymbol{\beta}} = (\mathbf{X}'\mathbf{X})^{-1}\mathbf{X}'\mathbf{y}$, assuming $\mathbf{X}'\mathbf{X}$ is invertible.

(ii) Note that $\mathrm{Cov}(\mathbf{Y}) = \sigma^2\mathbf{I}$. Thus,

$$\mathrm{Cov}(\widehat{\boldsymbol{\beta}}) = (\mathbf{X}'\mathbf{X})^{-1}\mathbf{X}'\mathrm{Cov}(\mathbf{Y})\mathbf{X}(\mathbf{X}'\mathbf{X})^{-1} = \sigma^2(\mathbf{X}'\mathbf{X})^{-1} = \frac{\sigma^2}{1-\rho^2}\begin{pmatrix} 1 & -\rho \\ -\rho & 1 \end{pmatrix}.$$

(iii) In the two-dimensional case with the given $\mathbf{X}'\mathbf{X}$,

$$(\mathbf{X}'\mathbf{X})^{-1} = \frac{1}{1-\rho^2}\begin{pmatrix} 1 & -\rho \\ -\rho & 1 \end{pmatrix}.$$

Therefore, the variance of $\widehat{\boldsymbol{\beta}}$ is very large when ρ is close to -1 or 1.

(iv) Note that

$$\begin{aligned}(\mathbf{X}'\mathbf{X} + \lambda\mathbf{I})^{-1} &= \frac{1}{(1+\lambda)^2 - \rho^2}\begin{pmatrix} 1+\lambda & -\rho \\ -\rho & 1+\lambda \end{pmatrix} \\ &= \frac{1-\rho^2}{(1+\lambda)^2-\rho^2}(\mathbf{X}'\mathbf{X})^{-1} + \frac{\lambda}{(1+\lambda^2)-\rho^2}\mathbf{I}.\end{aligned}$$

Therefore,

$$\begin{aligned}\widetilde{\boldsymbol{\beta}} &= (\mathbf{X}'\mathbf{X} + \lambda\mathbf{I})^{-1}\mathbf{X}'\mathbf{y} \\ &= \frac{1-\rho^2}{(1+\lambda)^2-\rho^2}(\mathbf{X}'\mathbf{X})^{-1}\mathbf{X}'\mathbf{y} + \frac{\lambda}{(1+\lambda)^2-\rho^2}\mathbf{X}'\mathbf{y} \\ &= \underbrace{\frac{1-\rho^2}{(1+\lambda)^2-\rho^2}}_{u}\widehat{\boldsymbol{\beta}} + \underbrace{\frac{\lambda}{(1+\lambda)^2-\rho^2}}_{v}\mathbf{X}'\mathbf{y}.\end{aligned}$$

(v) If $\rho = \pm 1$, then $u = 0$. In this case, the covariates would be linearly dependent. In fact, there would be just a single covariate.

Further Reading

Historical Background. Collinearity or multicollinearity refers to the problem in regression analysis where the columns of the model matrix are nearly linearly dependent. One of the original solutions to this problem was ridge regression (Hoerl and Kennard, 1970), though other approaches exist including lasso regression and principal component analysis.

Points of Interest.

1. Ridge regression was developed to solve the imprecision of least squares estimators for collinear covariates. It provides more precise estimators because the variance and mean squared error for ridge estimators are typically smaller than for ordinary least squares estimators. Chapter 13 of Christensen (2002) discusses ridge regression and other approaches that overcome the problem of highly correlated covariates.

2. Several other methods exist that deal with collinearity, including latent variable methods and tree-based models. A good review of such methods is Dormann et al. (2013), wherein the authors benchmark various statistical methods that account for collinearity.

S5.1.4* – Variable selection with the Hopfield network

(i) The linear model can be written as $\mathbf{y} = \mathbf{X}_0\,\mathbf{D}\,\boldsymbol{\beta} + \sigma\boldsymbol{\epsilon}$, where $\boldsymbol{\epsilon}$ is a vector of independent standard normal random variables. Let $\mathbf{X} = \mathbf{X}_0\mathbf{D}$ to obtain the desired likelihood. Note that $\mathbf{D}\boldsymbol{\beta}$ is the vector with elements $(z_j\,\beta_j)$.

(ii) The given prior for $\boldsymbol{\beta}$ conditional on $\mathbf{z}$ implies that $\boldsymbol{\beta} = \mathbf{A}\,\boldsymbol{\eta}$, where $\boldsymbol{\eta}$ is a vector of independent standard normal random variables and $\mathbf{A} = \sigma\sqrt{g}\,(\mathbf{X}'\mathbf{X})^{-1/2}$. Thus, $\mathbf{y} = \mathbf{X}\,\mathbf{A}\boldsymbol{\eta} + \sigma\boldsymbol{\epsilon}$, which is marginally normal with mean $\mathbf{0}$ and covariance matrix

$$\boldsymbol{\Omega} = \sigma^2(\mathbf{I} + g\,\mathbf{X}(\mathbf{X}'\mathbf{X})^{-1}\mathbf{X}').$$

Then, $p(\mathbf{y} \mid \mathbf{z})$ is the density function for $\mathbf{y}$.

(iii) With the uniform prior, the posterior for $\mathbf{z}$ only depends on its appearance in the term $\mathbf{X}(\mathbf{X}'\mathbf{X})^{-1}\mathbf{X}'$ in $p(\mathbf{y} \mid \mathbf{z})$. Note that $\mathbf{X}$ denotes the $n \times p'$ matrix that contains the jth column of $\mathbf{X}_0$ if $z_j = 1$; here, p' is the number of $\{z_j\}$ equal to 1. Thus, the log-posterior for $\mathbf{z}$ is

$$\log p(\mathbf{z} \mid \mathbf{y}) = \kappa - \tfrac{1}{2}\log|\boldsymbol{\Omega}| - \tfrac{1}{2}\mathbf{y}'\boldsymbol{\Omega}^{-1}\mathbf{y},$$

where κ is constant with respect to $\mathbf{z}$.

(iv) Now, $p(z_j \mid \mathbf{z}_{-j}, \mathbf{y}) \propto p(z_j, \mathbf{z}_{-j} \mid \mathbf{y})$ can be calculated from part (iii), which is all that is required. Then, $p(z_j = 1 \mid \mathbf{z}_{-j}, \mathbf{y}) - p(z_j = 0 \mid \mathbf{z}_{-j}, \mathbf{y})$, and select $z_j = 1$ or 0 depending on whether the difference is positive or negative, respectively. This is equivalent to considering $p(z_j = 1, \mathbf{z}_{-j} \mid \mathbf{y}) - p(z_j = 0, \mathbf{z}_{-j} \mid \mathbf{y})$, which can be calculated from part (iii) using two forms of $\boldsymbol{\Omega}$: the former with $\mathbf{X}$ including column j and the latter excluding column j. The other columns include those indicated by $\mathbf{z}_{-j}$.

(v) By definition, $p(\widehat{z}_j \mid \mathbf{z}_{-j}, \mathbf{y}) \geq p(z_j \mid \mathbf{z}_{-j}, \mathbf{y})$; multiply both sides of this inequality by $p(\mathbf{z}_{-j} \mid \mathbf{y})$ to obtain

$$p(z_1, \ldots, z_{j-1}, \widehat{z}_j, z_{j+1}, \ldots, z_p \mid \mathbf{y}) \geq p(z_1, \ldots, z_{j-1}, z_j, z_{j+1}, \ldots, z_p \mid \mathbf{y}).$$

Further Reading

Historical Background. Hopfield networks were first described in Nakano (1972), Amari (1972) and Little (1974). However, Hopfield (1982) popularized the algorithm and detailed a sequence of conditional maximizations on a function of binary outcomes because he was concerned with how neuron activity (which is binary) relates to memory. Even in high-dimensional settings, this algorithm can experience rapid convergence. See Bruck (1990) for details on the convergence of the Hopfield network.

Points of Interest.

1. The closest related algorithm in the statistics literature is the EM algorithm (Dempster et al., 1977), which is an iterative optimization algorithm that is often used when there are missing values. There are two steps within each iteration of the EM algorithm: an expectation (E) step and a maximization (M) step. When the maximization step is over multiple variables, a common approach is to use a sequence of conditional maximization steps; this essentially is a single iteration of a Hopfield network. Such an algorithm is known as an ECM algorithm, where "CM" stands for conditional maximization (He and Liu, 2012).

2. In general, it is quite straightforward for the Bayesian to find the posterior mode for the indicator variables (i.e., which covariates are most likely to be active). In fact, there are many algorithms available, including MCMC, that explore the posterior distribution by exploiting the location of the mode.

Demonstration. Consider a Hopfield network with dimension $p = 1000$. The aim is to maximize $S = \mathbf{z}'\mathbf{W}\mathbf{z}$ over $\mathbf{z}$, where $\mathbf{z}$ is a p-dimensional vector of $\{-1, +1\}$ elements, $\mathbf{W} = (w_{ij})$, $w_{ii} = 0$, and $w_{ij} = v_i v_j$ for $i \neq j$. The $\{v_i\}$ are also in $\{-1, +1\}$ and are generated randomly and independently with $\mathrm{P}(v_i = +1) = 3/4$. Starting with a random configuration of the $\{z_i\}$, change $z_i \to -z_i$ if the value of S is increased with the switch. For just a single iteration for each component, there is an almost quadratic increase in S; see Fig. 5.3. In this particular demonstration, there were 378 switches. Note that the solution is easy to check because $\mathbf{z} = -\mathbf{v}$.

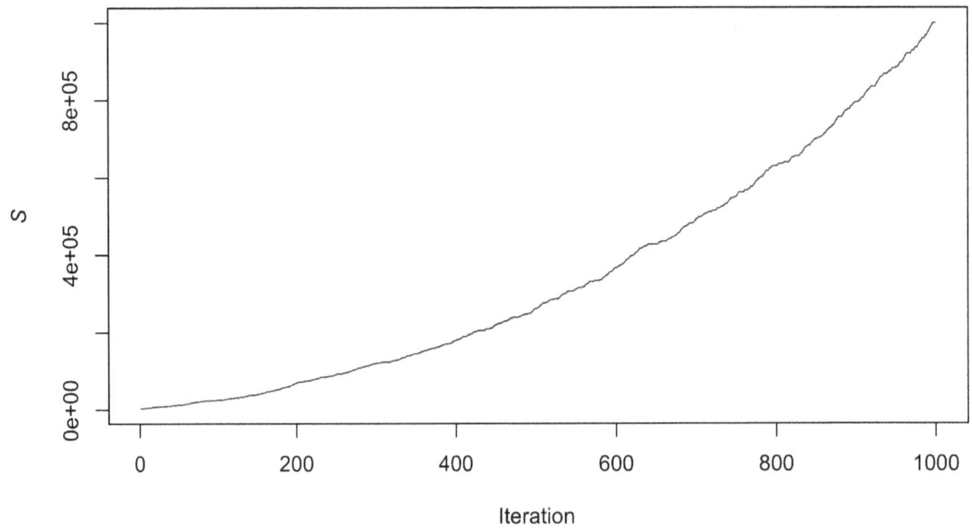

FIGURE 5.3
The value of $S = \mathbf{z}'\mathbf{W}\mathbf{z}$ increases quickly with the Hopfield network.

S5.1.5 – Variable selection with the value of the linear model

(i) The least squares estimator for β is given by

$$\widehat{\beta} = \arg\min_{\beta}(\mathbf{y} - \mathbf{X}\beta)'(\mathbf{y} - \mathbf{X}\beta),$$

which is obtained by setting $\frac{\partial}{\partial\beta}(\mathbf{y} - \mathbf{X}\beta)'(\mathbf{y} - \mathbf{X}\beta)$ to 0:

$$\frac{\partial}{\partial\beta}(\mathbf{y} - \mathbf{X}\beta)'(\mathbf{y} - \mathbf{X}\beta) = -2\mathbf{X}'\mathbf{y} + 2\mathbf{X}'\mathbf{X}\beta \overset{\text{set}}{=} 0.$$

Therefore, $\widehat{\beta} = (\mathbf{X}'\mathbf{X})^{-1}\mathbf{X}'\mathbf{y}$. Note that $\mathbf{X}'\mathbf{X}$ is invertible because $\mathbf{X}$ has full rank.

(ii) Let $\text{tr}(\mathbf{H})$ denote the trace of matrix $\mathbf{H}$. By the cyclic property of the trace operator,

$$\begin{aligned}\text{tr}(\mathbf{H}) &= \text{tr}\{\mathbf{X}(\mathbf{X}'\mathbf{X})^{-1}\mathbf{X}'\} \\ &= \text{tr}\{(\mathbf{X}'\mathbf{X})^{-1}\mathbf{X}'\mathbf{X}\} \\ &= \text{tr}(\mathbf{I}_d) = d,\end{aligned}$$

where $\mathbf{I}_d$ denotes the $d \times d$ identity matrix.

(iii) Note that the expected value of $\mathbf{z}'\mathbf{H}\mathbf{z}$ is a scalar, and $\text{tr}(c) = c$ for any scalar c. Thus,

$$\text{E}(\mathbf{z}'\mathbf{H}\mathbf{z}) = \text{E}\{\text{tr}(\mathbf{z}'\mathbf{H}\mathbf{z})\} = \text{E}\{\text{tr}(\mathbf{z}\mathbf{z}'\mathbf{H})\} = \text{tr}\{\text{E}(\mathbf{z}\mathbf{z}')\mathbf{H}\} = \text{tr}(\mathbf{H}) = d,$$

which uses the cyclic property of the trace operator to obtain the second equality. The third equality follows because $\text{E}\{\text{tr}(\mathbf{A})\} = \text{tr}\{\text{E}(\mathbf{A})\}$ for any random square matrix $\mathbf{A}$. To obtain the fourth equality, note that $\mathbf{z}$ is a vector of independent standard normal random variables, so $\text{E}(\mathbf{z}\mathbf{z}') = \mathbf{I}_d$.

(iv) Write out $\widehat{\beta}$ to obtain

$$\begin{aligned}
\mathrm{E}(\widehat{\beta}'\mathbf{X}'\mathbf{X}\widehat{\beta}) &= \mathrm{E}\{\mathbf{y}'\mathbf{X}(\mathbf{X}'\mathbf{X})^{-1}\mathbf{X}'\mathbf{X}(\mathbf{X}'\mathbf{X})^{-1}\mathbf{X}'\mathbf{y}\} \\
&= \mathrm{E}(\mathbf{y}'\mathbf{H}\mathbf{y}). \\
&= \beta^{*\,\prime}\mathbf{X}'\mathbf{H}\mathbf{X}\beta^* + \mathrm{tr}(\sigma^2\mathbf{H}) \\
&= \beta^{*\,\prime}\mathbf{X}'\mathbf{X}\beta^* + \sigma^2 d,
\end{aligned}$$

where the third equality follows because $\mathrm{E}(\mathbf{v}'\mathbf{A}\mathbf{v}) = \mu'\mathbf{A}\mu + \mathrm{tr}(\mathbf{A}\boldsymbol{\Sigma})$ for a symmetric matrix $\mathbf{A}$ and a random vector $\mathbf{v}$ with mean μ and covariance $\boldsymbol{\Sigma}$.

(v) Using the result in part (iv), an unbiased estimator for $\beta^{*\,\prime}\mathbf{X}'\mathbf{X}\beta^*$ is $\widehat{\beta}'\mathbf{X}'\mathbf{X}\widehat{\beta} - \sigma^2 d$.

Further Reading

Historical Background. The value of the model is a useful measure for model evaluation and selection. It highlights the significance of certain covariates and the trade-off between model fit and complexity. In general, the larger the value, the more the covariates in the design matrix $\mathbf{X}$ contribute to the explanation of the variance in the response $\mathbf{y}$. The comparison of model values underpins many popular model selection procedures. For example, this concept is the motivation behind the AIC variable selection procedure (Akaike, 1974), which is applied here for a known σ; see Question 5.1.6 for a slightly different approach to AIC.

Points of Interest.

1. Let d denote the total number of observed covariates. There is often interest in how many of these covariates are predictive for the response. Consider the model with p covariates and p parameters, where $p \le d$. The value of this model with design matrix $\mathbf{X}_p$ is

$$V(\mathbf{X}_p) = \beta_p^{*\,\prime}\mathbf{X}_p'\mathbf{X}_p\beta_p^*,$$

where β_p^* is the true parameter value for the p covariates. It is straightforward to show that

$$V(\mathbf{X}_p) = \beta^{*\,\prime}\mathbf{X}'\mathbf{H}_p\mathbf{X}\beta^*,$$

where β^* is the true parameter value, $\mathbf{H}_p$ is the hat matrix from $\mathbf{X}_p$, and $\mathbf{X}$ is the $n \times d$ fully-observed design matrix. A principled approach to variable selection is to select the subset of covariates that result in the highest value V.

2. Consider an initial design matrix $\mathbf{X}_p$. If a column containing a true covariate is added to $\mathbf{X}_p$, then $V(\mathbf{X}_{p+1}) > V(\mathbf{X}_p)$. However, if the new column does not contain a true covariate, then $V(\mathbf{X}_{p+1}) = V(\mathbf{X}_p)$. Thus, an unbiased estimator for $V(\mathbf{X}_p)$ is helpful for variable selection, and this exercise shows how to obtain such an estimator.

Demonstration. To demonstrate the point of the exercise, take $d = 5$ and $\beta^* = (2, 0, -4, 0, 4)'$. The $\mathbf{X}$ matrix is of size $n \times d$, and each element is independently generated as a standard normal random variable. It is now possible to compute $V(\mathbf{X}_p)$ for all p; here, this is done for $p = 1, \ldots, 5$. The function will increase for $p = 1$, $p = 3$ and $p = 5$ while remaining 0 for $p = 2$ and $p = 4$; see Fig. 5.4. Hence, finding an unbiased estimator for $V(\mathbf{X}_p)$ for any part of the design matrix with p columns is going to be helpful for variable selection.

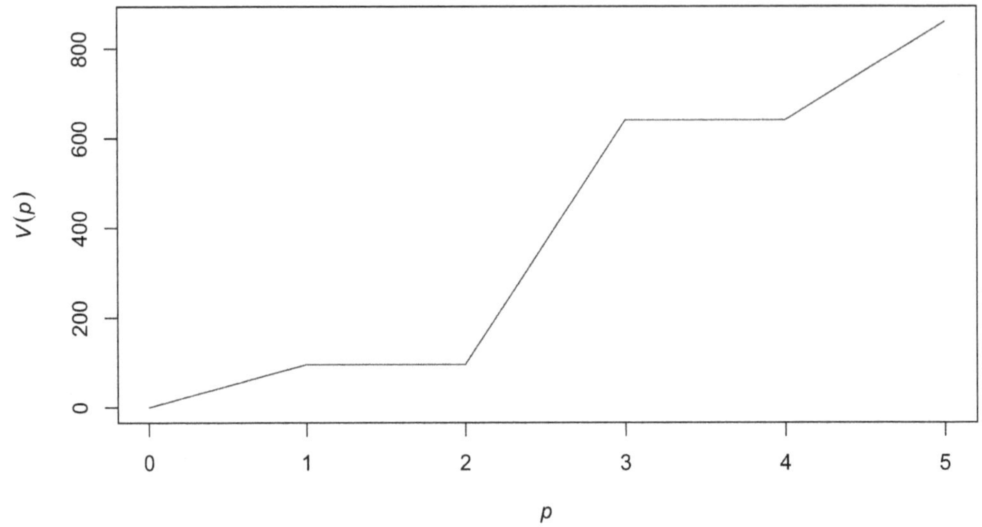

FIGURE 5.4
The value of $V(p) = V(\mathbf{X}_p)$ for $p = 0, 1, \ldots, 5$.

S5.1.6 – Variable selection: A form of AIC

(i) One simulation yields the following values of $L(d)$ in order: 97.97, 166.68, 342.61, 342.61 and 342.61.

(ii) The value of $L(d)$ increases as the number of true covariates get included; see the first $d_0 = 3$ values in part (i). However, adding extraneous covariates no longer improves $L(d)$. A penalty term on the number of covariates may be appropriate because it is typically desirable to select parsimonious models over models with irrelevant covariates. An ideal penalty decreases $L(d)$ as $d > d_0$. Note that the penalty should not be so large that it disturbs the increase from $d = 0$ to $d = d_0$.

(iii) The true model is

$$\mathbf{y} = \mathbf{X}_0\boldsymbol{\beta}^* + \sigma^* \boldsymbol{\epsilon}.$$

Hence,

$$\begin{aligned} \mathrm{E}\left(\mathbf{Y}'\mathbf{H}_d\mathbf{Y}\right) &= \boldsymbol{\beta}^{*\prime}\mathbf{X}_0'\mathbf{H}_d\mathbf{X}_0\boldsymbol{\beta}^* + \mathrm{tr}\{(\sigma^*)^2\mathbf{H}_d\} \\ &= \boldsymbol{\beta}^{*\prime}\mathbf{X}_0'\mathbf{H}_d\mathbf{X}_0\boldsymbol{\beta}^* + (\sigma^*)^2 d, \end{aligned}$$

where the first equality follows because $\mathrm{E}(\mathbf{v}'\mathbf{A}\mathbf{v}) = \boldsymbol{\mu}'\mathbf{A}\boldsymbol{\mu} + \mathrm{tr}(\mathbf{A}\boldsymbol{\Sigma})$ for a symmetric matrix $\mathbf{A}$ and a random vector $\mathbf{v}$ with mean $\boldsymbol{\mu}$ and covariance $\boldsymbol{\Sigma}$. The second equality follows because $\mathrm{tr}(\mathbf{H}_d) = d$; see the solution to part (ii) of Question 5.1.5 for more details.

(iv) Using the result from part (iii),

$$\begin{aligned}\mathrm{E}(\mathbf{Y}'\mathbf{I}_n\mathbf{Y}) &= \boldsymbol{\beta}^{*\,\prime}\mathbf{X}_0'\mathbf{I}_n\mathbf{X}_0\boldsymbol{\beta}^* + \mathrm{tr}\{(\sigma^*)^2\mathbf{I}_n\} \\ &= \boldsymbol{\beta}^{*\,\prime}\mathbf{X}_0'\mathbf{X}_0\boldsymbol{\beta}^* + (\sigma^*)^2\, n.\end{aligned}$$

Therefore, $\mathrm{E}(\mathbf{Y}'\mathbf{Y})/n > (\sigma^*)^2$ because $\boldsymbol{\beta}^{*\,\prime}\mathbf{X}_0'\mathbf{X}_0\boldsymbol{\beta}^* > 0$.

(v) Consider the quantity $\mathbf{y}'\mathbf{H}_d\mathbf{y} - d\,\mathbf{y}'\mathbf{y}/n$, which has expected value

$$\begin{aligned}\mathrm{E}(\mathbf{Y}'\mathbf{H}_d\mathbf{Y} - d\,\mathbf{Y}'\mathbf{Y}/n) &= \boldsymbol{\beta}^{*\,\prime}\mathbf{X}_0'\mathbf{H}_d\mathbf{X}_0\boldsymbol{\beta}^* + (\sigma^*)^2\, d - d\left(\boldsymbol{\beta}^{*\,\prime}\mathbf{X}_0'\mathbf{X}_0\boldsymbol{\beta}^*/n + (\sigma^*)^2\right) \\ &= L(d) - d\,\underbrace{\boldsymbol{\beta}^{*\,\prime}\mathbf{X}_0'\mathbf{X}_0\boldsymbol{\beta}^*/n}_{\gamma},\end{aligned}$$

where $\gamma > 0$. Thus, one possible variable selection criterion is maximizing $\mathbf{y}'\mathbf{H}_d\mathbf{y} - d\,\mathbf{y}'\mathbf{y}/n$.

Further Reading

Historical Background. The approach here is based on the Akaike information criterion (Akaike, 1974). However, standard AIC does not incorporate a penalty; the apparent penalty is actually a term that ensures unbiasedness, not a term that penalizes the addition of unnecessary covariates. Thus, the AIC is still prone to overestimating the number of covariates. For instance, even stochastic noise can increase the AIC, despite being an insignificant predictor of the response.

Points of Interest.

1. The combination of $L(d)$ with the penalty term γd ensures that the criterion value increases if a true covariate is added and decreases if an extraneous covariate is added. An unbiased estimator is given by $\mathbf{y}'\mathbf{H}_d\mathbf{y} - (\sigma^*)^2\, d$, which already appears to include a penalty term. However, the $(\sigma^*)^2\, d$ term is only included to provide an unbiased estimator. If σ^* is unknown, a penalty of the form γd is needed, where $\gamma > 0$ can be computed from the data. As shown in this question, a simple unbiased estimator for $L(d) - \gamma d$ is $\mathbf{y}'\mathbf{H}_d\mathbf{y} - d\,\mathbf{y}'\mathbf{y}/n$. Note that one simply needs to change the hat matrix $\mathbf{H}_d$ for each subset of covariates of interest.

2. If σ^* is known, then the decision rule is already known: maximize $L(d) - (\sigma^*)^2\, d$. Note that $L(d)$ is written as the value of the model $V(\mathbf{X}_d)$ in Question 5.1.5, where $\mathbf{X}_d$ is a fixed design matrix.

S5.1.7 – On the lasso estimator

(i) If $|y| < \lambda$, then

$$\tfrac{1}{2}y^2 < \tfrac{1}{2}(y-\theta)^2 + \lambda|\theta|$$

for any θ because

$$-y\theta + \tfrac{1}{2}\theta^2 + \lambda|\theta| > 0.$$

Therefore, if $|y| < \lambda$, the minimizer is $\widehat{\theta} = 0$. Otherwise, consider the derivative of $l(\theta)$ with respect to θ: $l'(\theta) = \theta - y + \lambda \cdot \text{sgn}(\theta)$, where $\text{sgn}(\theta)$ is $+1$ if $\theta > 0$ and -1 if $\theta < 0$. Therefore, if $y > \lambda$ or $y < -\lambda$, the minimizers are $\widehat{\theta} = y - \lambda$ and $\widehat{\theta} = y + \lambda$, respectively.

(ii) The expected value of $\widehat{\theta}$ is

$$\mathrm{E}(\widehat{\theta}) = \tfrac{1}{2}(e^{\theta^*} - e^{-\theta^*})\, e^{-\lambda},$$

which can be obtained using standard integration. In particular, evaluate

$$\mathrm{E}(\widehat{\theta}) = \tfrac{1}{2}\int_{-\infty}^{-\lambda} (y+\lambda)\, e^{y-\theta^*}\, dy + \tfrac{1}{2}\int_{-\lambda}^{\lambda} 0\, e^{-|y-\theta^*|} dy + \tfrac{1}{2}\int_{\lambda}^{\infty} (y-\lambda)\, e^{-y+\theta^*}\, dy,$$

which is obtained by breaking up $\widehat{\theta}$ into the three cases described in part (i).

(iii) From part (ii), $\lambda > \theta^*$, so $e^{-\lambda} < e^{-\theta^*}$. The bound of interest arises by noting that

$$\mathrm{E}(\widehat{\theta}) = \tfrac{1}{2}(e^{\theta^*} - e^{-\theta^*})\, e^{-\lambda} < \tfrac{1}{2}(e^{\theta^*} - e^{-\theta^*})\, e^{-\theta^*} = \tfrac{1}{2}(1 - e^{-2\theta^*}).$$

Equality is obtained if $\theta^* = \lambda$.

(iv) This follows from part (i) with y replaced by $\theta_{\text{LSE}} = \sum_{i=1}^n y_i x_i / \sum_{i=1}^n x_i^2$ and λ replaced by $n\lambda / \sum_{i=1}^n x_i^2$. Thus,

$$\widehat{\theta} = \theta_{\text{LSE}} \max\left\{0, 1 - \frac{n\lambda}{|\sum_{i=1}^n x_i y_i|}\right\}.$$

(v) Although θ_{LSE} converges to θ^*, it is the $\max\{\cdot\}$ term that determines the final estimator. Thus, the asymptotic behavior of the estimator in part (iv) depends on the behavior of $\sum_{i=1}^n x_i^2 / n$ as $n \to \infty$.

Further Reading

Historical Background. The lasso was introduced in Tibshirani (1996). Unlike most variable selection criteria such as AIC and BIC, the lasso is fast. It can also be compared to the ridge regression solution, which does not achieve variable selection. In particular, estimators are not naturally set to 0 with the ridge estimator, but they are with the lasso.

Points of Interest.

1. The main penalized regression approaches can be viewed as

$$\frac{1}{2}(\mathbf{y} - \mathbf{X}\boldsymbol{\theta})'(\mathbf{y} - \mathbf{X}\boldsymbol{\theta}) + \lambda||\boldsymbol{\theta}||_k,$$

   where $k = 0$ specifies an information criteria, $k = 1$ specifies the lasso, and $k = 2$ specifies ridge regression. The choices of $k = 0$ and $k = 1$ lead to variable selection because some components of the estimator of $\boldsymbol{\theta}$ can be 0.

2. The convergence rates for the lasso estimators of the non-zero regression coefficients can be worse than the convergence rates for the least squares estimator. However, the lasso estimates the zero components exactly for all large n (Chatterjee and Lahiri, 2011).

3. A popular recent develop in regularization and variable selection is the "elastic net," which uses a combination of the L_1 (lasso regression) and L_2 (ridge regression) penalties (Zou and Hastie, 2005).

6

Asymptotics

Asymptotics, or large sample theory, is the study of the properties of estimators, statistical tests, and other procedures. The usual aim in asymptotics is to demonstrate the consistency of estimation processes, particularly to ensure that the estimators converge in some sense to the true value. Though some people decry the notion of large sample studies and the existence of a true model, an estimator must estimate something. Typically, the estimand is defined as having an underlying truth. Otherwise, there would not be a sense of whether the estimator is good or bad. The main idea is that, if the estimation procedure does not perform well for large samples, there cannot be confidence in its performance for moderate or small samples. Thus, the study of asymptotics is essential to understanding and constructing estimation procedures of high quality. In fact, most classical procedures are based on the central limit theorem, the law of large numbers, or asymptotically demonstrable χ^2 tests – all of which are asymptotic results that are regularly relied on in practice. For example, the χ^2 approximation to the likelihood-ratio test is often used in hypothesis testing.

Classical asymptotics often use several types of convergence: almost sure convergence, convergence in distribution, convergence in probability, and convergence in mean. Question 6.1.1 explores the connection between the latter three types of convergence. Almost sure convergence is featured in Question 6.1.2, where a proof for the law of large numbers is constructed. According to many texts, proving the law of large numbers is a convoluted affair. The approach in Question 6.1.2 uses backward martingales, which is straightforward and elegant given the martingale convergence theorem.

Large sample properties of the likelihood function are of primary interest in classical asymptotics. For example, a basic objective is to show that the MLE converges in some way to the true value; however, given the list of required conditions for this to be true, proving such a convergence is not trivial. An asymptotic distribution for the MLE is often obtained via the central limit theorem, which is most accessible if the estimator can be expressed as a sum of independent random variables; certainly, the log-likelihood yields such a representation. Question 6.1.3 obtains the approximate distribution for the MLE of the exponential family by exploiting the convexity of the negative log-likelihood. This result simplifies the usual conditions required for convergence, which often assume a local convex property about the true value.

The study of asymptotics is not exclusive to classical statistics. In the Bayesian framework, asymptotic results generally concern the amount of mass of a parameter's posterior distribution that is outside of an arbitrarily small neighborhood centered on the true value; ideally, this posterior mass goes to 0, indicating convergence of the posterior distribution to the truth. Thus, distances between parametric values become crucial. The use of many common distance functions, such as Euclidean distance, are typically not feasible because they only make point-wise comparisons (Walker, 1969). Rather, distance functions that explicitly compare two probability distributions, such as the Hellinger distance, simplify the study of Bayesian asymptotics and yield elementary and natural conditions for consistency. Question 6.2.1 makes use of the Hellinger distance to study the consistency of the

DOI: 10.1201/9781003493471-6

posterior distribution for the mean of a normal model. Question 6.2.2 uses a similar approach to establish consistency for the rate parameter of a Poisson model. Question 6.2.3 generalizes this approach for an arbitrary parametric family.

Question 6.2.4 discusses Bayesian bootstrapping and relates it to objective priors; interestingly, the original prior-free objective Bayesian procedure was the Bayesian bootstrap (Rubin, 1981). Question 6.2.5 investigates the asymptotic properties of marginal likelihood functions, which are commonly used for Bayesian testing. Finally, Question 6.2.6 establishes posterior consistency results for a discrete parameter space; here, a notable simplification occurs.

Q6.1 Questions – Classical Framework

Q6.1.1 – Different types of convergence

Introduction. The convergence of random variables is an important topic in probability theory. Different metrics of convergence emphasize different characteristics of the behavior of random variables. Several well-known metrics include convergence in distribution, convergence in mean, and convergence in probability. This exercise explores the equivalence between these different types of convergence. In the question, the assumption is that the sequence $\{X_n\}$ represents a sequence of estimators of increasing sample size n.

When considering non-random deterministic sequences, say $\{a_n\}$, there is only the one sequence to consider; it either converges or does not. For a random sequence $\{X_n\}$, there are associated sequences, such as the sequence of probabilities on certain events $\{P_n(A_n)\}$ and the sequence of distribution functions $\{F_{X_n}(x)\}$. Investigating these other sequences yields the different types of convergence associated with $\{X_n\}$.

Question. Denote a sequence of real random variables as $\{X_n\}$.

(i) Prove the Markov inequality: if Z is a nonnegative random variable with finite expectation, then $a\,\mathrm{P}(Z \geq a) \leq \mathrm{E}(Z)$ for all $a > 0$.

(ii) If $\mathrm{E}(|X_n|) \to 0$ as $n \to \infty$, show that $\mathrm{P}(|X_n| > \epsilon) \to 0$ for all $\epsilon > 0$ (i.e., X_n converges to 0 in probability).

(iii) If X_n converges to 0 in probability, show that X_n converges to 0 in distribution:

$$\mathrm{P}(X_n \leq x) \to \begin{cases} 0 & x < 0 \\ 1 & x \geq 0 \end{cases}.$$

(iv) Let $A_n(\epsilon) = 1(|X_n| > \epsilon)$, where $1(\cdot)$ is the indicator function. If X_n converges to 0 in probability, what happens to $\mathrm{E}\{A_n(\epsilon)\}$ as $n \to \infty$?

(v) If $|X_n| \leq 1$ for all n and X_n converges to 0 in probability, show that $\mathrm{E}(|X_n|) \to 0$ by writing X_n in terms of $A_n(\epsilon)$ and the complement of $A_n(\epsilon)$.

Q6.1.2 – Strong law of large numbers from a backward martingale

Introduction. The law of large numbers is a fundamental asymptotic result that states that a sample mean converges to the corresponding population mean as the sample size increases. There are two versions of this theorem: [1] the strong law refers to almost sure convergence and [2] the weak law refers to convergence in probability. Because almost sure convergence implies convergence in probability, proving the strong law is often more difficult. However, this question constructs an elegant proof for the strong law of large numbers using backward martingales. The proof relies on Doob's martingale convergence theorems (Doob, 1953).

Question. Suppose $X_1, X_2, \ldots$ are independent and identically distributed with $\mathrm{E}(|X_1|) < \infty$. Define

$$S_n = \sum_{i=1}^{n} X_i$$

and $\mathcal{G}_m = \sigma(S_m, S_{m+1}, \ldots)$, which is the σ-algebra on $\{S_m, S_{m+1}, \ldots\}$.

(i) Explain why $\mathrm{E}(X_n \mid \mathcal{G}_m) = \mathrm{E}(X_n \mid S_m)$ for all $n \leq m$.

(ii) Explain why $\mathrm{E}(X_n \mid S_m) = \mathrm{E}(X_i \mid S_m)$ for all $i, n \leq m$.

(iii) Prove that $\mathrm{E}(X_n \mid \mathcal{G}_m) = S_m/m$ for all $n \leq m$.

(iv) A process $Y = \{Y_n : n \in N\}$ is a backward martingale with respect to the decreasing sequence of sub σ-algebras $\{F_n : n \in N\}$ if $\mathrm{E}(Y_n \mid F_m) = Y_m$ for all $n \leq m$ and $\mathrm{E}(|Y_n|) < \infty$ for all n. The backward martingale convergence theorem states that, if $Y = \{Y_n : n \in N\}$ is a backward martingale with respect to $\{F_n : n \in N\}$, then there exists a random variable Y_∞ such that $Y_n \to Y_\infty$ with probability 1, $Y_n \to Y_\infty$ in mean, and $Y_\infty = \mathrm{E}(Y_1 \mid F_\infty)$. Show that $\{S_m/m\}$ is a backward martingale with respect to $\mathcal{G}_m$.

(v) Prove the strong law of large numbers: $S_m/m \to \mathrm{E}(X_1)$ with probability 1.

Q6.1.3 – Approximate distributions for MLEs

Introduction. The central limit theorem (CLT) establishes that a sum of suitably scaled independent random variables converges to a normal random variable, even if the original variables themselves are not normally distributed. The theorem is fundamental in statistics because it implies that methods developed for normal distributions can be used in problems involving other types of distributions. In this question, the CLT is applied to the negative log-likelihood function to obtain an asymptotic distribution at an arbitrary point; this leads to an approximate distribution for the maximum likelihood estimator.

Question. Let $f(x \mid \theta)$ be a family of density functions indexed by $\theta \in \Theta$ and $x \in \mathcal{X} \subset (-\infty, +\infty)$. Let $l(\theta; x) = -\log f(x \mid \theta)$ denote the negative log-likelihood, and suppose that $l(\theta; x)$ is strictly convex in θ for all x. Additionally, let $l'(\theta; x) = \partial l(\theta; x)/\partial\theta$.

(i) Consider a sample $\{x_1, \ldots, x_n\}$ of size n from $f(\cdot \mid \theta)$, and let $\widehat{\theta}$ be the maximum likelihood estimator of θ. Show that $\mathrm{P}(\widehat{\theta} \leq z) = \mathrm{P}(l'(z) \geq 0)$, where

$$l'(z) = \sum_{i=1}^{n} l'(z; x_i).$$

(ii) Assuming the central limit theorem holds for $l'(z)/n$ for all z, find an approximate distribution for $\widehat{\theta}$.

(iii) Suppose $f(x \mid \theta) = \theta\, e^{-x\theta}$, where $x > 0$ and $\theta > 0$. Show that $-\log f(x \mid \theta)$ is convex for all θ.

(iv) Find the normal approximation to $l'(z)/n$, where $f(x \mid \theta) = \theta\, e^{-x\theta}$.

(v) Find an approximate distribution for the MLE of θ in part (iii).

Q6.1.4 – Glivenko–Cantelli theorem

Introduction. The Glivenko–Cantelli theorem is a fundamental result concerning the convergence properties of the empirical distribution function. The strong result is that

$$\sup_x |F_n(x) - F^*(x)| \to 0$$

with probability 1, where F^* is the true distribution generating observations $\{x_i\}$ and F_n is the usual empirical distribution function based on a sample of size n. This exercise attempts to explain the theorem from foundational results that rely on continuity and the law of large numbers.

Question. Let $F^*(x)$ be a continuous distribution function on the real line and $\{x_i\}$ be independent samples from F^*.

(i) What is the result that provides $F_n(x) \to F^*(x)$ almost surely for all x?

(ii) For a fixed $\epsilon > 0$, show that there exists $z_1 < \cdots < z_K$ for a finite K (that depends on ϵ) for which

$$F^*(z_{j+1}) - F^*(z_j) \leq \epsilon.$$

(iii) Find a suitable upper and lower bound for

$$F_n(x) - F^*(x)$$

in terms of F_n and F evaluated at z_j and z_{j+1} with $z_j < x < z_{j+1}$.

(iv) Prove the Glivenko–Cantelli theorem.

(v) What changes occur, if any, when F^* has a single discontinuity at $x = 0$?

S6.1 Solutions – Classical Framework

S6.1.1 – Different types of convergence

(i) Note that $a\,1\{a \leq Z\} \leq Z\,1\{a \leq Z\} \leq Z$. Integrate both sides of $a\,1\{a \leq Z\} \leq Z$ with respect to the density function for Z (i.e., P) to obtain the result:

$$a\,\mathrm{P}(a \leq Z) = \int a\,1(a \leq Z)\,dP \leq \int Z\,dP = \mathrm{E}(Z).$$

(ii) Let $a = \epsilon$ and $Z = |X_n|$ in part (i). Then, apply Markov's inequality to obtain

$$\begin{aligned} \epsilon \cdot \mathrm{P}(|X_n| \geq \epsilon) &\leq \mathrm{E}(|X_n|), \\ \mathrm{P}(|X_n| \geq \epsilon) &\leq \mathrm{E}(|X_n|)/\epsilon \to 0 \end{aligned}$$

for all $\epsilon > 0$ as $n \to \infty$. Because $\mathrm{P}(|X_n| \geq \epsilon)$ is nonnegative, it must be that $\mathrm{P}(|X_n| \geq \epsilon) \to 0$ for all $\epsilon > 0$.

(iii) Write $\mathrm{P}(X_n \leq x) = 1 - \mathrm{P}(X_n > x)$ and note that $\mathrm{P}(|X_n| > \epsilon) \to 0$ implies $\mathrm{P}(X_n > \epsilon) \to 0$. If $x > 0$, there exists some positive $\epsilon < x$ such that $\mathrm{P}(X_n > x) < \mathrm{P}(X_n > \epsilon) \to 0$. Thus, when $x > 0$, $\mathrm{P}(X_n \leq x) = 1 - \mathrm{P}(X_n > x) \to 1$. If $x \leq 0$, the converse is true: $\mathrm{P}(X_n \leq x) \to 0$.

(iv) Solving this problem directly,

$$\mathrm{E}\{A_n(\epsilon)\} = \mathrm{E}\{1(|X_n| > \epsilon)\} = \mathrm{P}(|X_n| > \epsilon) \to 0$$

as $n \to \infty$.

(v) Write X_n in terms of $A_n(\epsilon)$ and $A_n^C(\epsilon)$:

$$X_n = X_n \cdot 1(|X_n| > \epsilon) + X_n \cdot 1(|X_n| \leq \epsilon).$$

Break this problem into two cases because there are two indicator functions; this will generally decrease the complexity of similar problems. First, consider the case when $|X_n| > \epsilon$:

$$\begin{aligned} \mathrm{E}(|X_n|) &= \mathrm{E}\{|X_n \cdot 1(|X_n| > \epsilon) + X_n \cdot 1(|X_n| \leq \epsilon)|\} \\ &= \mathrm{E}\{|X_n| \cdot |1(|X_n| > \epsilon)|\} \\ &\leq \mathrm{E}\{1(|X_n| > \epsilon)\} \quad \text{(by assumption that } |X_n| \leq 1) \\ &= \mathrm{P}(|X_n| > \epsilon) \to 0. \end{aligned}$$

Then, consider the case when $|X_n| \leq \epsilon$:

$$\begin{aligned} \mathrm{E}(|X_n|) &= \mathrm{E}\{|X_n \cdot 1(|X_n| > \epsilon) + X_n \cdot 1(|X_n| \leq \epsilon)|\} \\ &= \mathrm{E}\{|X_n| \cdot |1(|X_n| \leq \epsilon)|\} \\ &\leq \epsilon\,\mathrm{E}\{1(|X_n| \leq \epsilon)\} \to 0. \end{aligned}$$

Therefore, $\mathrm{E}(|X_n|) \to 0$.

Further Reading

Historical Background. The types of convergence discussed in this exercise are fundamental to large sample theory, which assesses properties of random variables (i.e., estimators and statistical tests). For example, estimators are typically motivated by demonstrating convergence to the true value; in this case, convergence is typically in probability or with probability 1. Thus, a fundamental understanding of the different types of convergence is important within statistics. Excellent reads on the topic include Grimmett and Stirzaker (1982), Chandra (2012), and Chapter 17 in Jacod and Protter (2000). For a historical context, see Prokhorov (1956), which discusses the various types of convergence, as well as their implications for studying random variables. Finally, for a measure theoretic approach to convergences in probability theory, see Athreya and Lahiri (2006).

Points of Interest.

1. Consider the sequence of random variables $\{X_n\}$, where $X_n = Z_n + 1/n$ and the $\{Z_n\}$ are a sequence of independent standard normal random variables. Then, $\mathrm{P}(X_n \le x) = \Phi(x - 1/n) \to \Phi(x)$. In words, X_n converges in distribution to a standard normal random variable.

2. Now, consider an example of the Kolmogorov 0-1 law, which concerns tail events associated with independent σ-algebras (Stroock, 2010). Suppose that $X_n = Z_n$, where the $\{Z_n\}$ are a sequence of independent Bernoulli random variables with mean $1/n$. Then, X_n converges in probability to 0 because $\mathrm{P}(X_n > \epsilon) < 1/(n\epsilon) \to 0$ as $n \to \infty$ for all $\epsilon > 0$. However, X_n does not converge to 0 with probability one. To see this, note that

$$\mathrm{E}\left(\sum_{n=1}^{\infty} Z_n\right) = \infty.$$

From the second Borel–Cantelli lemma, it is that

$$\mathrm{P}(Z_n = 1 \text{ infinitely often}) = 1.$$

On the other hand, if the $\{Z_n\}$ had means of $1/n^{1+a}$ for any $a > 0$, then the Borel–Cantelli lemma implies

$$\mathrm{P}(Z_n = 1 \text{ infinitely often}) = 0$$

because $\sum_n \mathrm{P}(Z_n = 1) < \infty$. In this case, X_n converges to 0 with probability one.

3. The Kolmogorov 0-1 law for the Bernoulli can be proved with relatively simple algebra as opposed to measure theory. Let $\{Z_n\}$ be independent Bernoulli random variables with $\mathrm{P}(Z_n = 1) = \epsilon_n > 0$, where $\epsilon_1 = 1$ and all other $\epsilon_n < 1$. Now, define the events $A_n = \{Z_n = 1, Z_m = 0, m > n\}$, which are mutually disjoint events for all $n \ge 1$. Then, the probability that Z_n converges to 0 almost surely is

$$P = \sum_{n=1}^{\infty} \mathrm{P}(A_n) = \sum_{n=1}^{\infty} \epsilon_n \prod_{k=n+1}^{\infty} (1 - \epsilon_k).$$

Here, $P = 0$ or $P = 1$. To show this, define

$$\pi = \prod_{k=2}^{\infty} (1 - \epsilon_k).$$

If $\pi = 0$, then $P = 0$ because each term must be 0:

$$\prod_{k=n}^{\infty}(1-\epsilon_k) = \frac{\pi}{\prod_{j=2}^{n-1}(1-\epsilon_j)} = 0$$

for all n. If $\pi > 0$, then define

$$P_N = \sum_{n=1}^{N} \epsilon_n \prod_{k=n+1}^{\infty}(1-\epsilon_k) = \pi\left[\epsilon_1 + \sum_{k=2}^{N}\frac{\epsilon_k}{\sum_{j=2}^{k}(1-\epsilon_j)}\right].$$

The sum inside the large square brackets can be written as

$$\frac{\epsilon_N + \epsilon_{N-1}(1-\epsilon_N) + \cdots + \epsilon_1(1-\epsilon_2)\dots(1-\epsilon_N)}{\prod_{n=2}^{N}(1-\epsilon_n)}.$$

The numerator can be shown to be equal to 1 because $\epsilon_1 = 1$, and the denominator is converging to π as $N \to \infty$, so $P_N \to 1$. If $\epsilon_n = 1/n$, then $\pi = 0$. If $\epsilon_n = 1/n^2$, then $\pi = \frac{1}{2}$.

S6.1.2 – Strong law of large numbers from a backward martingale

(i) Because the $\{X_i\}$ are independent, $S_{m+1}, S_{m+2}, \dots$ does not provide more information about X_m than S_m does. Formally, $\mathcal{G}_m = \sigma(S_m, S_{m+1}, S_{m+2}, \dots) = \sigma(S_m, X_{m+1}, X_{m+2}, \dots)$. Thus, by independence,

$$\mathrm{E}(X_n \mid \mathcal{G}_m) = \mathrm{E}(X_n \mid S_m, X_{m+1}, \dots) = \mathrm{E}(X_n \mid S_m).$$

(ii) The independence of the $\{X_i\}$ implies exchangeability, meaning that the joint law of $\{X_1, \dots, X_n\}$ is invariant under permutations. Therefore, X_n can be exchanged for any X_i and the conditional expectations would remain the same.

(iii) Note that S_m is $\mathcal{G}_m$-measurable. Using the results from parts (i) and (ii),

$$\begin{aligned} S_m = \mathrm{E}(S_m \mid \mathcal{G}_m) &= \sum_{i=1}^{m}\mathrm{E}(X_i \mid \mathcal{G}_m) \overset{\text{(i)}}{=} \sum_{i=1}^{m}\mathrm{E}(X_i \mid S_m) \\ &\overset{\text{(ii)}}{=} \sum_{i=1}^{m}\mathrm{E}(X_n \mid S_m) = m\mathrm{E}(X_n \mid S_m) \overset{\text{(i)}}{=} m\mathrm{E}(X_n \mid \mathcal{G}_m), \end{aligned}$$

where $n \le m$.

(iv) The aim is to show that $\mathrm{E}(S_n/n \mid \mathcal{G}_m) = S_m/m$ for all $n \le m$. Without loss of generality, let $m = n+1$. Then,

$$\begin{aligned} \mathrm{E}(S_n/n \mid \mathcal{G}_{n+1}) &= \frac{1}{n}\mathrm{E}(S_n \mid \mathcal{G}_{n+1}) = \frac{1}{n}\mathrm{E}\left(S_{n+1} - X_{n+1} \mid \mathcal{G}_{n+1}\right) \\ &= \frac{1}{n}\mathrm{E}\left(S_{n+1} \mid \mathcal{G}_{n+1}\right) - \frac{1}{n}\mathrm{E}\left(X_{n+1} \mid \mathcal{G}_{n+1}\right) \\ &\overset{\text{(iii)}}{=} \frac{1}{n}S_{n+1} - \frac{1}{n(n+1)}S_{n+1} = \frac{1}{n+1}S_{n+1}, \end{aligned}$$

so the process $\{S_m/m\}$ is a $\mathcal{G}_m$-backward martingale.

(v) By the backward martingale convergence theorem, there exists a random variable S_∞ such that $S_m/m \to S_\infty$ almost surely, where $S_\infty = \mathrm{E}(S_1/1 \mid \mathcal{G}_\infty) = \mathrm{E}(X_1 \mid \mathcal{G}_\infty)$. Thus, $S_m/m \to \mathrm{E}(X_1)$ almost surely, proving the strong law of large numbers.

Further Reading

Historical Background. Traditional and comprehensible proofs for the strong law of large numbers require a finite variance for X_1. The more difficult proofs also require a finite value of $\mathrm{E}|X_1|$ (Etemadi, 1981). The proof using backward martingales only requires this latter result (Doob, 1953). Note that $\mathrm{E}|X_1| < \infty$ is a necessary condition for the law of large numbers (Grimmett and Stirzaker, 1982).

Points of Interest.

1. Backward martingales can also be used to demonstrate the strong law of large numbers for exchangeable sequences, which are conditionally independent and identically distributed sequences (Aldous, 1985). This is perhaps not surprising given that exchangeable sequences can be so easily expressed as backward martingales.

2. Berti et al. (2004) showed that a conditionally identically distributed sequence satisfies the strong law of large numbers under the usual condition that $\mathrm{E}(|X_1|)$ is finite. A sequence $\{X_n\}$ is conditionally identically distributed with respect to a filtration $\mathcal{G} = \{\mathcal{G}_n\}_{n\geq 0}$ if
$$\mathrm{E}\{f(X_k) \mid \mathcal{G}_n\} = \mathrm{E}\{f(X_{n+1}) \mid \mathcal{G}_n\}$$
almost surely for all $k > n \geq 0$ and all bounded measurable f.

S6.1.3 – Approximate distributions for MLEs

(i) The MLE is given by
$$\widehat{\theta} = \min_\theta \sum_{i=1}^n l(\theta; x_i).$$
Because $l(\theta; x)$ is strictly convex in θ for all x, it is that $l''(\theta; x) \geq 0$. Because $\widehat{\theta}$ is the minimizer of a convex function,
$$\widehat{\theta} \leq z \Longleftrightarrow \sum_{i=1}^n l'(z; x_i) \geq 0.$$
Therefore, $\mathrm{P}(\widehat{\theta} \leq z) = \mathrm{P}(l'(z) \geq 0)$.

(ii) Using the CLT,
$$l'(z)/n = n^{-1} \sum_{i=1}^n l'(z; x_i) \overset{d}{\approx} \mathrm{N}\left(\mu(z), \sigma^2(z)/n\right),$$

where $\mu(z) = \mathrm{E}\{l'(z;x_i)\}$ and $\sigma^2(z) = \mathrm{Var}\{l'(z;x_i)\}$. Assume that ϵ is a standard normal random variable, then

$$\begin{aligned}
\mathrm{P}(\widehat{\theta} \leq z) &= \mathrm{P}(l'(z) \geq 0) \\
&\approx \mathrm{P}\left(\mu(z) + \epsilon\sigma(z)/\sqrt{n} \geq 0\right) \\
&= \mathrm{P}\left(\epsilon \geq -\sqrt{n}\,\mu(z)/\sigma(z)\right) \\
&= 1 - \Phi\left(\sqrt{n}\,\mu(z)/\sigma(z)\right),
\end{aligned}$$

where Φ denotes the standard normal cumulative distribution function.

(iii) Let $f(x \mid \theta) = \theta e^{-x\theta}$. Then,

$$l(\theta;x) = -\log\theta + x\theta, \quad l'(\theta;x) = -1/\theta + x, \quad l''(\theta;x) = 1/\theta^2 \geq 0.$$

Therefore, $l(\theta;x)$ is convex.

(iv) Note that $\mathrm{E}(X) = 1/\theta$ and $\mathrm{Var}\,(X) = 1/\theta^2$. Then,

$$\begin{aligned}
\mu(z) &= \mathrm{E}\{l'(z;X_i)\} = \mathrm{E}\left(-1/z + X_i\right) = -1/z + 1/\theta, \\
\sigma^2(z) &= \mathrm{Var}\{l'(z,X_i)\} = \mathrm{Var}\left(-1/z + X_i\right) = \mathrm{Var}\,(X_i) = 1/\theta^2.
\end{aligned}$$

Thus, the approximate distribution for $l'(z)/n$ is $\mathrm{N}(1/\theta - 1/z, 1/\theta^2)$.

(v) The approximate cumulative distribution for the MLE is

$$F_{\widehat{\theta}}(z) = \Phi\left(\sqrt{n}(1 - \theta/z)\right).$$

Note that this converges to 0 and 1 if $z < \theta$ and $z > \theta$, respectively. Therefore, the distribution becomes a point mass distribution at θ as $n \to \infty$.

Further Reading

Historical Background. The central limit theorem is one of the most useful tools to show convergence in distribution for sequences of random variables. In its most common form, the random variables are independent and identically distributed. However, convergence of the mean to the normal distribution may also occur for non-identical or non-independent variables under certain conditions (Dudley, 2014).

The asymptotic convergence of random variables can be shown using limit theorems, such as the central limit theorem and the continuous mapping theorem. The latter theorem states that continuous functions preserve limits even if their arguments are sequences of random variables; this was first proved by Mann and Wald (1943). See Chapter 2 in van der Vaart (2000) and Chapter 5 in Wasserman (2004) for more details.

Points of Interest.

1. In this question, the asymptotic distribution for $\widehat{\theta}$ in part (v) coincides with that of $1/\overline{x}$, where $\overline{x}$ is approximately normal with mean $1/\theta$ and variance $1/(n\theta^2)$. However, this is not true in general. The usual approximate distribution for $\widehat{\theta}$ is normal, whereas the asymptotic distribution based on $l'(\theta;x)$ may not be normal. Knowing if $l(\theta;x)$ is convex can lead to improved asymptotic distributions for the MLE.

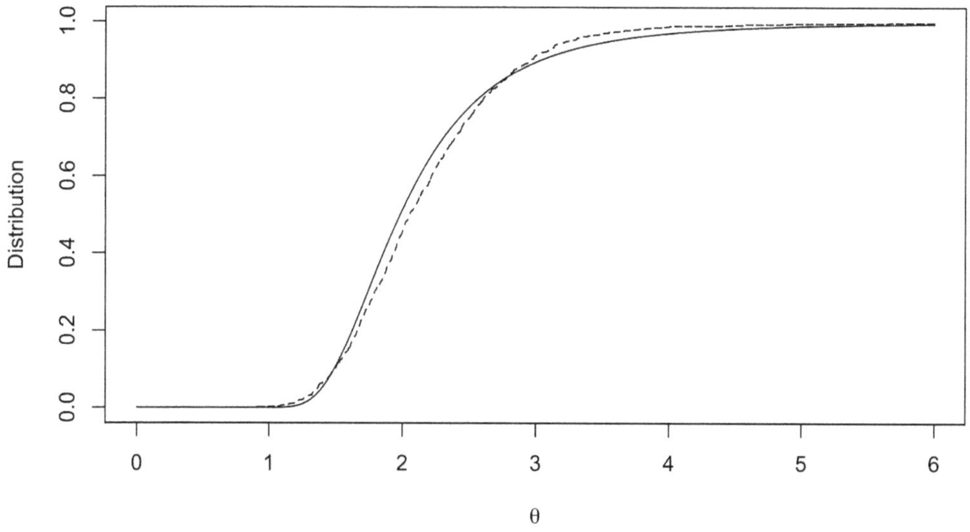

FIGURE 6.1
Comparing the true distribution function (full line) to the approximated distribution function (dashed line) using the MLE for the Fisk density.

Demonstration. Consider approximating the distribution of the MLE for the Fisk density

$$f(x \mid \theta) = \frac{\theta\, x^{\theta-1}}{(1+x^{\theta})^2},$$

where $x > 0$ and $\theta > 0$. It is easy to show log-concavity, taking

$$l(\theta; x) = \log(\theta) + (\theta - 1)\log(x) - 2\log(1 + x^{\theta}).$$

Let the true value be $\theta^* = 2$. The true distribution of $\widehat{\theta}$ is approximated using an empirical distribution function constructed from 1000 MLEs coming from a sample of size $n = 10$. The approximate distribution of the MLE is given by

$$\widehat{F}_{\widehat{\theta}}(z) = 1 - \Phi\left(\sqrt{n}\frac{D(z,\theta^*)}{\sqrt{V(z,\theta^*) - D^2(z,\theta^*)}}\right),$$

where

$$D(\theta,\theta^*) = \int l'(\theta; x)\, f(x \mid \theta^*)\, dx \quad \text{and} \quad V(\theta,\theta^*) = \int \{l'(\theta; x)\}^2 f(x \mid \theta^*)\, dx.$$

The true and approximated cumulative distribution functions for $\widehat{\theta}$ are depicted in Fig. 6.1 as a full line and a dashed line, respectively.

S6.1.4 – Glivenko–Cantelli theorem

(i) The law of large numbers indicates that

$$n^{-1} \sum_{i=1}^{n} 1(x_i \leq x) \to F^*(x)$$

with probability one for each x. Recall that the expectation of a Bernoulli variable is the probability of observing 1.

(ii) Take $z_1 = Q^*(\epsilon)$, where Q^* is the true quantile function. Hence, $F^*(z_1) = \epsilon$. In general, take $F^*(z_j) = j\epsilon$; only a $K = 1/\epsilon$ amount of $\{z_j\}$ will be required.

(iii) Now,

$$F_n(z_j) - F^*(z_{j+1}) \leq F_n(x) - F^*(x) \leq F_n(z_{j+1}) - F^*(z_j)$$

whenever $z_j < x < z_{j+1}$.

(iv) Apply the law of large numbers to z_j and z_{j+1}. For large enough n, $|F_n(x) - F(x)| \leq \epsilon$ for all x. Hence, the theorem follows.

(v) The result will still hold; this can be seen by splitting the x axis into two regions ($x < 0$ and $x \geq 0$) and applying the Glivenko–Cantelli theorem separately on the two regions.

Further Reading

Historical Background. The theorem is named after Valery Glivenko and Francesco Paolo Cantelli, who discovered the result in 1933. The class of function $g \in \mathcal{G}$ for which

$$\sup_{g \in \mathcal{G}} \left| \int g(x)\,(dF_n(x) - dF^*(x)) \right| \to 0$$

almost surely is known as a Glivenko–Cantelli class (Dudley et al., 1991). The trivial case is $g(x) = 1(x \leq y)$, which ensures that the supremum goes to 0; here, y specifies the class $g(x)$. Today, there is interest in different (non-trivial) formulations of g for which the limit is still 0.

Points of Interest.

1. The Glivenko–Cantelli theorem is a special case of uniform convergence being an immediate consequence of pointwise convergence. This is because distribution functions are bounded to lie in $[0,1]$ and are monotone. In general, to show the uniform convergence of a sequence of functions f_n to f^* from pointwise convergence, consider the triangle inequality

$$|f_n(x) - f^*(x)| \leq |f_n(x) - f_n(z_j)| + |f_n(z_j) - f^*(z_j)| + |f^*(z_j) - f^*(x)|.$$

Certain properties for f^* will provide the finite $z_1, \ldots, z_K$ for which $|f^*(x) - f^*(z_j)| < \epsilon$ for all x; one way to ensure this is if the space of x is compact and f^* is continuous. The second term on the right side of the inequality will go to 0 from pointwise convergence, and the first term will impose restrictions on the sequence $\{f_n\}$ such that it converges to 0.

Q6.2 Questions – Bayesian Framework

Q6.2.1* – Posterior consistency for a normal model

Introduction. This question establishes consistency for one of the simplest models: a normal model with unknown mean and known variance. It is not trivial to show that the posterior for the mean accumulates around the true parameter value. Most approaches that show posterior consistency necessarily involve distances. The use of Euclidean distance between parameter values is generally problematic, though it is feasible for normal models. However, this question establishes consistency using Hellinger distance, which is a distance between density functions that can be used for most parametric models.

Question. Consider the normal model

$$f(y_i \mid \theta) = \frac{1}{\sigma\sqrt{2\pi}} \exp\left\{-\tfrac{1}{2}(y_i - \theta)^2/\sigma^2\right\}, \quad i = 1, \ldots, n,$$

where $\sigma = 1$ is known. Define the likelihood ratio

$$\mathcal{L}_n(\theta) = \prod_{i=1}^{n} \frac{f(y_i \mid \theta)}{f(y_i \mid \theta^*)},$$

where $\theta^* = 0$ is the true parameter value. Consider a normal prior for θ with mean 0 and variance 1, denoted $\pi(\theta)$, and write the posterior density as

$$\pi(\theta \mid y) = \frac{\mathcal{L}_n(\theta)\, \pi(\theta)}{\int \mathcal{L}_n(\theta)\, \pi(\theta)\, d\theta}$$

for a sample $y = \{y_1, \ldots, y_n\}$ of size n.

(i) Find half the squared Hellinger distance $\frac{1}{2} d_H^2(\theta, \theta^*) = 1 - \int \sqrt{f(y \mid \theta) f(y \mid \theta^*)} dy$.

(ii) Explicitly show that $I_n = \int \mathcal{L}_n(\theta)\, \pi(\theta)\, d\theta$ satisfies $n^{-1} \log I_n \stackrel{a.s.}{\to} 0$; this implies that $I_n > \exp(-nc)$ almost surely for all large n and any $c > 0$.

(iii) Show that $n^{-1} \log \mathcal{L}_n(\widehat{\theta}_n) \stackrel{a.s.}{\to} 0$, where $\widehat{\theta}_n$ is the maximum likelihood estimator.

(iv) Write the posterior for a set A as

$$\pi(A \mid y) = I_n^{-1} \int_A \prod_{i=1}^{n} \frac{f(y_i \mid \theta)}{f(y_i \mid \theta^*)}\, \pi(\theta)\, d\theta.$$

Find an upper bound for the integral over the set A using the MLE.

(v) Show that the posterior weight on the set $A_\epsilon = \{\theta : \ |\theta| > \epsilon\}$ goes to 0 as $n \to \infty$ for any $\epsilon > 0$.

Q6.2.2 – Posterior consistency for a Poisson model

Introduction. This question establishes posterior consistency for the rate parameter in the Poisson model. It uses the same approach as in Question 6.2.1. However, Euclidean distance cannot be used to show posterior consistency in the Poisson model. Thus, this question better highlights the need for Hellinger distance in establishing posterior consistency.

Question. Suppose $x = \{x_1, \ldots, x_n\}$ are independent and identically distributed from the Poisson distribution

$$p_\theta(x) = \frac{\theta^x}{x!} e^{-\theta}, \quad x \in \{0, 1, 2, \ldots\}$$

with $\theta > 0$. Consider the prior $\pi(\theta) = e^{-\theta}$, and let θ^* denote the true parameter value.

(i) If $l_n(\theta)$ denotes the likelihood function and

$$I_n = \int_{\theta>0} \frac{l_n(\theta)}{l_n(\theta^*)} \pi(\theta)\, d\theta,$$

show that

$$n^{-1} \log I_n \overset{a.s.}{\to} 0$$

using Stirling's approximation: as $z \to \infty$, $\log \Gamma(z) \approx z \log(z) - z + O(\log(z))$.

(ii) Show that $I_n > e^{-nc}$ almost surely for all large n and any $c > 0$.

(iii) Find the half squared Hellinger distance between the Poisson distribution with parameter θ and the Poisson distribution with parameter θ^*:

$$\tfrac{1}{2} d_H^2(\theta, \theta^*) = 1 - \sum_{x=0}^{\infty} \sqrt{p_\theta(x)\, p_{\theta^*}(x)}.$$

Explain why it is strictly smaller than 1 if $\theta \neq \theta^*$.

(iv) Show that $n^{-1} \log\{l_n(\widehat{\theta}_n)/l_n(\theta^*)\} \overset{a.s.}{\to} 0$.

(v) Show the posterior is consistent: $\pi(A_\epsilon \mid \boldsymbol{x}) \overset{a.s.}{\to} 0$, where $A_\epsilon = \{\theta : |\sqrt{\theta} - \sqrt{\theta^*}| > \epsilon\}$.

Q6.2.3 – Posterior consistency with respect to the Hellinger distance

Introduction. This question continues to show posterior consistency using the Hellinger distance. Whereas the previous two questions concerned normal and Poisson models, this question establishes posterior consistency results for a wide array of parametric families. In particular, one can use the Hellinger distance to show that a likelihood ratio with the true parameter in the denominator and an arbitrary value in the numerator converges to 0 almost surely.

Question. Consider the sample $x = \{x_1, \ldots, x_n\}$, which is independent and identically distributed from some density function $g(x \mid \theta^*)$. For shorthand, write $g^*(x) = g(x \mid$

θ^*) and $g_\theta(x) = g(x \mid \theta)$. The maximum likelihood estimator $\widehat{\theta}$ is assumed to exist and assumed to satisfy $n^{-1} \log \mathcal{L}_n(\widehat{\theta}) \to 0$, where

$$\mathcal{L}_n(\theta) = \prod_{i=1}^{n} \frac{g(x_i \mid \theta)}{g(x_i \mid \theta^*)}, \quad \theta \in \Theta.$$

In this question, all convergence results and inequalities are taken to be almost surely.

(i) What is the expected value of $\mathcal{L}_n(\theta)$?

(ii) Half the squared Hellinger distance between densities g^* and g_θ is

$$\tfrac{1}{2} d_H^2(g^*, g_\theta) = 1 - \int \sqrt{g(x \mid \theta^*)\, g(x \mid \theta)}\, dx.$$

Prove that

$$\mathrm{E}\left\{ \mathcal{L}_n(\theta)^{1/2} \right\} = \left\{ 1 - \tfrac{1}{2}\, d_H^2(g^*, g_\theta) \right\}^n,$$

where expectation is with respect to the $\{x_i\}$.

(iii) A Bayesian assigns the prior $\pi(\theta)$ to θ. Using $\mathcal{L}_n(\theta) \le \mathcal{L}_n(\widehat{\theta})$, prove that

$$\int_{A_\epsilon} \mathcal{L}_n(\theta)\, \pi(\theta)\, d\theta \le e^{\frac{1}{2} nc}\, e^{-\frac{1}{2} n\epsilon^2}$$

for large n and any $c > 0$, where $A_\epsilon = \{\theta : d_H(g^*, g_\theta) > \epsilon\}$. *Hint*: consider a sequence of non-negative random variables Z_n. If $\mathrm{E}\,(Z_n) < e^{-n\delta}$, then $Z_n < e^{-\frac{1}{2} n\delta}$ almost surely for all large n and $\delta > 0$.

(iv) Suppose that $\int_\Theta \mathcal{L}_n(\theta)\, \pi(\theta)\, d\theta > e^{-nd}$ for all large n and any $d > 0$. Prove that

$$\pi(A_\epsilon \mid \boldsymbol{x}) \to 0 \quad \text{as} \quad n \to \infty,$$

where

$$\pi(A_\epsilon \mid \boldsymbol{x}) = \frac{\int_{A_\epsilon} \mathcal{L}_n(\theta)\, \pi(\theta)\, d\theta}{\int_\Theta \mathcal{L}_n(\theta)\, \pi(\theta)\, d\theta}$$

is the posterior mass assigned to A_ϵ.

(v) Explain the significance of the result in part (iii).

Q6.2.4 – Bootstrap, objective, and proper posteriors

Introduction. Bootstrapping is a numerical approach that uses resampled or simulated data to estimate the sampling distribution of maximum likelihood estimators; this allows for estimation of the sampling distribution of nearly any statistic via the use of random sampling methods. This question shows that, for the normal model, the parametric bootstrap posterior is equivalent to an objective posterior and converges to a proper posterior as $n \to \infty$.

Question. Suppose $\boldsymbol{x} = \{x_1, \ldots, x_n\}$ are independent and identically distributed from a normal density with unknown mean θ and known variance σ^2. Let $\widehat{\theta}$ be the sample mean and maximum likelihood estimator. The parametric bootstrap considers $x_1^*, \ldots, x_n^*$ to be independent and identically distributed from the normal density with mean $\widehat{\theta}$ and variance σ^2 and sets θ_b as the sample mean of the $\{x_i^*\}$. This process is repeated B times to get a bootstrapped sample $\{\theta_b\}_{b=1}^B$.

(i) What is the common density function for θ_b?

(ii) Show that the density function from part (i) is equivalent to an objective posterior for θ. What is the objective prior?

(iii) Consider a (proper) normal prior for θ with mean 0 and variance σ^2/λ, where $\lambda > 0$ is known. Find the posterior density of θ.

(iv) Find the squared Hellinger distance between the proper posterior density found in part (iii) and the density found in part (i). Note that half the squared Hellinger distance between $\mathrm{N}(\mu_1, \sigma_1^2)$ and $\mathrm{N}(\mu_2, \sigma_2^2)$ is

$$1 - \sqrt{\frac{2\sigma_1\sigma_2}{\sigma_1^2 + \sigma_2^2}} \exp\left\{ -\frac{(\mu_1 - \mu_2)^2}{4(\sigma_1^2 + \sigma_2^2)} \right\}.$$

(v) Show that the distance in part (iv) goes to 0 as $n \to \infty$, and find the convergence rate.

Q6.2.5* – Asymptotic properties of a marginal likelihood ratio

Introduction. Marginal likelihoods are used in many Bayesian problems, including posterior consistency analysis and the study of information merging, which is the idea that statisticians should use models that agree with each other for large samples. Hence, it is crucial to understand the large sample properties of marginal likelihoods. This question investigates the asymptotic properties of marginal likelihood functions and demonstrates the use of marginal likelihoods in the merging of information.

Question. Consider the Bayesian marginal likelihood

$$m(\boldsymbol{x}) = \int \prod_{i=1}^{n} f(x_i \mid \theta)\, \pi(\theta)\, d\theta$$

for some parametric family of density functions $f(x \mid \theta)$ and prior $\pi(\theta)$. The observations $\boldsymbol{x} = (x_1, \ldots, x_n)'$ are assumed to be independent and identically distributed from $f(\cdot \mid \theta^*)$ for some $\theta^* \in \Theta$. Additionally, assume that $\pi(\theta : d_K(\theta^*, \theta) < \epsilon) > 0$ for all $\epsilon > 0$, where $d_K(\theta^*, \theta) = \int f(x \mid \theta^*) \log\{f(x \mid \theta^*)/f(x \mid \theta)\} dx$ is the Kullback–Leibler divergence between $f(x \mid \theta^*)$ and $f(x \mid \theta)$.

(i) Show that it is possible to write

$$\mathcal{L}_n = \frac{m(\boldsymbol{x})}{f(\boldsymbol{x} \mid \theta^*)} \geq \pi(A_\epsilon) \int \exp\left\{ -\sum_{i=1}^{n} \log \frac{f(x_i \mid \theta^*)}{f(x_i \mid \theta)} \right\} \pi_\epsilon(\theta)\, d\theta,$$

where $A_\epsilon = \{\theta : d_K(\theta^*, \theta) < \epsilon\}$ and $\pi_\epsilon(\theta)$ is the prior $\pi(\theta)$ restricted to the support A_ϵ and normalized.

(ii) Show that

$$\mathcal{L}_n \geq \pi(A_\epsilon)\, \exp\left\{-n\, n^{-1} \sum_{i=1}^{n} H(x_i)\right\}$$

for some function H.

(iii) Using the law of large numbers, show that $\mathcal{L}_n \geq \pi(A_\epsilon)\, e^{-2n\epsilon}$ almost surely for all large n.

(iv) Show that $\mathcal{L}_n < e^{2n\epsilon}$ almost surely for all large n and any $\epsilon > 0$.

(v) Show that $n^{-1} \log \mathcal{L}_n \overset{a.s.}{\to} 0$.

Q6.2.6 – Bayesian consistency for a discrete parameter space

Introduction. This question concerns the posterior consistency given a prior over a discrete parameter space. The aim is to discover a condition on the prior for which the posterior mass on the true parameter value converges to 1 almost surely.

Question. For the family of density functions $f(x \mid \theta)$, define a discrete prior such that $\pi_j = \mathrm{P}(\theta = \theta_j)$. Let the true value be θ_0 with $\pi_0 > 0$. For all other $\{\theta\}_{j \neq 0}$, let

$$\int \sqrt{f(x \mid \theta_0)\, f(x \mid \theta_j)}\, dx \leq 1 - \epsilon$$

for some $\epsilon > 0$. Given a sample $\boldsymbol{x} = \{x_1, \ldots, x_n\}$, define the posterior on the complement of θ_0 as

$$Q_n = \frac{\sum_{j=1}^{\infty} \prod_{i=1}^{n} \frac{f(x_i|\theta_j)}{f(x_i|\theta_0)} \pi_j}{\sum_{j=0}^{\infty} \prod_{i=1}^{n} \frac{f(x_i|\theta_j)}{f(x_i|\theta_0)} \pi_j}.$$

(i) Explain why

$$Q_n \leq \frac{\sum_{j=1}^{\infty} \sqrt{\prod_{i=1}^{n} \frac{f(x_i|\theta_j)}{f(x_i|\theta_0)}}\, \pi_j^{1/2}}{\sqrt{\sum_{j=0}^{\infty} \prod_{i=1}^{n} \frac{f(x_i|\theta_j)}{f(x_i|\theta_0)} \pi_j}}.$$

(ii) Define

$$J_n = \sum_{j=1}^{\infty} \sqrt{\prod_{i=1}^{n} \frac{f(x_i \mid \theta_j)}{f(x_i \mid \theta_0)}}\, \pi_j^{1/2}.$$

Show that $\mathrm{E}(J_{n+1} \mid \boldsymbol{x}) \leq (1 - \epsilon)\, J_n$.

(iii) Find a suitable upper bound for $\mathrm{E}(J_n)$.

(iv) Suppose $\sum_{j=1}^{\infty} \pi_j^{1/2} = \gamma < \infty$. Show that $J_n < \gamma\, e^{-n\epsilon/2}$ almost surely for all large n.

(v) Show that $Q_n \overset{a.s.}{\to} 0$ for all large n. Interpret this result.

S6.2 Solutions – Bayesian Framework

S6.2.1* – Posterior consistency for a normal model

(i) Half the squared Hellinger distance is given by

$$\tfrac{1}{2}\, d_H^2(\theta,\theta^*) = 1 - \int \sqrt{\frac{1}{\sigma\sqrt{2\pi}} \exp\left\{-\frac{1}{2\sigma^2}(y-\theta)^2\right\} \frac{1}{\sigma\sqrt{2\pi}} \exp\left\{-\frac{1}{2\sigma^2}(y-\theta^*)^2\right\}}dy.$$

Letting $\sigma = 1$ and $\theta^* = 0$,

$$\begin{aligned}
\tfrac{1}{2}\, d_H^2(\theta,\theta^*) &= 1 - \int \sqrt{2\pi}^{-1} \exp\left\{-\frac{1}{4}(y^2 - 2\theta y + \theta^2 + y^2)\right\} dy \\
&= 1 - \exp\{-\theta^2/8\} \int \sqrt{2\pi}^{-1} \exp\left\{-\tfrac{1}{2}(y-\theta/2)^2\right\} dy \\
&= 1 - \exp\{-\theta^2/8\},
\end{aligned}$$

where the integral in the second line is over a $\mathrm{N}(\cdot \mid \theta/2, 1)$ density.

(ii) The denominator for the posterior is given by

$$I_n = \int \prod_{i=1}^{n} \frac{f(y_i \mid \theta)}{f(y_i \mid \theta^*)} \pi(\theta) d\theta.$$

Note that

$$\begin{aligned}
\int \prod_{i=1}^{n} f(y_i \mid \theta)\pi(\theta)d\theta &= \int \prod_{i=1}^{n} \left[(2\pi)^{-\frac{1}{2}} \exp\left\{-\tfrac{1}{2}(y_i-\theta)^2\right\}\right] \cdot (2\pi)^{-\frac{1}{2}} \exp\{-\tfrac{1}{2}\theta^2\} d\theta \\
&= \int (2\pi)^{-(n+1)/2} \exp\left[-\tfrac{1}{2}\left\{(n+1)\theta^2 - 2\theta\sum_{i=1}^{n} y_i + \sum_{i=1}^{n} y_i^2\right\}\right] d\theta \\
&= \int (2\pi/(n+1))^{-1/2} \exp\left\{-\frac{n+1}{2}\left(\theta - \frac{\sum_{i=1}^{n} y_i}{n+1}\right)^2\right\} d\theta \\
&\quad \cdot (2\pi)^{-n/2}(n+1)^{-1/2} \exp\left[\tfrac{1}{2}\left\{\frac{(\sum_{i=1}^{n} y_i)^2}{n+1} - \sum_{i=1}^{n} y_i^2\right\}\right] \\
&= (2\pi)^{-n/2}(n+1)^{-1/2} \exp\left[\tfrac{1}{2}\left\{\frac{(\sum_{i=1}^{n} y_i)^2}{n+1} - \sum_{i=1}^{n} y_i^2\right\}\right],
\end{aligned}$$

where the integral in the third equality is over a $\mathrm{N}(\cdot \mid \sum_{i=1}^{n} y_i/(n+1), 1/(n+1))$ density. Additionally, because $\theta^* = 0$,

$$\prod_{i=1}^{n} f(y_i \mid \theta^*) = (2\pi)^{-n/2} \exp\left(-\tfrac{1}{2}\sum_{i=1}^{n} y_i^2\right).$$

Therefore, $I_n = (n+1)^{-1/2} \exp\{(\sum_{i=1}^n y_i)^2/(2(n+1))\}$ and

$$n^{-1}\log I_n = -\frac{1}{2n}\log(n+1) + \frac{(\sum_{i=1}^n y_i)^2}{2n(n+1)}.$$

By the strong law of large numbers, $\overline{y} \overset{a.s.}{\to} \theta^* = 0$. By the continuous mapping theorem, $\overline{y}^2 = (\sum_{i=1}^n y_i)^2/n^2 \overset{a.s.}{\to} 0$, so

$$\frac{(\sum_{i=1}^n y_i)^2}{n(n+1)} \overset{a.s.}{\to} 0.$$

Additionally, $n^{-1}\log(n+1) \to 0$. Therefore, $n^{-1}\log I_n \overset{a.s.}{\to} 0$.

(iii) Write out

$$\begin{aligned} n^{-1}\log \mathcal{L}_n(\widehat{\theta}_n) &= n^{-1}\left\{\sum_{i=1}^n \log f(y_i \mid \widehat{\theta}_n) - \sum_{i=1}^n \log f(y_i \mid \theta^*)\right\} \\ &= -\frac{1}{2n}\sum_{i=1}^n\{(y_i - \widehat{\theta}_n)^2 - (y_i - \theta^*)^2\} \\ &= \frac{1}{2n}\left\{2(\widehat{\theta}_n - \theta^*)\sum_{i=1}^n y_i + n(\theta^{*2} - \widehat{\theta}_n^2)\right\}. \end{aligned}$$

The MLE is $\widehat{\theta}_n = \overline{y}$, so

$$n^{-1}\log \mathcal{L}_n(\widehat{\theta}_n) = (\overline{y} - \theta^*)\overline{y} + \tfrac{1}{2}(\theta^{*2} - \overline{y}^2).$$

By the strong law of large numbers, $\overline{y} \overset{a.s}{\to} \theta^*$. Thus, $n^{-1}\log \mathcal{L}_n(\widehat{\theta}_n) \overset{a.s.}{\to} 0$.

(iv) By definition of the MLE, $\sqrt{\mathcal{L}_n(\theta)} \le \sqrt{\mathcal{L}_n(\widehat{\theta}_n)}$ for all θ. Therefore, the posterior for a set A has an upper bound

$$\begin{aligned} \pi(A \mid \boldsymbol{y}) &= I_n^{-1}\int_A \mathcal{L}_n(\theta)\,\pi(\theta)\,d\theta \\ &\le I_n^{-1}\sqrt{\mathcal{L}_n(\widehat{\theta}_n)}\int_A \sqrt{\mathcal{L}_n(\theta)}\,\pi(\theta)\,d\theta. \end{aligned}$$

From part (iii), $n^{-1}\log \mathcal{L}_n(\widehat{\theta}_n) \to 0$, so for any $c > 0$ and all large n, $n^{-1}\log \mathcal{L}_n(\widehat{\theta}_n) < c$. Thus, $\sqrt{\mathcal{L}_n(\widehat{\theta}_n)} < \exp(nc/2)$ almost surely for all large n. From part (ii), $I_n > \exp(-nc)$ almost surely for all large n and any $c > 0$, so $I_n^{-1} < \exp(nc)$ almost surely. Putting all this together,

$$\pi(A \mid \boldsymbol{y}) < \exp(3nc/2)\int_A \sqrt{\mathcal{L}_n(\theta)}\pi(\theta)d\theta \tag{6.1}$$

almost surely for any $c > 0$ and all large n.

(v) Consider the expectation

$$\begin{aligned} \mathrm{E}\left\{\int_{A_\epsilon}\sqrt{\mathcal{L}_n(\theta)}\pi(\theta)d\theta\right\} &= \int\int_{A_\epsilon}\pi(\theta)\sqrt{\prod_{i=1}^n \frac{f(y_i \mid \theta)}{f(y_i \mid \theta^*)}}\prod_{i=1}^n f(y_i \mid \theta^*)\,d\theta\,d\boldsymbol{y} \\ &= \int_{A_\epsilon}\left\{1 - \tfrac{1}{2}d_H^2(\theta, \theta^*)\right\}^n \pi(\theta)d\theta \\ &\le \exp(-\epsilon^2 n/8)\pi(A_\epsilon) \\ &\le \exp(-\epsilon^2 n/8), \end{aligned}$$

where the first inequality follows from part (i) and $|\theta| > \epsilon$. Using this upper bound and the Markov inequality,

$$\mathrm{P}\left(\int_{A_\epsilon} \sqrt{\mathcal{L}_n(\theta)}\pi(\theta)d\theta \geq e^{-nc'}\right) \leq \exp(nc' - n\epsilon^2/8)$$

for any $c' > 0$. Set $c' = \epsilon^2/16$ to ensure that the exponentiated term in the right side of the inequality is negative, so

$$\sum_n \mathrm{P}\left(\int_{A_\epsilon} \sqrt{\mathcal{L}_n(\theta)\pi(\theta)d\theta} \geq e^{-nc}\right) < \infty.$$

Thus, by the Borel–Cantelli lemma,

$$\int_{A_\epsilon} \sqrt{\mathcal{L}_n(\theta)}\pi(\theta)d\theta < \exp(-n\epsilon^2/16)$$

almost surely for all large n. Returning to (6.1) and choosing $c < \epsilon^2/16$, it follows that $\pi(A_\epsilon \mid y) \overset{a.s}{\to} 0$.

Further Reading

Historical Background. Walker (1969) defined the original sufficient conditions to establish posterior consistency for parametric models. In particular, this paper investigated how necessary these conditions were to ensure posterior consistency. Notably, theory developed for nonparametric models may be applied to parametric models, providing a different set of sufficient conditions.

Points of Interest.

1. This exercise introduces the fundamental idea that the sequence of posterior distributions accumulates in neighborhoods of the true parameter. A natural approach to defining such neighborhoods may use Euclidean distance with the aim to show

$$\pi(A_\epsilon \mid y) \overset{a.s.}{\to} 0,$$

where $A_\epsilon = \{\theta : ||\theta - \theta^*|| > \epsilon\}$. However, the rather stringent conditions in Walker (1969) suggest that Euclidean distance is not the best approach.

2. In short, all that is needed to ensure posterior consistency for any correctly specified parametric model is the existence of the MLE. The general condition that the denominator is suitably bounded below by e^{-nc} ensures that the prior puts positive mass on all Kullback–Leibler neighborhoods of the true density function. This theory is detailed in Schwartz (1965).

3. Much of the recent work on Bayesian consistency concerns nonparametric models for which the MLE does not often exist. Classical approaches typically use sieves if the MLE does not exist; sieves are increasing sequences of sets of density functions that converge to the entire space as the sample size increases to infinity. See Geman and Hwang (1982) for more details on the use of sieves, though the original idea is often attributed to Swedish statistician Ulf Grenander (Grenander, 1981). Perhaps for this reason, the pioneering work in Bayesian nonparametric consistency also used sieves (Ghosal et al., 1999; Barron et al., 1999).

S6.2.2 – Posterior consistency for a Poisson model

(i) The likelihood function is $l_n(\theta) \propto \theta^{n\overline{x}}\, e^{-n\theta}$, so

$$I_n = \int \frac{l_n(\theta)}{l_n(\theta^*)}\, \pi(\theta)\, d\theta = \int \frac{\theta^{n\overline{x}} e^{-n\theta}}{(\theta^*)^{n\overline{x}} e^{-n\theta^*}}\, e^{-\theta}\, d\theta = e^{n\theta^*}\, (\theta^*)^{-n\overline{x}} \frac{\Gamma\,(1+n\overline{x})}{(1+n)^{1+n\overline{x}}},$$

where the integrand in the second equality is the kernel of a $\text{Ga}(\cdot \mid 1+n\overline{x}, 1+n)$ distribution. Then,

$$n^{-1}\log I_n = \theta^* - \overline{x}\log\theta^* + n^{-1}\log\Gamma(1+n\overline{x}) - (\overline{x}+1/n)\log(1+n).$$

By Stirling's approximation, $n^{-1}\log\Gamma(1+n\overline{x})$ is approximately $\overline{x}\log(n\overline{x}) - \overline{x}$ for large n. Finally, as $n \to \infty$, the result follows.

(ii) For any $c > 0$, if $n^{-1}\log I_n \overset{a.s.}{\to} 0$, then $I_n > e^{-nc}$ almost surely for all large n.

(iii) Half the squared Hellinger distance is given by

$$\begin{aligned}
\tfrac{1}{2}\, d_H^2(\theta,\theta^*) &= 1 - \sum_{x=0}^{\infty} \sqrt{p_\theta(x)\, p_{\theta^*}(x)} \\
&= 1 - \sum_{x=0}^{\infty} \sqrt{\frac{\theta^x}{x!}\, e^{-\theta}\, \frac{(\theta^*)^x}{x!}\, e^{-\theta^*}} \\
&= 1 - \sum_{x=0}^{\infty} \sqrt{(\theta\theta^*)^x e^{-(\theta+\theta^*)} \frac{1}{(x!)^2}} \\
&= 1 - \sum_{x=0}^{\infty} \frac{(\theta\theta^*)^{x/2}}{x!} e^{-(\theta+\theta^*)/2} \\
&= 1 - \exp\{-(\theta+\theta^*)/2\} \sum_{x=0}^{\infty} \frac{(\theta\theta^*)^{x/2}}{x!} \\
&= 1 - \exp\{-(\theta+\theta^*)/2 + \sqrt{\theta\theta^*}\},
\end{aligned}$$

where the last equality uses the series expansion of $\exp(\sqrt{\theta\theta^*})$. This distance is bounded above by one because an arithmetic average is greater than or equal to its geometric average.

(iv) Note that $\widehat{\theta}_n = \overline{x}$ and $\overline{x} \overset{a.s.}{\to} \theta^*$ by the strong law of large numbers. Thus,

$$n^{-1}\,\log\{l_n(\widehat{\theta}_n)/l_n(\theta^*)\} = \overline{x}\log(\overline{x}) - \overline{x} + \theta^* - \overline{x}\log\theta^*$$

converges to 0 almost surely.

(v) First, write

$$\pi(A_\epsilon \mid \boldsymbol{x}) = \frac{\int_{A_\epsilon} l_n(\theta)\pi(\theta)\, d\theta}{\int l_n(\theta)\pi(\theta)\, d\theta} = \frac{\int_{A_\epsilon} \frac{l_n(\theta)}{l_n(\theta^*)}\pi(\theta)\, d\theta}{\int \frac{l_n(\theta)}{l_n(\theta^*)}\pi(\theta)\, d\theta}.$$

From part (ii),

$$\frac{\int_{A_\epsilon} \frac{l_n(\theta)}{l_n(\theta^*)}\pi(\theta)\,d\theta}{\int \frac{l_n(\theta)}{l_n(\theta^*)}\pi(\theta)\,d\theta} \le e^{nc}\int_{A_\epsilon} \frac{l_n(\theta)}{l_n(\theta^*)}\pi(\theta)\,d\theta$$

for any $c > 0$. Now, the integral in the numerator must be bounded to prove convergence. This follows in exactly the same way as in part (v) of Question 6.2.1, except it is that $|\sqrt{\theta} - \sqrt{\theta^*}| > \epsilon$, which implies

$$\tfrac{1}{2}(\theta + \theta^*) - \sqrt{\theta\,\theta^*} = \tfrac{1}{2}(\sqrt{\theta} - \sqrt{\theta^*})^2 > \tfrac{1}{2}\epsilon^2.$$

Further Reading

Historical Background. A natural approach to measuring the distance between parameters is to use Euclidean distance. However, for some models such as the Poisson model, Euclidean distance cannot be used to show posterior consistency. Additionally, the use of Euclidean distance typically requires conditions that are stronger than necessary (Walker, 1969).

Points of Interest.

1. The Hellinger distance between θ_1 and θ_2 does not follow precisely from the Euclidean distance between θ_1 and θ_2 because knowledge of $|\sqrt{\theta_1} - \sqrt{\theta_2}|$ does not lead to knowledge of $|\theta_1 - \theta_2|$. In fact,

$$|\theta_1 - \theta_2| = |\sqrt{\theta_1} - \sqrt{\theta_2}|\,|\sqrt{\theta_1} + \sqrt{\theta_2}|,$$

so it is necessary to know the value of $|\sqrt{\theta_1} + \sqrt{\theta_2}|$.

2. Much of the literature on Bayesian consistency concerns nonparametric models because it is these models that might be so large that asymptotic challenges arise. Typically, identifiability issues are the source of consistency problems. A key sufficient condition for all types of consistency is that, for $\theta_1 \neq \theta_2$, the corresponding density and distribution functions are also different: $f(x \mid \theta_1) \neq f(x \mid \theta_2)$ and $F(x \mid \theta_1) \neq F(x \mid \theta_2)$, respectively. If this were not true, the posterior could accumulate at more than one true distribution.

3. A surprising Bayesian consistency result is that, with a suitable prior support that contains the true value, the posterior is guaranteed to accumulate at the correct distribution function. However, at extreme parameter values, it may be that additional distributions (or sets of distributions for multi-dimensional parameters) exist, which raises identifiability concerns. Moreover, if the distributions at these extreme values do not possess density functions, then one must ensure that, if the extreme distribution coincides with the true one, the posterior is forced to the distribution that has a density function. To see this, consider the family of density functions

$$f(x \mid \theta) = \frac{1 + \cos(x\theta)}{1 + \theta^{-1}\sin\theta}, \quad 0 \le x \le 1,$$

where $\theta \ge 0$. The uniform distribution function arises at $\theta = 0$ and $\theta = \infty$, but the uniform density function only exists at $\theta = 0$. A Bayesian hopes that the posterior for

θ accumulates mass at $\theta = 0$ rather than $\theta = \infty$; typically, a condition on the prior is needed to ensure that this happens. However, there is only a concern if (i) a distribution function exists at ∞, (ii) there is not a density function at ∞, and (iii) this distribution is in fact the true distribution. It is rare for all three of these conditions to be satisfied, so this is typically not a concern for many parametric models.

Miscellaneous. Stirling's formula is often used for the asymptotic approximation of factorials and gamma functions. The formula is named after Scottish mathematician James Stirling, who constructed a more precise result of an approximation obtained by French mathematician Abraham de Moivre (Le Cam, 1986).

S6.2.3 – Posterior consistency with respect to the Hellinger distance

(i) Using the independence of the $\{x_i\}$, the expected value of $\mathcal{L}_n(\theta)$ is 1.

(ii) Again, using the independence of the $\{x_i\}$,

$$\begin{aligned}
\mathrm{E}\left\{\mathcal{L}_n(\theta)^{1/2}\right\} &= \mathrm{E}\left\{\left(\prod_{i=1}^{n}\frac{g(x_i \mid \theta)}{g(x_i \mid \theta^*)}\right)^{1/2}\right\} \\
&= \prod_{i=1}^{n}\mathrm{E}\left[\left\{\frac{g(x_i \mid \theta)}{g(x_i \mid \theta^*)}\right\}^{1/2}\right] \\
&= \prod_{i=1}^{n}\int\left\{\frac{g(x_i \mid \theta)}{g(x_i \mid \theta^*)}\right\}^{1/2} g(x_i \mid \theta^*)dx_i \\
&= \left\{\int \sqrt{g(x \mid \theta)g(x \mid \theta^*)}dx\right\}^n \\
&= \left\{1 - \tfrac{1}{2}\, d_H^2(g^*, g_\theta)\right\}^n .
\end{aligned}$$

(iii) Because $(1-a)^n \le e^{-an}$ for all $0 < a < 1$ and $n \in \{1, 2, \ldots\}$, it follows that

$$\left\{1 - \tfrac{1}{2}\, d_H^2(g^*, g_\theta)\right\}^n \le \exp\left\{-\tfrac{1}{2}\, n d_H^2(g^*, g_\theta)\right\}.$$

Using the provided hint,

$$\mathcal{L}_n(\theta)^{1/2} \le \exp\left\{-\tfrac{1}{2}\, n d_H^2(g^*, g_\theta)\right\}$$

almost surely for all large n. Using the decomposition with upper bound $\mathcal{L}_n(\theta)$,

$$\begin{aligned}
\int_{A_\epsilon} \mathcal{L}_n(\theta)\pi(\theta)d\theta &\le \int_{A_\epsilon} \mathcal{L}_n(\widehat{\theta})^{1/2}\mathcal{L}_n(\theta)^{1/2}\pi(\theta)d\theta \\
&\le e^{cn/2}\int_{A_\epsilon} \exp\left\{-\tfrac{1}{2}\, n\, d_H^2(g^*, g_\theta)\right\}\pi(\theta)d\theta \\
&\le e^{cn/2}\int_{A_\epsilon} \exp\{-\epsilon^2 n/2\}\pi(\theta)d\theta \\
&\le \exp\left\{cn/2 - \epsilon^2 n/2\right\}.
\end{aligned}$$

The first inequality follows because $\mathcal{L}_n(\theta) \leq \mathcal{L}_n(\widehat{\theta})$, and the second inequality follows because $n^{-1} \log \mathcal{L}_n(\widehat{\theta}) \to 0$ implies that $\mathcal{L}_n(\widehat{\theta}) \leq \exp(cn)$ for all large n and any $c > 0$. The third inequality follows as the integral is over A_ϵ for which $d_H(g^*, g_\theta) > \epsilon$.

(iv) This part requires some manipulation after plugging in previous results:

$$\begin{aligned}\pi(A_\epsilon \mid \boldsymbol{x}) &= \frac{\int_{A_\epsilon} \mathcal{L}_n(\theta)\pi(\theta)d\theta}{\int_\Theta \mathcal{L}_n(\theta)\pi(\theta)d\theta} \\ &\leq \frac{\exp\{cn/2 - \epsilon^2 n/2\}}{\exp\{-nd\}} \\ &= \exp\left\{-n(\epsilon^2/2 - c/2 - d)\right\}.\end{aligned}$$

The first inequality holds for all n and for any c and d chosen. For all $\epsilon > 0$, take c and d such that $c/2 + d < \epsilon^2/2$. Therefore, $\epsilon^2/2 - c/2 - d \geq 0$, which implies that $\pi(A_\epsilon \mid \boldsymbol{x}) \to 0$ as $n \to \infty$.

(v) Note that $\pi(A_\epsilon \mid \boldsymbol{x})$ is the Bayesian posterior distribution with mass assigned to A_ϵ. The result in part (iii) states that the posterior mass decays to 0 outside of any Hellinger ball about the true density function. Consequently, all the mass must end up accumulating at the true density function.

Further Reading

Historical Background. Posterior consistency for parametric models was first tackled rigorously by Walker (1969). The sufficient conditions to ensure the convergence of posterior distributions to the true parameter value are solely related to the model. On the other hand, sufficient conditions for Bayesian nonparametric consistency concern both the prior and how posterior distributions accumulate in neighborhoods of the true density or distribution function. This question makes use of nonparametric tools and ideas to improve the results for parametric models (Ghosal et al., 1999).

Points of Interest.

1. Under certain conditions, which are provided in the question, the posterior mass $\pi(A_\epsilon \mid x)$ goes to 0 as $n \to \infty$. Because this holds for any $\epsilon > 0$, the posterior mass ends up accumulating in Hellinger neighborhoods of the true density g^*. One condition is that the likelihood ratio $\mathcal{L}_n(\theta)$ behaves suitably at the maximum likelihood estimator (i.e., $n^{-1} \log \mathcal{L}_n(\widehat{\theta}) \to 0$), which deals with the numerator of the Bayesian posterior at A_ϵ. The second condition deals with the denominator using an assumption on $\int_\Theta \mathcal{L}_n(\theta)\, \pi(\theta)\, d\theta$; the assumption provided in this question is a consequence of the prior distribution meeting certain requirements. For example, the posterior can only accumulate in suitable neighborhoods of g^* if the prior includes g^* in the support. A more strict requirement is in terms of Kullback–Leibler neighborhoods: $f(\theta : D(g^*, g_\theta) < \epsilon) > 0$ for all $\epsilon > 0$, where

$$D(g^*, g_\theta) = \int g^*(x)\, \log\{g^*(x)/g_\theta(x)\}\, dx$$

is the Kullback–Leibler divergence between g^* and g_θ.

2. The Hellinger distance is equivalent to the L_1 distance, which is often used in Bayesian consistency results because its form is easily obtained by square rooting the likelihood ratio. For more on Hellinger consistency from the classical perspective, see van de Geer (1993); for a Bayesian treatment, see Barron et al. (1999). Additionally, Ghosal and van der Vaart (2017) provide a general summary of Bayesian consistency.

3. The consistency of the Bayesian posterior is closely associated with a test of the form $H_0 : \theta = \theta_0$ vs $H_1 : \theta \neq \theta_0$. The classical test would rely on the likelihood ratio. In the Bayesian framework, the posterior is a monotone transform of the Bayes factor

$$B = \frac{\int_{A_\epsilon} \mathcal{L}_n(\theta)\, \pi(\theta)\, d\theta}{\int_{A'_\epsilon} \mathcal{L}_n(\theta)\, \pi(\theta)\, d\theta},$$

which forms the basis for assessing the hypotheses.

4. If the maximum likelihood estimator is suitably behaved (i.e., $n^{-1} \log \mathcal{L}_n(\widehat{\theta}) \to 0$) and the denominator has the given lower bound via the prior support condition, then the posterior is consistent. Note that the condition for the denominator is achieved if the prior places mass around the true value θ^*.

5. The square-rooted likelihood function

$$L(f) = \prod_{i=1}^{n} f^{1/2}(x_i)$$

is automatically strongly consistent (provided the suitable prior support condition is satisfied) with respect to the Hellinger distance; this should be easy to prove given all the exercises in this chapter. In fact, strong consistency holds under a likelihood function much closer to the correct one. For every Nth observation, where N is arbitrarily large, it is possible to have the power of $f(x_i)$ raised to $1 - \epsilon$:

$$L_{N,\epsilon}(f) = \prod_{i=1}^{n} f(x_i)^{1-\epsilon\, 1\{\text{mod}(i,N)=0\}}$$

for an arbitrary small ϵ.

S6.2.4 – Bootstrap, objective, and proper posteriors

(i) Because $x_1^*, \ldots, x_n^* \sim \mathrm{N}(\widehat{\theta}, \sigma^2)$ (i.e., independent and identically distributed),

$$\theta_b = n^{-1} \sum_{i=1}^{n} x_i^* \sim \mathrm{N}(\widehat{\theta}, \sigma^2/n)$$

with density function

$$f(\theta_b) = (2\pi\sigma^2/n)^{-\frac{1}{2}} \exp\left\{-\frac{n}{2\sigma^2}(\theta_b - \widehat{\theta})^2\right\}.$$

(ii) With some prior $\pi(\theta)$ and a likelihood based on $x_1, \dots, x_n \sim \mathrm{N}(\theta, \sigma^2)$ independently, the posterior distribution is

$$\begin{aligned}
\pi(\theta \mid \boldsymbol{x}) &\propto \prod_{i=1}^{n} \exp\left\{-\frac{1}{2\sigma^2}(x_i - \theta)^2\right\} \cdot \pi(\theta) \\
&\propto \exp\left\{-\frac{1}{2\sigma^2}\sum_{i=1}^{n}(x_i - \theta)^2\right\} \cdot \pi(\theta) \\
&\propto \exp\left\{-\frac{1}{2\sigma^2}\left(n\theta^2 - 2\theta\sum_{i=1}^{n} x_i\right)\right\} \cdot \pi(\theta) \\
&\propto \exp\left\{-\frac{n}{2\sigma^2}(\theta - \widehat{\theta})^2\right\} \cdot \pi(\theta).
\end{aligned}$$

With the flat objective prior $\pi(\theta) \propto 1$, the (objective) posterior distribution is equivalent to the distribution in part (i):

$$\pi(\theta \mid \boldsymbol{x}) = \mathrm{N}(\theta \mid \widehat{\theta}, \sigma^2/n).$$

(iii) Assuming the proper prior $\mathrm{N}(\theta \mid 0, \sigma^2/\lambda)$, the posterior distribution is

$$\begin{aligned}
\pi(\theta \mid \boldsymbol{x}) &\propto \exp\left\{-\frac{n}{2\sigma^2}(\theta^2 - 2\theta\widehat{\theta})\right\} \cdot \exp\left(-\frac{\lambda}{2\sigma^2}\theta^2\right) \\
&\propto \exp\left[-\tfrac{1}{2}\{(n+\lambda)\theta^2/\sigma^2 - 2n\theta\widehat{\theta}/\sigma^2\}\right],
\end{aligned}$$

which is the kernel of a normal distribution. Therefore,

$$\pi(\theta \mid \boldsymbol{x}) = \mathrm{N}\left(\theta \,\middle|\, \frac{n\widehat{\theta}}{n+\lambda}, \frac{\sigma^2}{n+\lambda}\right).$$

(iv) Half the squared Hellinger distance between $\mathrm{N}(\widehat{\theta}, \sigma^2/n)$ and $\mathrm{N}(n\widehat{\theta}/(n+\lambda), \sigma^2/(n+\lambda))$ is

$$\tfrac{1}{2}\, d_H^2 = 1 - \sqrt{\frac{2\frac{\sigma}{\sqrt{n}}\frac{\sigma}{\sqrt{n+\lambda}}}{\frac{\sigma^2}{n} + \frac{\sigma^2}{n+\lambda}}} \exp\left\{-\frac{(\widehat{\theta} - \frac{n\widehat{\theta}}{n+\lambda})^2}{4(\frac{\sigma^2}{n} + \frac{\sigma^2}{n+\lambda})}\right\}.$$

(v) The term inside the square root becomes

$$\frac{2n(n+\lambda)}{n\sqrt{1+\lambda/n}\,(2n+\lambda)} = \frac{\sqrt{1+\lambda/n}}{1+\lambda/(2n)},$$

which converges to 1 as $1 - \lambda^2/(4n)$. Additionally, the exponential term is of the form $\exp(-c\widehat{\theta}^2/n)$ for some $c > 0$; this converges to 1 as $1 - c\widehat{\theta}^2/n$. Therefore,

$$\tfrac{1}{2}\, d_H^2 = 1 - (1 - \lambda^2/(4n))(1 - c\widehat{\theta}^2/n) + o(1/n),$$

which converges to 0 at a rate of $1/n$.

Further Reading

Historical Background. The parametric bootstrap was first introduced in Efron (2012). Notably, it does not neatly fit into the classical nor the Bayesian framework because it is a parametric version of the nonparametric bootstrap that aims to approximate a posterior distribution.

Points of Interest.

1. As seen in this question, the squared Hellinger distance between an objective and proper posterior density is of the order $1/n$. In fact, this is exactly the difference between two proper posteriors with different variances. An objective posterior can be closer to a proper posterior than another proper posterior if the variances of the proper priors are sufficiently different. Here, the goal was to derive a posterior distribution that is sufficiently motivated and satisfies a Hellinger distance of $o(1/n)$ (known as the $1/n$ Hellinger rule).

2. One of the original prior-free objective Bayesian procedures is the Bayesian bootstrap introduced in Rubin (1981). This procedure generates random distribution functions that have certain Dirichlet weights attached to the data points that act as atoms; it is essentially a randomization of the empirical distribution function. Newton and Raftery (1994) adapted this approach for parametric models. In their adaptation, the log-likelihood function is randomly weighted by the same Dirichlet weights, and a sample from the objective posterior (known as the weighted likelihood bootstrap) is the random parameter that maximizes the log-likelihood.

3. The parametric bootstrap is the parametric analogue of the original bootstrap. However, instead of taking n samples from the empirical distribution, the samples arise from the parametric model with the MLE. Donald Rubin pointed out that both approaches generate random distribution functions, so there is a similar connection between the parametric bootstrap and the weighted likelihood bootstrap. Additionally, both bootstraps generate random parameters, one by maximizing the likelihood from the n samples taken from the MLE density estimator and the other by maximizing a randomized log-likelihood. The $1/n$ Hellinger rule can be used to justify both approaches.

4. The idea behind the parametric bootstrap posterior is similar to the concept of fiducial inference, which is feasible if $\widehat{\theta} = g(z, \theta^*)$, where g is some known function, z is an independent variable with known distribution, and θ^* is the true parameter value. A fiducial posterior attempts to find the distribution of θ^* conditional on $\widehat{\theta}$; typically, this is a difficult problem. However, the parametric bootstrap posterior generates a sample via $\theta = g(z, \widehat{\theta})$, where the credible intervals for θ and confidence intervals for θ^* can be made identical.

Demonstration. The parametric bootstrap may be an adequate substitute for an objective Bayesian posterior distribution. To see this, consider a Poisson model with sample size $n = 50$ and mean $\lambda = 5$. The MLE is computed to be 4.95. Then, take 1000 independent samples of size n from the Poisson distribution with mean 4.95; Fig. 6.2 depicts the histogram of these bootstrapped samples. Overlaid is the posterior gamma distribution for λ, which used a gamma prior with $\alpha = \frac{1}{2}$ and $\beta = \frac{1}{2}$. The parametric bootstrap posterior well-approximates the proper posterior.

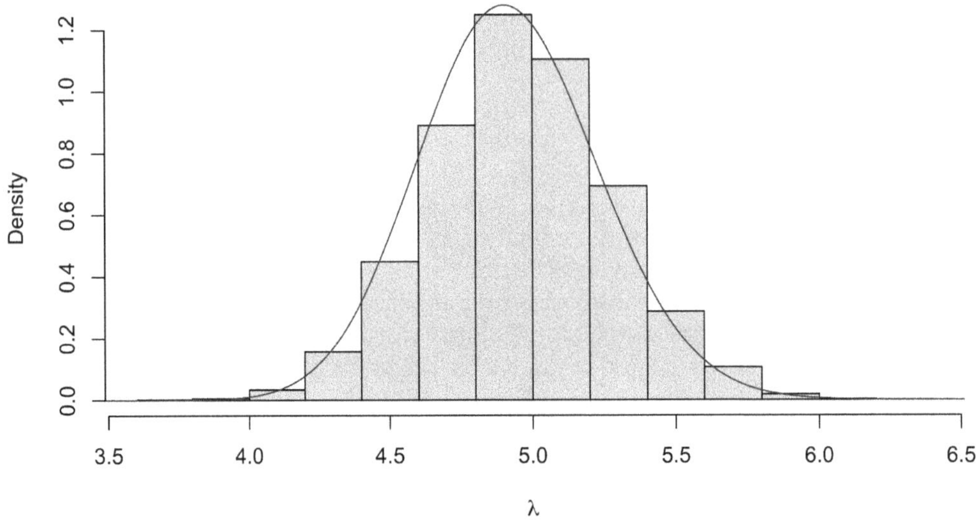

FIGURE 6.2
Histogram of the parametric bootstrap posterior with a proper gamma posterior overlaid.

S6.2.5* – Asymptotic properties of a marginal likelihood ratio

(i) If $\pi_\epsilon(\theta)$ is $\pi(\theta)$ restricted to A_ϵ and normalized, then $\int_{A_\epsilon} \pi(\theta)d\theta = \pi(A_\epsilon)\int_\Theta \pi_\epsilon(\theta)d\theta$ with normalizing constant $\pi(A_\epsilon)$. Therefore,

$$
\begin{aligned}
\mathcal{L}_n = \frac{m(\boldsymbol{x})}{f(\boldsymbol{x} \mid \theta^*)} &= \frac{\int_\Theta \prod_{i=1}^n f(x_i \mid \theta)\pi(\theta)d\theta}{\prod_{i=1}^n f(x_i \mid \theta^*)} \\
&= \int_\Theta \prod_{i=1}^n \frac{f(x_i \mid \theta)}{f(x_i \mid \theta^*)}\pi(\theta)d\theta \\
&= \int_\Theta \exp\left[-\sum_{i=1}^n \log\left\{\frac{f(x_i \mid \theta^*)}{f(x_i \mid \theta)}\right\}\right]\pi(\theta)d\theta \\
&\geq \int_{A_\epsilon} \exp\left[-\sum_{i=1}^n \log\left\{\frac{f(x_i \mid \theta^*)}{f(x_i \mid \theta)}\right\}\right]\pi(\theta)d\theta \\
&= \pi(A_\epsilon)\int \exp\left[-\sum_{i=1}^n \log\left\{\frac{f(x_i \mid \theta^*)}{f(x_i \mid \theta)}\right\}\right]\pi_\epsilon(\theta)d\theta,
\end{aligned}
$$

where the inequality follows because the terms inside the integral are non-negative and $A_\epsilon \subset \Theta$.

(ii) By Jensen's inequality,

$$\begin{aligned}
\mathcal{L}_n &\geq \pi(A_\epsilon) \int \exp\left[-\sum_{i=1}^{n} \log\left\{\frac{f(x_i \mid \theta^*)}{f(x_i \mid \theta)}\right\}\right] \pi_\epsilon(\theta) d\theta \\
&\geq \pi(A_\epsilon) \exp\left[\int -\sum_{i=1}^{n} \log\left\{\frac{f(x_i \mid \theta^*)}{f(x_i \mid \theta)}\right\} \pi_\epsilon(\theta) d\theta\right] \\
&= \pi(A_\epsilon) \exp\left[-nn^{-1} \sum_{i=1}^{n} \int \log\left\{\frac{f(x_i \mid \theta^*)}{f(x_i \mid \theta)}\right\} \pi_\epsilon(\theta) d\theta\right].
\end{aligned}$$

Therefore, $H(x_i) = \int \log\{f(x_i \mid \theta^*)/f(x_i \mid \theta)\}\, \pi_\epsilon(\theta)\, d\theta$.

(iii) By the law of large numbers, $n^{-1} \sum_{i=1}^{n} H(x_i) \overset{a.s.}{\to} \mathrm{E}\{H(x_i)\}$ as $n \to \infty$, where

$$\begin{aligned}
\mathrm{E}\{H(x_i)\} &= \int \int \log \frac{f(x_i \mid \theta^*)}{f(x_i \mid \theta)} \pi_\epsilon(\theta) d\theta f(x_i \mid \theta^*) dx_i \\
&= \int \left\{ \int \log \frac{f(x_i \mid \theta^*)}{f(x_i \mid \theta)} f(x_i \mid \theta^*) dx_i \right\} \pi_\epsilon(\theta) d\theta \\
&= \int d_K(\theta^*, \theta) \pi_\epsilon(\theta) d\theta \\
&\leq \epsilon \int \pi_\epsilon(\theta) d\theta = \epsilon.
\end{aligned}$$

Thus, $n^{-1} \sum_{i=1}^{n} H(x_i) \leq 2\epsilon$ almost surely for all large n, so

$$\mathcal{L}_n \geq \pi(A_\epsilon) \exp\left[-n\left\{n^{-1} \sum_{i=1}^{n} H(x_i)\right\}\right] \geq \pi(A_\epsilon) e^{-2n\epsilon}$$

almost surely for all large n.

(iv) By Markov's inequality, $\mathrm{P}(\mathcal{L}_n \geq e^{2n\epsilon}) \leq e^{-2n\epsilon} \mathrm{E}(\mathcal{L}_n)$, where

$$\begin{aligned}
\mathrm{E}(\mathcal{L}_n) &= \mathrm{E}\left\{\int_\Theta \prod_{i=1}^{n} \frac{f(x_i \mid \theta)}{f(x_i \mid \theta^*)} \pi(\theta) d\theta\right\} \\
&= \int \cdots \int \frac{\int \prod_{i=1}^{n} f(x_i \mid \theta) \pi(\theta) d\theta}{\prod_{i=1}^{n} f(x_i \mid \theta^*)} \prod_{i=1}^{n} f(x_i \mid \theta^*) d\boldsymbol{x} \\
&= 1.
\end{aligned}$$

Hence, $\mathrm{P}(\mathcal{L}_n \geq e^{2n\epsilon}) \leq e^{-2n\epsilon}$, so $\sum_n \mathrm{P}(\mathcal{L}_n \geq e^{2n\epsilon}) < \infty$. By the Borel–Cantelli lemma, $\mathcal{L}_n < e^{2n\epsilon}$ almost surely for all large n and any $\epsilon > 0$.

(v) Using the result in part (iii),

$$n^{-1} \log \mathcal{L}_n \geq n^{-1} \log\{\pi(A_\epsilon) e^{-2n\epsilon}\} = n^{-1} \log\{\pi(A_\epsilon)\} - 2\epsilon.$$

As $n \to \infty$, $n^{-1} \log\{\pi(A_\epsilon)\} \to 0$, so

$$n^{-1} \log \mathcal{L}_n \geq n^{-1} \log\{\pi(A_\epsilon)\} - 2\epsilon \to -2\epsilon.$$

From part (iv), $\mathcal{L}_n < e^{2n\epsilon}$ almost surely, where the value of ϵ is arbitrarily small. Therefore, $n^{-1} \log \mathcal{L}_n \overset{a.s.}{\to} 0$.

Further Reading

Historical Background. The marginal likelihood is used in a number of Bayesian procedures. Therefore, its asymptotic behavior is important to study. The asymptotic properties of marginal likelihoods are mostly applied to the large sample behavior of Bayes factors. Additional examples of their use include the approximation of integrals using Gaussian-type integrals; these Laplace approximations are commonly used for parametric models. A general asymptotic result is that the log of the marginal likelihood divided by the true likelihood and the sample size converges to 0 (i.e., $n^{-1}\log\mathcal{L}_n \to 0$ almost surely). This is an important result in information theory (Barron, 1988).

Points of Interest.

1. To see why dividing by the sample size matters, consider a sample of size n from the standard normal density function. Then, consider a normal model with unknown mean θ and known variance 1, and a normal prior for θ with mean 0 and precision parameter λ. Then, the log of the marginal likelihood divided by the true joint density is given by

$$\mathcal{L}_n = \tfrac{1}{2}\log\{\lambda/(n+\lambda)\} + \tfrac{1}{2}n^2\overline{x}^2/(n+\lambda).$$

   Note that $\overline{x} \to 0$ but $n\overline{x}^2$ does not converge to 0. Let $\overline{x} = z/\sqrt{n}$, where z is a standard normal random variable, so $n\overline{x}^2 = z^2$. Dividing $\mathcal{L}_n$ by any sequence that converges to infinity faster than $\log n$ will force it to 0. Therefore, dividing by n guarantees convergence for all models. However, if the designated true model is not true, then $\overline{x}$ will converge to a constant and $\mathcal{L}_n$ will converge to $-\infty$.

2. In Bayesian statistics, the merging of information suggests that, with an arbitrarily large sample size, all (reasonable) Bayesians should reach the same conclusions, regardless of their specified priors. Two posteriors π_n and ν_n merge in information if

$$n^{-1}d_K(\pi_n,\nu_n) \overset{a.s.}{\to} 0$$

   as $n \to \infty$, where $d_K(\pi_n,\nu_n)$ represents the Kullback–Leibler divergence between π_n and ν_n. Note that

$$d_K(\pi_n,v_n) = \mathrm{E}_{\pi_n}\{\log(\pi/\nu)\} + \log(m_\nu/m_\pi),$$

   where π and ν represent the prior distributions and m_π and m_ν are their corresponding marginal likelihoods. The expectation can be upper-bounded by using the MLE $\widehat{\theta}$:

$$n^{-1}d_K(\pi_n,\nu_n) = n^{-1}\log\{f(\boldsymbol{x} \mid \widehat{\theta})/m(\boldsymbol{x})\}\, d_K(\pi,\nu) + n^{-1}\log(m_\nu/m_\pi).$$

   If the marginal likelihoods and the MLE density models with the π-marginal likelihoods merge in information, then the posteriors merge in information.

3. If $f(\boldsymbol{x} \mid \theta^*)$ merges in information with the marginal likelihood $m(\boldsymbol{x})$, then the sequence of Cesàro averaged predictive density functions

$$\overline{p}_n(x) = n^{-1}\sum_{i=1}^{n} p_i(x \mid x_{1:i-1})$$

   converges to $f(\boldsymbol{x} \mid \theta^*)$ with respect to the expected Kullback–Leibler divergence. The proof is provided in Barron (1988).

S6.2.6 – Bayesian consistency for a discrete parameter space

(i) Note that $Q_n \leq 1$, so $Q_n \leq \sqrt{Q_n}$. For the numerator, use the result

$$\sqrt{\sum_{j=1}^{\infty} a_j} \leq \sum_{j=1}^{\infty} \sqrt{a_j}$$

provided that $a_j > 0$ for all j.

(ii) Take the expectation of J_{n+1} conditional on $\boldsymbol{x}$:

$$\begin{aligned}
\mathrm{E}(J_{n+1} \mid \boldsymbol{x}) &= \int \sum_{j=1}^{\infty} \sqrt{\prod_{i=1}^{n+1} \frac{f(x_i \mid \theta_j)}{f(x_i \mid \theta_0)}} \pi_j^{1/2} f(x_{n+1} \mid \theta_0) dx_{n+1} \\
&= \sum_{j=1}^{\infty} \sqrt{\prod_{i=1}^{n} \frac{f(x_i \mid \theta_j)}{f(x_i \mid \theta_0)}} \pi_j^{1/2} \int \sqrt{f(x_{n+1} \mid \theta_j) f(x_{n+1} \mid \theta_0)} dx_{n+1} \\
&\leq (1-\epsilon) J_n,
\end{aligned}$$

where the inequality follows because $\int \sqrt{f(x \mid \theta_j)\, f(x \mid \theta_0)}\, dx \leq 1-\epsilon$.

(iii) Once again using $\int \sqrt{f(x \mid \theta_j)\, f(x \mid \theta_0)}\, dx \leq 1-\epsilon$,

$$\begin{aligned}
\mathrm{E}(J_n) &= \int \cdots \int \sum_{j=1}^{\infty} \sqrt{\prod_{i=1}^{n} \frac{f(x_i \mid \theta_j)}{f(x_i \mid \theta_0)}} \pi_j^{1/2} \prod_{i=1}^{n} f(x_i \mid \theta_0) d\boldsymbol{x} \\
&= \sum_{j=1}^{\infty} \sqrt{\pi_j} \left[\int \cdots \int \prod_{i=1}^{n} \sqrt{f(x_i \mid \theta_j) f(x_i \mid \theta_0)} d\boldsymbol{x} \right] \\
&\leq (1-\epsilon)^n \sum_{j=1}^{\infty} \sqrt{\pi_j}.
\end{aligned}$$

(iv) By Markov's inequality,

$$\mathrm{P}(J_n \geq e^{-nc}) \leq e^{nc} \, (1-\epsilon)^n \, \gamma.$$

If $c = \frac{1}{2}\epsilon$ and $\gamma < \infty$, then

$$\sum_{n=1}^{\infty} \mathrm{P}(J_n \geq e^{-n\epsilon/2}) < \infty.$$

Thus, by the Borel–Cantelli lemma, $J_n < \gamma \, e^{-n\epsilon/2}$ almost surely for all large n.

(v) The denominator for the upper bound of Q_n is lower bounded by $\sqrt{\pi_0}$, so

$$Q_n < \gamma \, e^{-n\epsilon/2} / \sqrt{\pi_0}$$

almost surely for all large n. Hence, $Q_n \stackrel{a.s.}{\to} 0$ as $n \to \infty$, and the posterior mass accumulates on the correct parameter value. In other words, the Bayesian will eventually know the true value.

Further Reading

Historical Background. The consistency of the Bayesian posterior is an active area of research; see Ghosal and van der Vaart (2017) for a recent summary. If the prior is discrete and puts positive mass on the true parameter value, then the sequence of posteriors accumulates at the true value if the sum of square roots of the prior probabilities has a finite sum. This might seem like a strange condition; perhaps it is best understood in terms of the Hellinger distance, which introduces the use of square root terms.

Points of Interest.

1. Conditions for parametric consistency for continuous parameter spaces appear in Walker (1969), though these strict conditions apply to the model itself rather than the prior. For a discrete parameter space, the aim is to show that the posterior puts zero mass for large n on the complement of a Euclidean ball about θ_0. For a continuous parameter space, the aim is to show that the posterior puts zero mass for large n on the complement of a Hellinger neighborhood about θ_0.

2. To see the connection between the prior condition and Hellinger distance, consider the equality
$$\frac{f_{n,A}(x_{n+1})}{f(x_{n+1} \mid \theta_0)} = \frac{J_{n+1,A}}{J_{n,A}} \tag{6.2}$$
for some parameter set A that satisfies $d_H(f_0, f_{n,A}) > \epsilon$ for all n, where d_H denotes Hellinger distance. Here,
$$f_{n,A}(x) = \int_A f(x \mid \theta)\, \pi_A(\theta \mid \boldsymbol{x})\, d\theta,$$
where $\pi_A(\cdot \mid \boldsymbol{x})$ is the posterior restricted and normalized to the set A. Hence, $f_{n,A}$ is the predictive density restricted to the set A. Then, define
$$J_{n,A} = \int_A \prod_{i=1}^{n} \frac{f(x_i \mid \theta)}{f(x_i \mid \theta_0)}\, \pi(\theta)\, d\theta,$$
so the posterior on the set A is given by $Q_n(A) = J_{n,A}/I_n$, where
$$I_n = \int \prod_{i=1}^{n} \frac{f(x_i \mid \theta)}{f(x_i \mid \theta_0)}\, \pi(\theta)\, d\theta.$$
A suitable support condition on the prior ensures that $I_n > e^{-nc}$ almost surely for all large n and any $c > 0$. The Hellinger distance can be recovered from (6.2) by taking the expectations of the square root of both sides of the equality. Note that
$$\mathrm{E}(\sqrt{J_{n+1}} \mid \boldsymbol{x}) = \{1 - \tfrac{1}{2}\, d_H^2(f_0, f_{n,A})\}\, \sqrt{J_n}.$$
Therefore, if $d_H(f_0, f_{n,A}) > \epsilon$ for all n, then A is a set of parameters that yields a convex set of density functions that are further than 2ϵ away from f_0 with respect to the Hellinger distance. Then, $J_{n,A} < e^{-n\delta}$ almost surely for all large n which, when combined with the result for the denominator I_n, gives $Q_n(A) \overset{a.s.}{\to} 0$. More details on this strategy for establishing consistency can be found in Walker (2004).

3. The combination of the Markov inequality and the Borel–Cantelli lemma arises routinely in Bayesian asymptotic calculations. Through these results, it is typically feasible to prove the almost sure convergence of a sequence of positive random variables at an exponential rate given that the expectation is also exponentially small.

7

Time Series

Time series models generalize the classic independent and identically distributed set up by defining a sequence of conditional distribution functions that explain how and to what extent a current observation depends on past observations. The most basic time series model is a Markov process (or first-order process), where dependence is only on the immediately preceding observation. More complex time series models can be constructed by increasing the order of the process (i.e., instilling dependence on more than just the most previous observation).

Due to the mathematical convenience of the Markov assumption, first-order models, denoted AR(1), receive more attention than other time-series models. Question 7.1.1 shows that an arbitrary first-order model can be constructed using a copula density, which allows for the modeling of arbitrary dependencies. Question 7.1.2 and Question 7.1.3 investigate normal and Poisson AR(1) models, respectively. Additionally, Question 7.1.4 constructs a discrete AR(1)-type model using the Dirichlet distribution; this DAR(1) model is then connected to the Pólya urn scheme. However, the existence of naturally occurring processes that produce first-order data is questionable. Thus, other mathematical simplifications may better represent naturally occurring processes. One such assumption is a linear dependence of the current observation on the previous p observations; this structure gives rise to the autoregressive model of order p, denoted AR(p).

A central concept in time series modeling is stationarity, which has a number of interpretations. Perhaps one of the most common interpretations is that each observation has the same marginal distribution, which is known as the stationary distribution. Strong stationarity implies that the joint distribution is the same for any sequence of observations of the same length, no matter when this sequence occurs. Weak stationarity implies that each observation has the same mean and variance, and the covariance between any two observations only depends on the time between them.

A common theme in the exercises is the construction of stationary Markov processes given marginal distributions. In particular, the aim is to find a kernel density $k(x' \mid x)$ for which $f(x') = \int k(x' \mid x)\, f(x)\, dx$ for a given $f(x)$. To meet this goal, Markov chains can be constructed for posterior sampling by defining the joint density $f(y, x) = f(y \mid x)\, f(x)$ for an almost arbitrary choice of $f(y \mid x)$. Then, a Gibbs sampler yields the transition kernel

$$k(x' \mid x) = \int f(x' \mid y)\, f(y \mid x)\, dy,$$

which has $f(x)$ as the stationary density. It is remarkable to note that many popular kernels have such simple derivations. For example, if $f(x)$ is standard normal and $f(y \mid x)$ is normal with mean x and variance 1, then $f(x \mid y)$ is normal with mean $\frac{1}{2}y$ and variance $\frac{1}{2}$, and $k(x' \mid x)$ is normal with mean $\frac{1}{2}x$ and variance $3/4$; this is the well-known AR(1) Gaussian model of the form $x' = \rho x + \sqrt{1-\rho^2}\, z$, where z is an independent standard normal random variable and $\rho = \frac{1}{2}$.

The last two questions explore different topics for time series modeling. Question 7.1.5 motivates a test for positive autocorrelation in a stationary time series, where

DOI: 10.1201/9781003493471-7

autocorrelation is a measure that quantifies the degree to which values in a time series are related to past values. Finally, Question 7.1.6 concerns hidden Markov models (HMMs), which are one of the more difficult models to fit to time series data. However, as demonstrated in the question, a HMM with Gaussian distributions is relatively simple to fit, and prediction can be done using the Kalman filter.

Q7.1 Questions – Time Series

Q7.1.1* – Constructing time series densities with copulas

Introduction. A convenient property of Markov processes is that the joint density function $k(y,x)$ has identical marginal density functions. For example, suppose that the stationary density is $f(x)$ and the transition density is $p(y \mid x)$. Then, $k(y,x) = p(y \mid x)\, f(x)$ has marginals $f(y)$ and $f(x)$. An alternative approach to constructing the joint density $k(y,x)$ uses a copula density function, as demonstrated in this exercise.

Question. Suppose $p(x_t \mid x_{t-1})$ is the conditional density for a first-order time-homogeneous series $\{x_t\}$. Further, suppose

$$p(y \mid x) = f(y)\, c(F(y), F(x)) \tag{7.1}$$

for some copula density $c(u,v)$ and some distribution function F with corresponding density function f. A copula density $c(u,v)$ is a density function on $(0,1)^2$ with marginals that are the uniform density on $(0,1)$. That is, $\int c(u,v)\, dv = 1$ for all $0 < u < 1$. If $C(u,v)$ is the copula distribution function, then

$$c(u,v) = \frac{\partial^2 C(u,v)}{\partial u \partial v}.$$

(i) Show that $f(x)$ is the stationary density for the process $\{x_t\}$.

(ii) Find the distribution function $P(x_t \le y \mid x_{t-1} = x)$ in terms of the copula function.

(iii) If

$$C(u,v) = \frac{uv}{1 - \rho(1-u)(1-v)}$$

for some $0 \le \rho < 1$, find the distribution function in part (ii).

(iv) If $\rho = 0$, what can be said about the process?

(v) If $\{x_t\}$ is a first-order homogeneous process with stationary density $f(x)$, show that it is possible to write the conditional transition density $p(y \mid x)$ as (7.1) for some copula density $c(u,v)$.

Q7.1.2 – Normal AR(1) models

Introduction. One of the simplest time series models is the normal AR(1) model, which is strictly stationary under certain constraints on the parameters. This question focuses on the estimation of the dependence parameter ρ and argues that its MLE converges to the true value almost surely.

Question. Consider the time series model

$$x_n = \rho x_{n-1} + \sigma z_n$$

for $n \in \mathbb{N}^+$, where $|\rho| < 1$ and the $\{z_n\}$ are independent standard normal random variables. Assume that x_0 is normal with mean 0 and variance σ^2.

(i) By considering the variance of x_n, explain why the model is non-stationary if $|\rho| \geq 1$.

(ii) Find the stationary distribution for $\{x_n\}$ when $|\rho| < 1$.

(iii) For an observed sample $\{x_0, \ldots, x_n\}$, write down the likelihood function and find the MLE for ρ, written as $\widehat{\rho}_n$.

(iv) Show that

$$\mathrm{E}\left(\widehat{\rho}_n - \rho \mid \boldsymbol{x}_{1:n-1}\right) = w_n\left(\widehat{\rho}_{n-1} - \rho\right),$$

where $0 < w_n < 1$ and ρ is the true value.

(v) Show that $\widehat{\rho}_n \overset{a.s.}{\to} \rho$ as $n \to \infty$.

Q7.1.3 – Poisson AR(1) models

Introduction. The previous exercise concerned the estimation of the dependence parameter in a normal AR(1) model. As seen in this question, parameter estimation for non-normal AR(1) models is more challenging. For a Poisson AR(1) model, $k(x' \mid x)$ must satisfy

$$p(x') = \sum_{x=0}^{\infty} k(x' \mid x)\, p(x),$$

where $p(x)$ is a Poisson probability mass function. It is not immediately obvious how to find $k(x' \mid x)$. One approach is to exploit the conjugacy of the Binomial–Poisson pair.

Question. Let $p(x)$ be a Poisson mass function with rate parameter $\lambda > 0$ for $x \in \{0, 1, 2, \ldots\}$:

$$p(x) = \mathrm{P}(X = x) = \frac{\lambda^x}{x!} e^{-\lambda}.$$

Additionally, let $p(y \mid n)$ be a binomial mass function with probability parameter p for $y \in \{0, 1, \ldots, n\}$:

$$p(y \mid n) = \mathrm{P}(Y = y \mid n) = \binom{n}{y} p^y (1-p)^{n-y}.$$

If $n = 0$, then $Y = 0$ with probability 1.

(i) If $p(x, y) = p(x)\, p(y \mid x)$, find $p(x \mid y)$.

(ii) Consider the kernel

$$k(x' \mid x) = \sum_y p(x' \mid y)\, p(y \mid x).$$

Show that the stationary mass function for this kernel is $p(x)$.

(iii) Show that it is possible to write $x' = A(x) + B$, where $A(x)$ is a random variable depending on x and B is an independent random variable not dependent on x.

(iv) If x_0 is from $p(x)$ and x_i is from $k(\cdot \mid x_{i-1})$ for $i = 1, \ldots, n$, write down the likelihood function for parameters p and λ.

(v) Describe an algorithm for estimating the parameters via maximum likelihood.

Q7.1.4 – The DAR(1) model and the Dirichlet distribution

Introduction. The discrete autoregressive model of order p, denoted DAR(p), is a time series model used for analyzing and forecasting discrete-valued sequential data. In its simplest form, the DAR(1) model is often connected to a Gibbs sampler involving a Dirichlet distribution and a discrete distribution. The extension to an arbitrary order p coincides with the Pólya urn scheme, which is related to Dirichlet distributions and exchangeable sequences. This question investigates the DAR(1) model and concludes with an exchangeability result.

Question. One version of the DAR(1) model is the sequence $\{x_n\}$ with $\mathrm{P}(X_1 = j) = w_j$ for $j = 1, \ldots, M$, and

$$\mathrm{P}(X_n = j \mid x_{n-1}) = \frac{cw_j + 1(x_{n-1} = j)}{c+1}, \quad n > 1$$

for some $c > 0$. Additionally, $\mathbf{p} = (p_1, \ldots, p_M)'$ has a Dirichlet distribution with parameters $(c\,w_1, \ldots, c\,w_M)$, written as $\mathbf{p} \sim \pi(\mathbf{p})$.

(i) Given $x_1 = j$, find the posterior probability mass function for $\mathbf{p}$, denoted $\pi(\mathbf{p} \mid x_1)$.

(ii) Let

$$\mathrm{P}(X_2 = j \mid \mathbf{p}) = p_j \quad \text{with} \quad \mathbf{p} \sim \pi(\mathbf{p} \mid x_1).$$

Show that $\mathrm{P}(X_2 = j \mid x_1)$ coincides with the expression given in the question.

(iii) Find the stationary probability mass function of the DAR(1) model.

(iv) The model is extended to

$$\mathrm{P}(X_n = j \mid \mathbf{x}_{1:n-1}) = \frac{cw_j + \sum_{i=1}^{n-1} 1(x_i = j)}{c + n - 1}.$$

What is the marginal probability mass function for x_n?

(v) Show that the sequence $\{x_n\}$ is exchangeable.

Q7.1.5 – Testing for dependence in a time series

Introduction. Autocorrelation, or serial correlation, is a statistical measure that quantifies the degree to which values in the series are related to past values of the same series. An understanding of autocorrelation for time series is useful to identify patterns and predict future values based on historical data. This question motivates a test for positive autocorrelation in a stationary time series.

Question. Suppose x_0 is standard normal and

$$x_n = \rho\, x_{n-1} + \sqrt{1-\rho^2}\, z_n, \quad n \geq 1,$$

where the $\{z_n\}$ are independent standard normal random variables and $0 < \rho < 1$.

(i) Show that the stationary distribution for the process is standard normal.

(ii) Show that the covariance between x_i and x_j is $\rho^{|i-j|}$.

(iii) If Σ is the covariance matrix for $(x_1, \ldots, x_n)$, show that the variance of

$$S = x_1 + \cdots + x_n$$

is given by

$$\text{Var}\,(S) = \sum_{1 \leq i,j \leq n} \Sigma_{i,j}.$$

(iv) Consider a test for the hypotheses $H_0 : \rho = 0$ vs $H_1 : \rho > 0$. What is the distribution of S under the null hypothesis, and what is an appropriate critical region for the test?

(v) If the true value is $\rho = \frac{1}{2}$, what is the Type II error for the test? Comment on this.

Q7.1.6 – The Kalman filter

Introduction. State space models, also known as hidden Markov models, are one of the more difficult time series models to estimate. They pose unique challenges because the likelihood function involves multiple intractable integrals. This question looks at a special case with Gaussian distributions for which estimation and prediction is relatively straightforward. The prediction process is known as the Kalman filter.

Question. An observed sequence $\{y_n\}_{n \geq 1}$ is based on the following model:

$$p(y_n \mid x_n) = \text{N}(y_n \mid x_n, \sigma^2), \quad p(x_n \mid x_{n-1}) = \text{N}(x_n \mid x_{n-1}, \tau^2),$$

where the $\{x_n\}$ are unobserved and the $\{y_n\}$ are independent among each other once conditioned on the $\{x_n\}$. Assume that $x_0 \sim \text{N}(0, \tau^2)$.

(i) The aim is to find a recursive expression for $p(x_n \mid \mathbf{y}_{1:n})$. To this end, show that

$$p(x_n \mid \mathbf{y}_{1:n}) \propto p(y_n \mid x_n)\, p(x_n \mid \mathbf{y}_{1:n-1}).$$

(ii) Expand $p(x_n \mid \mathbf{y}_{1:n-1})$ such that $p(x_n \mid \mathbf{y}_{1:n})$ can be expressed as a function of $p(x_{n-1} \mid \mathbf{y}_{1:n-1})$.

(iii) If $p(x_{n-1} \mid \mathbf{y}_{1:n-1}) = \mathrm{N}(x_{n-1} \mid \mu_{n-1}, \phi_{n-1}^2)$, find $p(x_n \mid \mathbf{y}_{1:n-1})$.

(iv) Find $p(x_n \mid \mathbf{y}_{1:n})$.

(v) Explain how a sample from $p(x_{n-1} \mid \mathbf{y}_{1:n-1})$ could be used to obtain a sample from $p(x_n \mid \mathbf{y}_{1:n})$.

S7.1 Solutions – Time Series

S7.1.1* – Constructing time series densities with copulas

(i) The aim is to show that $f(y) = \int p(y \mid x)\, f(x)\, dx$. Assume that

$$f(y) = \int f(y) c(F(y), F(x))\, f(x)\, dx,$$

so it must be that $\int c(F(y), F(x))\, f(x)\, dx = 1$. Consider the transformation $v = F(x)$, so $dv = f(x)\, dx$. By definition, $\int c(F(y), v)\, dv = 1$ regardless of the value of $F(y)$.

(ii) The distribution function is given by

$$\int_{s \le y} f(s)\, c(F(s), F(x))\, ds = \int_{u \le F(y)} c(u, F(x))\, du = \frac{\partial}{\partial v} C(F(y), F(x)),$$

where $u = F(s)$.

(iii) Using part (ii), the aim is to find $\partial C / \partial v$, which is given by

$$\frac{u\{1 - \rho(1-u)\}}{\{1 - \rho(1-u)(1-v)\}^2}.$$

By definition, this is a cumulative distribution function in u on $(0, 1)$. If $u = 0$, the distribution function in part (ii) is 0. If $u = 1$, it is 1. It is also easy to check that the function is increasing.

(iv) If $\rho = 0$, the distribution in part (iii) is u. Thus, the process is independent of previous observations, and the copula is $C(u, v) = uv$, which is known as the independence copula with $c(u, v) = 1$ for all u, v.

(v) Consider $k(y, x) = f(x)\, p(y \mid x)$ and define

$$P(X_{t+1} \le y, X_t \le x) = K(y, x),$$

where $k(y, x) = \partial^2 K(y, x)/\partial x \partial y$. Then, write $C(u, v) = K(F^{-1}(u), F^{-1}(v))$, so $K(y, x) = C(F(y), F(x))$. Therefore, $k(y, x) = f(x)\, f(y)\, c(F(y), F(x))$, implying that

$$p(y \mid x) = f(y)\, c(F(y), F(x)).$$

Further Reading

Historical Background. For general information on copulas, see Nelsen (2007). Here, the important result to complete the exercise is Sklar's theorem, which states that any joint distribution function can be represented using a copula distribution function (Sklar, 1959). That is, it is possible to write

$$P(X \le x, Y \le y) = C\left(F_X(x), F_Y(y)\right),$$

where F_X and F_Y are the marginal distribution functions for X and Y, respectively. This follows by simply defining

$$C(u,v) = \mathrm{P}(F_X^{-1}(u), F_Y^{-1}(v)).$$

The uniform marginals are a consequence of $\mathrm{P}(X \leq F_X^{-1}(u)) = u$ and the notion that, if X is sampled from F_X, then $F_X(X)$ is a uniform random variable.

An interesting book on the simulation of copulas is Mai and Scherer (2017). They explore theory relating to various copulas, including the Archimedean, elliptical, and Marshall–Olkin copulas. Much emphasis is placed on the construction of pair copulas and their simulation. The book concludes with a chapter exploring the applications and implementations of copulas for Monte Carlo methods.

Points of Interest.

1. Related to the copula construction of Markov stationary processes is the construction of conditionally identically distributed (CID) sequences $\{X_{1:\infty}\}$ (Berti et al., 2004). CID sequences are proving useful for Bayesian-style inference; see Fong et al. (2024) for a recent example. Formally, such sequences satisfy

   $$\mathrm{E}\{f(X_k) \mid x_{1:n}\} = \mathrm{E}\{f(X_{n+1}) \mid x_{1:n}\}$$

   for all $k > n$ and all bounded measurable functions $f : \mathbb{R} \to \mathbb{R}$. Given $x_{1:n}$, x_k has the same distribution as x_{n+1} for all $k > n$. This holds if the sequence of distribution functions forms a martingale; that is, if $X_{k+1} \sim F_k$, then

   $$\mathrm{E}(F_k \mid F_n) = F_n$$

   for all $k > n$. Interestingly, a martingale $\{F_k\}$ can be constructed with copulas:

   $$f_{k+1}(x) = f_k(x)\, c(F_k(x), F_k(x_{k+1})).$$

   It then follows that $\mathrm{E}\{f_{k+1}(x) \mid f_k\} = f_k(x)$, where the expectation is with respect to x_{n+1}.

2. A notable nonparametric copula that has connections with the empirical distribution function and the Dirichlet process is

   $$C(u,v) = (1-\alpha)\, u\, v + \alpha \, \min\{u,v\},$$

   where $\min\{u,v\}$ is the maximal copula (i.e., the largest a copula can be) and $0 < \alpha < 1$ converges to 0 as the sample size increases. In particular, this copula is a mixture between the independence copula and the maximal copula. If x_{k+1} comes from F_k, where

   $$F_{k+1}(x) = (1-\alpha_k)\, F_k(x) + \alpha_k\, 1(x_{k+1} \leq x),$$

   then the sequence $\{F_k\}$ is a martingale.

3. The AR(1) Gaussian stationary model has $y = \rho x + \sqrt{1-\rho^2} z$ as the transition density $p(y \mid x)$. The Gaussian copula generalizes this to an arbitrary distribution by taking

   $$\mathrm{P}(y \mid x) = \Phi\left(\frac{\Phi^{-1}(F(y)) - \rho\Phi^{-1}(F(x))}{\sqrt{1-\rho^2}}\right)$$

for some distribution function F, which becomes the stationary distribution. Note that it is possible to sample y from x by taking a draw $\xi \sim \text{Unif}(0,1)$ and setting

$$F(y) = \Phi\left(\rho\Phi^{-1}(F(x)) + \sqrt{1-\rho^2}\,\Phi^{-1}(\xi)\right).$$

Note that $\Phi^{-1}(\xi) = z$ is a standard normal random variable. Thus,

$$y = F^{-1}\left(\Phi\left(\rho\Phi^{-1}(F(x)) + \sqrt{1-\rho^2}\,z\right)\right)$$

becomes the Gaussian AR(1) model when $F \equiv \Phi$, the standard normal cumulative distribution function.

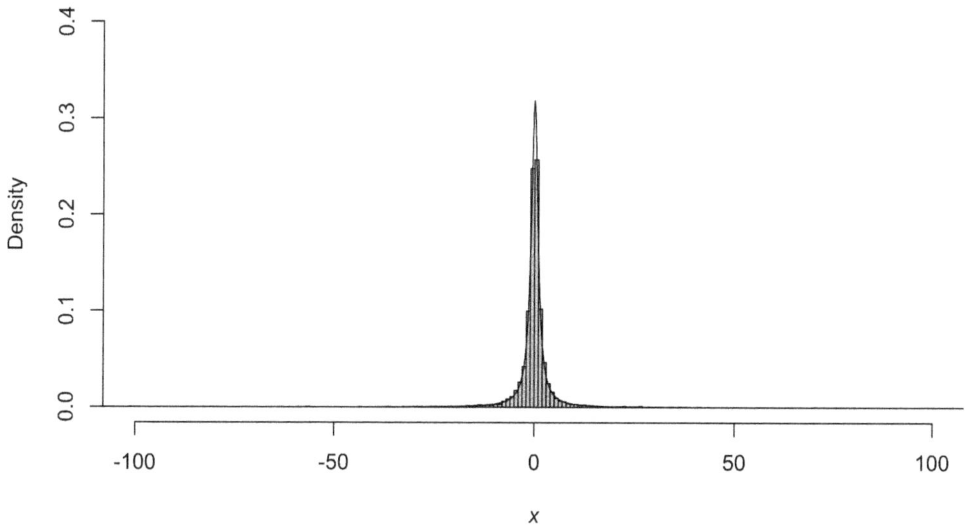

FIGURE 7.1
Histogram of samples from a Cauchy AR(1) model with the true Cauchy density function overlaid.

Demonstration. Fig. 7.1 presents a histogram of samples in $[-100, 100]$ generated from F, which is the standard Cauchy distribution with $\rho = 0.8$; overlaid is the true Cauchy density function. Some large values that are not shown in Fig. 7.1 are shown in Fig. 7.2, which is the plot of the sequence $\{x_i\}$ from the Cauchy AR(1) model for a sample of size 10,000.

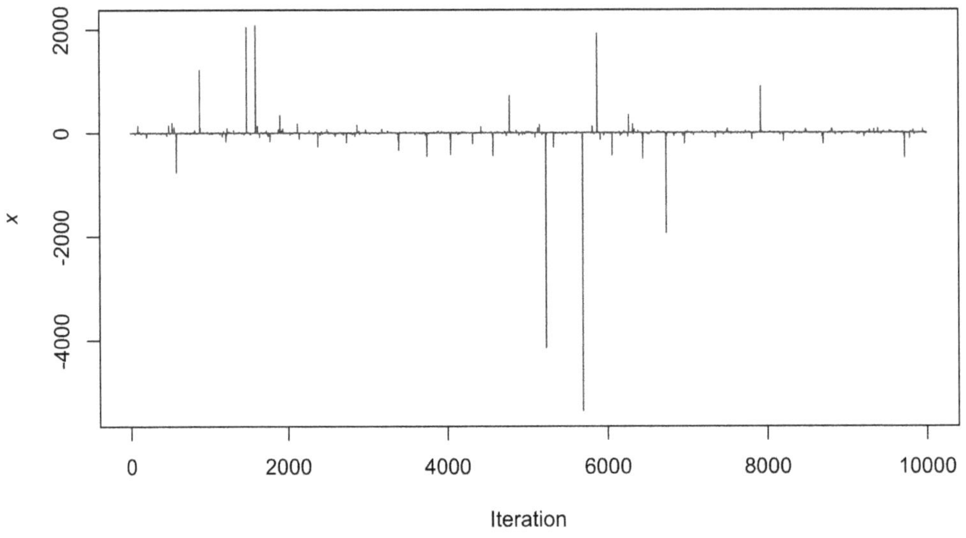

FIGURE 7.2
Plot of samples from a Cauchy AR(1) model.

S7.1.2 – Normal AR(1) models

(i) First, compute the variance for x_1:

$$\mathrm{Var}(X_1) = \mathrm{Var}(\rho X_0 + \sigma z_1) = \rho^2\sigma^2 + \sigma^2 = \sigma^2(1+\rho^2).$$

For x_2,

$$\mathrm{Var}(X_2) = \mathrm{Var}(\rho X_1 + \sigma z_2) = \sigma^2(1+\rho^2+\rho^4).$$

Using induction, it can be seen that

$$\mathrm{Var}(X_n) = \sigma^2 \sum_{k=0}^{n} \rho^{2k}.$$

To prove this, assume the equation holds for x_{n-1}. Then,

$$\begin{aligned}\mathrm{Var}(X_n) &= \mathrm{Var}(\rho X_{n-1} + \sigma z_n) \\ &= \sigma^2 \sum_{k=0}^{n-1} \rho^{2k+2} + \sigma^2 \\ &= \sigma^2 \sum_{k=0}^{n} \rho^{2k}.\end{aligned}$$

Therefore, if $|\rho| > 1$, then $\mathrm{Var}(X_n) \to \infty$ as $n \to \infty$, which means that the variance of the process will grow with n.

(ii) Start by noting that $x_2 = \rho x_1 + \sigma z_2 = \rho(\rho x_0 + \sigma z_1) + \sigma z_2$. By induction,

$$x_n = \sum_{k=0}^{n} \rho^k \sigma z_{n-k},$$

where $x_0 = \sigma z_0$. This linear combination of independent normal random variables suggests that x_n is Gaussian. Thus, to determine the stationary distribution, only the mean and the variance need to be determined. It is clear that $\mathrm{E}(X_n) = 0$ for all n. For the variance, note that

$$\mathrm{Var}(X_n) = \mathrm{Var}(\rho X_{n-1} + \sigma z_n) = \rho^2 \mathrm{Var}(X_{n-1}) + \sigma^2.$$

Because $\{x_n\}$ is stationary, $\mathrm{Var}(X_n) = \mathrm{Var}(X_{n-1})$, so it must be that $\mathrm{Var}(X_n) = \sigma^2/(1-\rho^2)$. Therefore, the stationary distribution is $\mathrm{N}(0, \sigma^2/(1-\rho^2))$.

(iii) The likelihood function for the sample is

$$l_n(\rho \mid \boldsymbol{x}_{0:n}) = (2\pi\sigma^2)^{-n/2} \exp\left\{ -\frac{1}{2\sigma^2} \sum_{k=1}^{n} (x_k - \rho x_{k-1})^2 \right\}.$$

To find the MLE, take the partial derivative of the log-likelihood function with respect to ρ:

$$\begin{aligned} \frac{\partial}{\partial \rho} L_n(\rho) &= -\frac{1}{2\sigma^2} \sum_{k=1}^{n} \frac{\partial}{\partial \rho} (x_k - \rho x_{k-1})^2 \\ &= \frac{1}{\sigma^2} \sum_{k=1}^{n} x_{k-1}(x_k - \rho x_{k-1}). \end{aligned}$$

Set this to zero to obtain the MLE:

$$\widehat{\rho}_n = \frac{\sum_{k=1}^{n} x_{k-1} x_k}{\sum_{k=1}^{n} x_{k-1}^2}.$$

(iv) The only term in

$$\mathrm{E}(\widehat{\rho}_n \mid \boldsymbol{x}_{1:n-1}) = \mathrm{E}\left(\frac{\sum_{k=1}^{n} x_{k-1} x_k}{\sum_{k=1}^{n} x_{k-1}^2} \mid \boldsymbol{x}_{1:n-1} \right)$$

that contains a random variable given $\boldsymbol{x}_{1:n-1}$ is $\mathrm{E}(X_n\, x_{n-1} \mid x_{n-1}) = \rho\, x_{n-1}^2$. Thus,

$$\mathrm{E}(\widehat{\rho}_n \mid \boldsymbol{x}_{1:n-1}) = w_n \widehat{\rho}_{n-1} + \rho \frac{x_{n-1}^2}{\sum_{k=1}^{n} x_{k-1}^2}$$

with weights

$$w_n = \frac{\sum_{k=1}^{n-1} x_{k-1}^2}{\sum_{k=1}^{n} x_{k-1}^2}.$$

The result follows by subtracting by ρ.

(v) Note that $\mathrm{E}(|\widehat{\rho}_n - \rho|)$ is a non-negative supermartingale. Thus, by the Doob martingale convergence theorem, it is that $|\widehat{\rho}_n - \rho|$ converges almost surely to a finite random variable; to show that the finite random variable is 0, use the variance of $\widehat{\rho}_n$.

A more straightforward solution is to write

$$\widehat{\rho}_n = \frac{n^{-1}\sum_{k=1}^{n} x_k\, x_{k-1}}{n^{-1}\sum_{k=1}^{n} x_{k-1}^2}.$$

By the strong law of large numbers, the numerator and denominator both converge almost surely to their respective means, which are $\rho\,\sigma^2/(1-\rho^2)$ and $\sigma^2/(1-\rho^2)$, respectively. Therefore, $\widehat{\rho}_n \overset{a.s.}{\to} \rho$.

Further Reading

Historical Background. Time series modeling has a long and extensive history. One of the most notable early contributions was the introduction of the Yule–Walker equations, which made statistical analysis for time series feasible. These equations were published in Yule (1927) and Walker (1931).

Points of Interest.

1. Normal autoregressive models of order p are of the form

$$x_n = \sum_{j=1}^{p} \rho_j x_{n-j} + \sigma\, z_n.$$

It is typically of interest to estimate the dependence parameters $\{\rho_j\}$. To do so, note that the data can provide estimators for $r(k) = \mathrm{E}(X_j\, X_{j-k})$ for each $k = 1, \ldots, p$. Multiply the expression for x_n by x_{n-k} for each $k = 1, \ldots, p$ to obtain the following equations:

$$\begin{aligned} r(1) &= \rho_1 r(0) + \rho_2 r(1) + \cdots + \rho_p r(p-1) \\ r(2) &= \rho_1 r(1) + \rho_2 r(0) + \rho_3 r(1) + \cdots + \rho_p r(p-2) \\ &\vdots \\ r(p) &= \rho_1 r(p-1) + \rho_2 r(p-2) + \cdots + \rho_p r(0). \end{aligned}$$

Write this system of equations as $\mathbf{r} = \mathbf{R}\boldsymbol{\rho}$, where $\boldsymbol{\rho} = (\rho_1, \ldots, \rho_p)'$ and $\mathbf{r} = (r(1), \ldots, r(p))'$. An estimator for $\boldsymbol{\rho}$ is $\widehat{\boldsymbol{\rho}} = \mathbf{R}^{-1}\mathbf{r}$; this is known as the Yule–Walker estimator (Yule, 1927; Walker, 1931).

2. The normal AR(1) model can be separated into two terms: the ρx_n term representing a thinning operation and the innovation term σz_n. In general, the model is written as

$$x_n = \rho_n \star x_{n-1} + z_n,$$

where z_n is the innovation variable that is independent of x_{n-1} and $\rho_n \star x$ is a random thinning operation given by

$$\rho \star x = \sum_{i=1}^{N(x)} w_i,$$

where the $\{w_i\}$ are independent and identically distributed and $N(x)$ is integer-valued. The model in this question arises when $N(x) = x$ and $w_1 = \rho$. Additionally, the model in the next exercise arises when $N(x) = x$ and w_1 is Bernoulli with parameter ρ. See Grunwald et al. (1996) for a unified theory of AR(1) linear models.

3. Langevin diffusion has recently received a lot of attention due to its connection with artificial intelligence and generative modeling; for example, it has been used to produce realistic images. This type of diffusion is a continuous-path Markov process, and its stochastic differential equation is given by

$$dX_t = dt\,\frac{d}{dx}\log p(X_t) + \sqrt{2}dW_t,$$

where the stationary density is p and W_t is standard Brownian motion. To obtain a Gaussian stationary density and the necessary discrete-time construction with step size h, the diffusion can be approximated as

$$X_{t+h} = X_t - h\,X_t + \sqrt{2\,h}\,z_t,$$

where the $\{z_t\}$ are independent standard normal random variables. Taking $h = 0.1$ provides a set of $\{X_t\}$, the histogram of which resembles a normal density shape.

Demonstration. In normal autoregressive models, there is often interest in testing whether $\rho = 0$ (i.e., whether there is dependence in the series). To test this, compute the mean and variance of $\widehat{\rho}_n$. For independent standard normal $\{x_i\}$, the mean is 0. Using the standard conditional variance formula,

$$\mathrm{Var}(\widehat{\rho}_n) = \mathrm{E}\left(\frac{X_{n-1}^2}{(X_0^2 + \cdots + X_{n-1}^2)^2}\right) + \mathrm{Var}\left(\frac{\sum_{k=1}^{n-1} X_k\,X_{k-1}}{X_0^2 + \cdots + X_{n-1}^2}\right).$$

It can be shown that the first term is $\{n(n-2)\}^{-1}$ and the second term is $(n-1)/\{n(n+2)\}$. To demonstrate this approximation, let $\rho = 0$ and $\sigma = 1$. Then, compute $\widehat{\rho}_n$ for 10,000 samples of size $n = 10$. Fig. 7.3 contains a histogram of these samples of $\widehat{\rho}_{10}$. Additionally, the normal density with mean 0 and variance $(n-1)/\{n(n+2)\} + 1/\{n(n-2)\}$ is overlaid.

S7.1.3 – Poisson AR(1) models

(i) By Bayes' theorem, $p(x \mid y) \propto p(x,y) = p(x)p(y \mid x)$, so

$$p(x \mid y) \propto \frac{\{\lambda(1-p)\}^{x-y}}{(x-y)!},$$

which is the kernel of a $\mathrm{Pois}(x - y \mid \lambda(1-p))$ distribution. Letting $z = x - y$, it can be seen that $x = y + z$, so x given y is a $y + \mathrm{Pois}(\lambda(1-p))$ random variable.

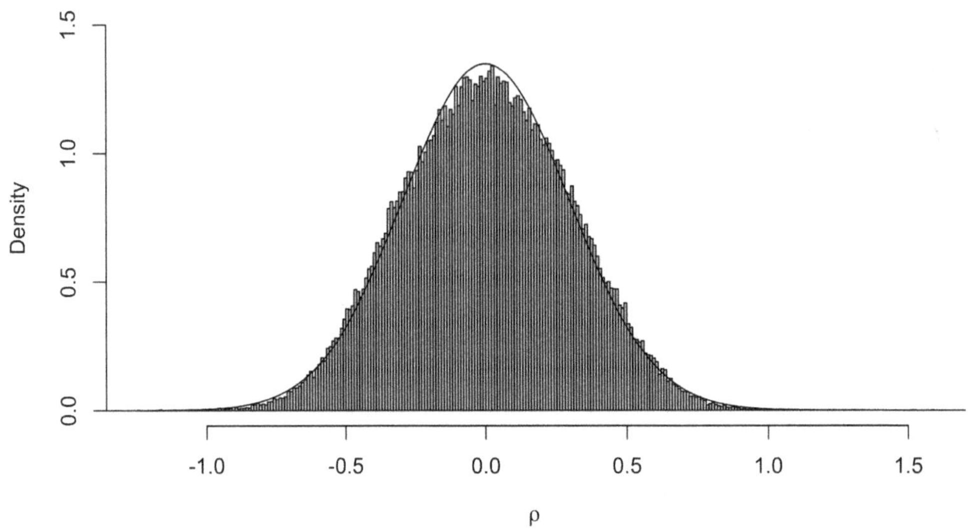

FIGURE 7.3
10,000 samples of $\widehat{\rho}_{10}$ with $\rho = 0$ and $\sigma = 1$. The corresponding normal density is overlaid.

(ii) To show that $p(x)$ is the stationary distribution, it is required to show

$$p(x') = \sum_{x=0}^{\infty} k(x' \mid x)p(x).$$

Note that

$$\begin{aligned}\sum_{x=0}^{\infty} k(x' \mid x)p(x) &= \sum_{x=0}^{\infty}\sum_{y=0}^{x} p(x' \mid y)\, p(y \mid x)\, p(x)\\ &= \sum_{y=0}^{\infty} p(x' \mid y)\, p(y),\end{aligned}$$

which is $p(x')$. The final equality was obtained by switching the order of summations and noting that $\sum_{x=y}^{\infty} p(y \mid x)p(x) = p(y)$.

(iii) Define $A(x)$ to be a $\text{Bin}(x, p)$ random variable and B to be a $\text{Pois}(\lambda(1-p))$ random variable that is independent of $A(x)$. From part (i), it is possible to write

$$x' = A(x) + B,$$

where $A(x) = y$ using the notation in part (i).

(iv) The likelihood function is

$$\begin{aligned}l(\lambda, p \mid x) &= p(x_0)\prod_{i=1}^{n} k(x_i \mid x_{i-1})\\ &= \frac{e^{-\lambda}\lambda^{x_0}}{x_0!}\prod_{i=1}^{n}\sum_{y_i=0}^{\min\{x_i, x_{i-1}\}} \text{Pois}(x_i - y_i \mid \lambda(1-p)) \times \text{Bin}(y_i \mid x_{i-1}, p),\end{aligned}$$

where $y_i \le \min\{x_i, x_{i-1}\}$ satisfies the support conditions for the Poisson and binomial mass functions.

(v) Consider the EM algorithm for estimating λ and p using maximum likelihood. Note that the $\{y_i\}$ in part (iv) can be treated as missing data. It is necessary to find the conditional probability of each y_i given the observed data, which can be written as $p(y_i \mid x_i, x_{i-1})$; this is a finite mass function on $\{0, 1, \ldots, \min(x_i, x_{i-1})\}$:

$$p(y_i \mid x_i, x_{i-1}) \propto \frac{\{\lambda(1-p)^2/p\}^{-y_i}}{(x_i - y_i)!(x_{i-1} - y_i)!y_i!} 1(0 \le y_i \le \min\{x_i, x_{i-1}\}).$$

Let $Q(\boldsymbol{\theta}' \mid \boldsymbol{\theta})$ denote the expectation of $L(\boldsymbol{\theta} \mid x, y)$ with respect to the distribution of $p(y \mid x, \boldsymbol{\theta})$, where $\boldsymbol{\theta} = (\lambda, p)'$. The EM algorithm updates $\boldsymbol{\theta}_t \to \boldsymbol{\theta}_{t+1}$ by maximizing $Q(\boldsymbol{\theta}' \mid \boldsymbol{\theta}_t)$ over $\boldsymbol{\theta}'$ for each iteration.

Further Reading

Historical Background. In the last two exercises, two different AR(1) models were considered. A general class of AR(1) models is illustrated in Joe (1996), which is based on previous work by McKenzie (1988), Lewis et al. (1989) and Al-Osh and Aly (1992).

Points of Interest.

1. The general construction for an AR(1) process is

$$x_n = A_n(x_{n-1}) + B_n,$$

where $A_n(x_{n-1})$ is a random variable depending on x_{n-1} and the $\{B_n\}$ are independent and identically distributed random variables that are independent of $\{A_n\}$. Note that the independence between the two terms holds if (i) x_n has distribution P_θ, (ii) the $\{B_n\}$ are independent and identically distributed from $P_{(1-\alpha)\theta}$ for some $0 < \alpha < 1$, and (iii) if x_{n-1} is from P_θ, then $A_n(x_{n-1})$ is from $P_{\alpha\theta}$. Thus, it is required that $P_{\alpha\theta} \bullet P_{(1-\alpha)\theta} = P_\theta$, where $P_1 \bullet P_2$ is the convolution operator

$$p_\theta(x) = \int p_{\alpha\theta}(y)\, p_{(1-\alpha)\theta}(x-y)\, dy.$$

Further, $A(x)$ has a distribution that coincides with $x_{\alpha\theta}$ if

$$x_{\alpha\theta} + x_{(1-\alpha)\theta} = x,$$

where $x_{\alpha\theta}$ and $x_{(1-\alpha)\theta}$ are independent and distributed as $P_{\alpha\theta}$ and $P_{(1-\alpha)\theta}$, respectively. If x has distribution P_θ, then $A(x)$ is $P_{\alpha\theta}$ based on the convolution property.

2. The general result for an AR(1) process in the previous point can be implemented as a Gibbs sampler. Consider the joint density

$$p(x, y) = p_\theta(x)\, p_{\alpha\theta}(y \mid x),$$

where a sample from $p_{\alpha\theta}(y \mid x)$ is precisely $A(x)$. Then, $p_{\alpha\theta}(x \mid y)$ is sampled as $y + x_{(1-\alpha)\theta}$. A Gibbs sampler transition kernel, in which $p_\theta(x)$ is the stationary density, is given by

$$k(x' \mid x) = \int p(x' \mid y)\, p(y \mid x)\, dy.$$

This can be written as $x' = A(x) + x_{(1-\alpha)\theta}$, where $A(x)$ is $x_{\alpha\theta}$. Note that $x' = x_\theta$ if $x = x_\theta$. For a more general Gibbs sampler based on AR(1) models, see Pitt et al. (2002).

3. Once the transition density has been set, inference for unknown parameters proceeds as usual. The likelihood function can be constructed and either Bayesian or classical analysis can be executed. In the Bayesian framework, exchangeability is often used to justify the existence of a prior if observations are assumed to be orderless. However, this assumption is clearly faulty for a time series of observations. It is then natural to ask what assumptions are required for a sequence $\{x_n\}$ to satisfy

$$f(x_1, \ldots, x_n) = \int \prod_{i=1}^{n} p(x_i \mid x_{1:i-1}, \theta)\, \pi(\theta)\, d\theta$$

for each n and some prior $\pi(\theta)$.

S7.1.4 – The DAR(1) model and the Dirichlet distribution

(i) Note that $\pi(\mathbf{p} \mid x_1) \propto \mathrm{P}(x_1 = j \mid \mathbf{p})\, \pi(\mathbf{p})$, which is proportional to

$$p_j \prod_{l=1}^{M} p_l^{cw_l - 1}.$$

This is a Dirichlet distribution with parameters

$$(cw_1, \ldots, cw_j + 1, \ldots, cw_M).$$

(ii) By the law of total probability,

$$\mathrm{P}(X_2 = j \mid x_1) = \int \mathrm{P}(X_2 = j \mid \mathbf{p}) \pi(\mathbf{p} \mid x_1) dx_1 = \int p_j \pi(\mathbf{p} \mid x_1) dx_1 = \mathrm{E}(p_j \mid x_1).$$

From the result in part (i), the expectation of the jth element of $\mathbf{p}$ is

$$\mathrm{E}(p_j \mid x_1) = \frac{cw_j + 1(x_1 = j)}{(\sum_{m=1}^{M} cw_m) + 1} = \frac{cw_j + 1(x_1 = j)}{c + 1},$$

where $\sum_{m=1}^{M} w_m = 1$.

(iii) The stationary probability for $\mathrm{P}(x_n = j)$ is w_j; this is based on the joint density function

$$\mathrm{P}(X_n = j \mid \mathbf{p}) = p_j, \quad \mathbf{p} \sim \mathrm{Dir}(c\mathbf{w}),$$

in which the marginal probabilities provide the stationary density.

(iv) A combination of recursion and induction leads to $\mathrm{P}(X_n = j) = w_j$; start from $\mathrm{P}(X_1 = j) = w_j$.

(v) The joint mass function for $(x_1, \ldots, x_n)$ is given by

$$p(x_1, \ldots, x_n) = \prod_{i=1}^{n} \frac{cw_{x_i} + \sum_{l<i} 1(x_l = x_i)}{c + i - 1}.$$

The denominator remains the same for all orderings. The numerator can be written as

$$\prod_{j=1}^{M} w_j\, (cw_j + 1) \cdots (cw_j + n_j - 1),$$

where $n_j = \sum_{i=1}^{n} 1(x_i = j)$; this equation also does not depend on the ordering of $\mathbf{x}_{1:n}$, so the sequence is exchangeable. Moreover,

$$\sum_{k=1}^{M} p(x_1, \ldots, x_i = k, \ldots, x_n) = p(\mathbf{x}_{-i}),$$

where $\mathbf{x}_{-i}$ is $\mathbf{x}_{1:n}$ without the ith component.

Further Reading

Historical Background. The DAR(1) family of models was developed to analyze discrete-valued time series. One of the original works on this family was McKenzie (1985), which introduced the well-known integer-valued autoregressive first-order model, denoted INAR(1). One key idea for such models is the thinning operator, which introduces randomness while maintaining the discrete nature of the data. Thinning is also a flexible approach to modeling the autocorrelation of a time series of counts.

Points of Interest.

1. Many time series models can be expressed using nonparametric-type models. In this exercise, the model is Dirichlet with an arbitrarily large number of dimensions. It is even possible to make the model infinite dimensional. The basic structure starts with the joint model
$$f(x, p) = f(x \mid p)\, f(p),$$
where $f(p)$ is a probability on p and p is a probability mass function or density function. Let x be a sample from p. Then, a first-order time series model can be constructed by
$$k(x' \mid x) = \int f(x' \mid p)\, f(dp \mid x),$$
where
$$f(p \mid x) = \frac{f(x \mid p)\, f(p)}{\int f(x \mid p)\, f(dp)}.$$
Indeed, many time series models can be understood in this way, particularly when $f(p \mid x)$ is conjugate with $f(p)$. See Pitt et al. (2002) for a number of such models.

2. To showcase the approach in the previous point, consider
$$f(x \mid \theta) = \mathrm{N}(x \mid \xi\theta, 1) \quad \text{and} \quad f(\theta) = \mathrm{N}(\theta \mid 0, 1),$$
which leads to
$$k(x' \mid x) = \mathrm{N}(x' \mid x\xi^2/(1+\xi^2), 1/(1+\xi^2))$$
and has stationary density $f(x) = \mathrm{N}(x \mid 0, 1/(1+\xi^2))$.

3. For a fully nonparametric model, consider the Dirichlet process model $f(p)$. Then, $\mathrm{E}(p) = q$ for some distribution q and

$$\mathrm{E}(p \mid x) = \frac{cq + 1_x}{c+1},$$

where 1_x is the single point mass function at x. Thus,

$$x' \begin{cases} \sim q & \text{with probability} \quad c/(1+c) \\ = x & \text{with probability} \quad 1/(1+c). \end{cases}$$

This is the formal definition of the DAR(1) model with stationary probability q; see Lawrance (1992) for this formal definition and similar models. Naturally, this construction is strongly connected with Bayesian models, which assume that observations $x_1, x_2, \ldots$ given p are independent and identically distributed from p. This leads to the exchangeability of the sequence of observations.

4. A time series model can also be constructed for the p:

$$k(p' \mid p) = \int \cdots \int f(p' \mid \mathbf{x}_{1:n}) \prod_{i=1}^{n} f(x_i \mid p)\, d\mathbf{x}.$$

The stationary model for p is precisely $f(p)$. An elegant transition model arises if p is a Dirichlet process, where the transitions $k(p' \mid p)$ are the number of atoms from p that get transferred to p'. For more on this model, see Fleming and Viot (1979), Ethier and Griffiths (1993), and Walker et al. (2007). Finally, Tavaré (1984) discussed the death process formed by the decrease in number of atoms from p_1 to p_t over t iterations.

S7.1.5 – Testing for dependence in a time series

(i) The stationary distribution can be derived recursively. First, consider $n = 1$:

$$x_1 = \underbrace{\rho x_0}_{\mathrm{N}(0,\rho^2)} + \underbrace{\sqrt{1-\rho^2} z_1}_{\mathrm{N}(0,1-\rho^2)},$$

where x_0 and z_1 are independent standard normal random variables. By the sum of independent normal random variables, it must be that $x_1 \sim \mathrm{N}(0, \rho^2 + 1 - \rho^2) = \mathrm{N}(0,1)$. Thus, x_1 has the same distribution as x_0: a standard normal distribution. Because the $\{z_n\}$ are independent and identically distributed, it is that marginally $x_n \sim \mathrm{N}(0,1)$ for all $n \geq 1$.

(ii) Denote the lag between observations i and j as $k = |i - j|$. Without loss of generality, assume that $j > i$, so $j = i + k$. The aim is to write x_j as a function of x_i such that the covariance between x_j and x_i can be written as a function of the variance of x_i. Thus,

$$\begin{aligned} x_j = x_{i+k} &= \rho x_{i+k-1} + \sqrt{1-\rho^2} z_{i+k} \\ &= \rho\left(\rho x_{i+k-2} + \sqrt{1-\rho^2} z_{i+k-1}\right) + \sqrt{1-\rho^2} z_{i+k} \\ &= \rho^2 x_{i+k-2} + \rho\epsilon_2 + \epsilon_1, \end{aligned}$$

where $\epsilon_l \equiv \sqrt{1-\rho^2}\, z_{i+k+1-l} \sim \mathrm{N}(0, 1-\rho^2)$ are independently distributed for $l = 1, \ldots, k$. Once again, use recursion to obtain

$$x_j = \rho^k x_i + \sum_{l=1}^{k} \rho^{l-1} \epsilon_l.$$

Thus, x_j is a function of x_i and

$$\begin{aligned} \mathrm{Cov}(X_i, X_j) &= \mathrm{Cov}\left(X_i, \rho^k X_i + \sum_{l=1}^{k} \rho^{l-1} \epsilon_l \right) \\ &= \mathrm{Cov}\left(X_i, \rho^k X_i \right) \\ &= \rho^k \,\mathrm{Var}(X_i), \end{aligned}$$

which uses the independence between the $\{\epsilon_l\}$ and x_i to obtain the second equality. Finally, $\mathrm{Cov}(X_i, X_j) = \rho^{|i-j|}$.

(iii) Using the definition of covariance:

$$\begin{aligned} \mathrm{Var}(S) &= \mathrm{Cov}\left(\sum_{i=1}^{n} X_i, \sum_{j=1}^{n} X_j \right) \\ &= \sum_{i=1}^{n} \sum_{j=1}^{n} \mathrm{Cov}(X_i, X_j) \\ &= \sum_{1 \le i,j \le n} \mathrm{Cov}(X_i, X_j) \\ &= \sum_{1 \le i,j \le n} \Sigma_{i,j}. \end{aligned}$$

(iv) The competing hypotheses imply a test for autocorrelation. The null hypothesis is that there is no autocorrelation, and the alternative hypothesis states there is (positive) autocorrelation. Under H_0, denote $S = \sum_{i=1}^{n} x_i$. Because $\rho = 0$,

$$x_i \sim \mathrm{N}(0,1), \quad i = 1, \ldots, n$$

independently, so $S \sim \mathrm{N}(0, n)$. Now, consider the test statistic $T = S/\sqrt{n} \sim \mathrm{N}(0,1)$. Thus, it is possible to leverage the readily available z-tables for standard normal random variables in hypothesis testing. Under H_1, denote $S_\rho = \sum_{i=1}^{n} x_i$. From part (iii), $\mathrm{Var}(S_\rho) = n + c_n$, where

$$c_n = \sum_{\substack{1 \le i,j \le n \\ i \ne j}} \Sigma_{i,j}.$$

From part (ii), $c_n > 0$ because $\Sigma_{i,j} > 0$ for all $i, j \in \{1, \ldots, n\}$, where $i \ne j$. Consider the statistic $T_\rho = S_\rho/\sqrt{n}$ to compare to T. In particular, compare the variances between T and T_ρ to test for autocorrelation, which are 1 and $1 + c_n/\sqrt{n}$, respectively. Although both T and T_ρ have zero means, the difference in magnitude of these statistics should reflect the discrepancy in variance. If $\rho > 0$, the magnitude of the observed test statistic should be large. Therefore, with an α level of significance, consider the critical region $\{T : |T| > z_{\alpha/2}\}$, where $z_{\alpha/2}$ is the critical value from the standard normal distribution for a two-tailed hypothesis with level of significance α.

(v) The Type II error is the probability of failing to reject H_0 when H_1 is true. Thus, to compute the Type II error, it is required to know the distribution of $T_{\rho=1/2}$. The variance of $S_{1/2}$ can be written as

$$\begin{aligned}
\text{Var}(S_{1/2}) &= n + \sum_{\substack{1\le i,j\le n \\ i\ne j}} \Sigma_{i,j} \\
&= n + 2 \sum_{1\le i<j\le n} (1/2)^{j-i} \\
&= n + 2 \sum_{k=1}^{n-1} (n-k)(1/2)^k \\
&= n + 2 \left[n \sum_{k=1}^{n-1} (1/2)^k - \sum_{k=1}^{n-1} k(1/2)^k \right] \\
&= n + 2 \left[n \left(1 - (1/2)^{n-1}\right) - \left(2 - (n+1)(1/2)^{n-1}\right)\right] \\
&= n + 2 \left[n - 2 + (1/2)^{n-1} \right] \\
&= 3n - 4 + (1/2)^{n-2},
\end{aligned}$$

which uses properties of well-known low-order polylogarithms to obtain the fifth equality. Thus, the variance of $T_{1/2}$ is

$$\begin{aligned}
\text{Var}(T_{1/2}) &= \text{Var}(S_{1/2})/n \\
&= 3 - 4/n + (1/2)^{n-2}/n,
\end{aligned}$$

so $T_{1/2} \sim \text{N}\left(0, 3 - 4/n + (1/2)^{n-2}/n\right)$. The Type II error when $\rho = \frac{1}{2}$ is the probability of failing to reject H_0 when H_1 is true:

$$\text{P}(-z_{\alpha/2} \le T_{1/2} \le z_{\alpha/2}) = \Phi\left(\frac{z_{\alpha/2}}{\sqrt{\text{Var}(T_{1/2})}}\right) - \Phi\left(\frac{-z_{\alpha/2}}{\sqrt{\text{Var}(T_{1/2})}}\right),$$

where Φ is the standard normal cumulative distribution function.

This is not a good test because it is typically desirable for the Type II error probability to go to 0; in this case, it does not.

Further Reading

Historical Background. Time series models assume a degree of autocorrelation. For example, AR(p) models assume that the current value depends on the previous p values in the series. Forecasting models, such as the autoregressive (integrated) moving average model, denoted AR(I)MA, rely on patterns detected through autocorrelation. Thus, whether a series has autocorrelation significantly affects model selection. A test for autocorrelation is crucial for identifying the appropriateness of these models (Hamilton, 2020).

Points of Interest.

1. There are better tests for autocorrelation than the one described in this exercise, such as the Durbin–Watson test (White, 1992) and the Ljung–Box Q test (Ljung and Box, 1978).

2. In general, autocorrelation in time series can be measured with the autocorrelation function (ACF), which considers the correlation between points separated by various time lags:
$$\rho_k = \frac{\sum_{t=1}^{T-k}(x_t - \overline{x})(x_{t+k} - \overline{x})}{\sum_{t=1}^{T}(x_t - \overline{x})^2},$$
where $t = 1, \ldots, T$, k is the time lag, and $\overline{x}$ is the mean of the observed time series (Shumway and Stoffer, 2000). The ACF at lag k can also be written as the covariance between x_t and x_{t-k} over the variance of x_t. A value of 1 implies perfect positive autocorrelation, and a value of -1 implies perfect negative autocorrelation; a value of 0 indicates no autocorrelation. Thus, the ACF is an alternative method to testing for autocorrelation, though it is not as principled as a formal hypothesis test, such as that found in this exercise. The ACF is typically calculated and plotted in a correlogram for a range of lags to visualize a trend in autocorrelation over time. The extent to which the ACF decreases as the lag increases indicates how long-lasting of an effect past values have on current values. A steep drop indicates that current values may only depend on the previous one or two values, and a gradual drop indicates that past values may have longer-lasting effects on the series. The Ljung–Box test is often used in conjunction with the ACF to test whether any group of autocorrelations are different from zero, thereby testing the general randomness of the series (Ljung and Box, 1978).

3. This exercise investigates a test for positive autocorrelation, which suggests that a positive (negative) value in the series will be followed by another positive (negative) value. However, one may be interested in testing for the presence of negative autocorrelation, where the sign of the current value in the series is flipped from the previous value (i.e., $\rho < 0$). The heuristic argument made in part (iv) would still apply; whether the autocorrelation is positive or negative, the impact that autocorrelation has is in the magnitude of the variance of S. A suitable follow-up question may concern a two-sided alternative hypothesis: $\rho \neq 0$. In such a case, the test statistic $T^2 = S^2/n$ may be considered. Under the null hypothesis, $T^2 \sim \chi_1^2$. Thus, the hypothesis test would concern the probability that the observed statistic falls in the tails of the χ_1^2 distribution. Interestingly, this is one of the rare occasions where a χ^2 hypothesis test naturally arises (Lobato et al., 2002).

4. The concept of autocorrelation extends well beyond the time series literature. Autocorrelation finds significant use in signal processing to analyze the properties of signals and to enhance signal detection (Zhang, 2022). Additionally, autocorrelation is studied with respect to the residuals of a model. Significant autocorrelation of these residuals may indicate that the model is not adequately capturing the dynamics of a series. Spatial autocorrelation is essential to spatial analysis and measures the degree to which a set of spatial features is correlated throughout space; a foundational principle in spatial statistics is Tobler's First Law of Geography, which suggests that things closer in space are more correlated (Cressie and Wikle, 2015). In image processing, spatial autocorrelation is used to understand the texture and identify regular patterns in images.

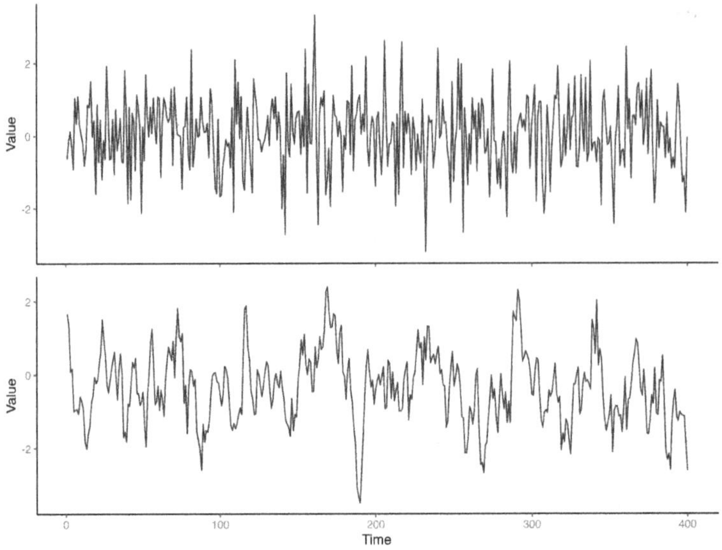

FIGURE 7.4
Trace plots of $\{x_n\}$ when $\rho = 0$ (upper) and $\rho = 0.8$ (lower).

Demonstration. Fig. 7.4 shows a realization of the process

$$x_n = \rho\, x_{n-1} + \sqrt{1-\rho^2}\, z_n, \quad n \geq 1,$$

where the $\{z_n\}$ and x_0 are independent standard normal random variables. In particular, two cases are shown: $\rho = 0$ (upper) and $\rho = 0.8$ (lower). The trace plots depict how the observations are independent when $\rho = 0$, but each realization is dependent on the previous one for $\rho = 0.8$.

S7.1.6 – The Kalman filter

(i) By Bayes' theorem,

$$\begin{aligned} p(x_n \mid \mathbf{y}_{1:n}) &\propto p(x_n, \mathbf{y}_{1:n}) \\ &\propto p(x_n, y_n, \mathbf{y}_{1:n-1}) \\ &\propto p(y_n \mid x_n, \mathbf{y}_{1:n-1}) p(x_n, \mathbf{y}_{1:n-1}) \\ &\propto p(y_n \mid x_n)\, p(x_n \mid \mathbf{y}_{1:n-1}), \end{aligned}$$

where the final line follows because y_n does not depend on $\mathbf{y}_{1:n-1}$ given x_n.

(ii) The aim is to rewrite $p(x_n \mid \mathbf{y}_{1:n-1})$ to set up a recursive equation for $p(x_n \mid \mathbf{y}_{1:n})$. To this end, consider

$$p(x_n \mid \mathbf{y}_{1:n-1}) = \int p(x_n \mid x_{n-1}) p(x_{n-1} \mid \mathbf{y}_{1:n-1}) dx_{n-1}.$$

From part (i),

$$p(x_n \mid \mathbf{y}_{1:n}) \propto p(y_n \mid x_n) \int p(x_n \mid x_{n-1}) p(x_{n-1} \mid \mathbf{y}_{1:n-1}) dx_{n-1},$$

which is a recursive expression for $p(x_n \mid \mathbf{y}_{1:n})$.

(iii) Note that $p(x_n \mid \mathbf{y}_{1:n-1})$ is a Gaussian distribution. Additionally,

$$\mathrm{E}(X_n \mid \mathbf{y}_{1:n-1}) = \mu_{n-1} \quad \text{and} \quad \mathrm{Var}(X_n \mid \mathbf{y}_{1:n-1}) = \phi_{n-1}^2 + \tau^2.$$

Therefore, $p(x_n \mid \mathbf{y}_{1:n-1}) = \mathrm{N}(x_n \mid \mu_{n-1}, \phi_{n-1}^2 + \tau^2)$.

(iv) By Bayes' theorem and a likelihood factorization,

$$\begin{aligned} p(x_n \mid \mathbf{y}_{1:n}) &\propto p(y_n \mid x_n)\, p(x_n \mid \mathbf{y}_{1:n-1}) \\ &\propto \mathrm{N}(y_n \mid x_n, \sigma^2)\, \mathrm{N}(x_n \mid \mu_{n-1}, \phi_{n-1}^2 + \tau^2), \end{aligned}$$

which yields $p(x_n \mid \mathbf{y}_{1:n}) = \mathrm{N}(x_n \mid \mu_n, \phi_n^2)$, where

$$\mu_n = \frac{y_n/\sigma^2 + \mu_{n-1}/(\tau^2 + \phi_{n-1}^2)}{1/\sigma^2 + 1/(\tau^2 + \phi_{n-1}^2)} \quad \text{and} \quad \phi_n^2 = \frac{1}{1/\sigma^2 + 1/(\tau^2 + \phi_{n-1}^2)}.$$

(v) Suppose $\{w_k, x_k^*\}_{k=1}^K$ represents a sample with weights from $p(x_{n-1} \mid \mathbf{y}_{1:n-1})$. Samples from $p(x_n \mid x_{n-1})$ can be obtained by sampling $\widetilde{x}_j$ from $\mathrm{N}(\cdot \mid x_k^*, \tau^2)$ with probability w_k for $j = 1, \ldots, K$. These $\{\widetilde{x}_j\}$ can then be weighted by $\widetilde{w}_j \propto \mathrm{N}(y_n \mid \widetilde{x}_j, \sigma^2)$. Thus, this process updates (w_k, x_k^*) to $(\widetilde{w}_j, \widetilde{x}_j)$, which is known as a particle filter.

Further Reading

Historical Background. This exercise explores prediction distributions for a hidden Markov model with Gaussian distributions. The original papers that developed this approach were Kalman (1960) and Kalman (1961). Thus, this process is named the Kalman filter.

Points of Interest.

1. Without the Gaussian assumption, the analytical estimation of the hidden Markov model is quite intractable. In such cases, both integration and the Bayes' theorem are needed to get the updates for $p(x_n \mid \mathbf{y}_{1:n})$. Hence, a non-conjugate framework will lead to difficulties. For example, Markov chain Monte Carlo methods struggle because they need to sample all hidden states; this is often done via a Metropolis algorithm, so the mixing can be poor. For such reasons, particle filters have become popular since the original work of Gordon et al. (1993). The idea is to estimate the predictive density at each juncture using a discrete distribution that can be updated quite easily. A common challenge is that the atoms are difficult to set, and only a few atoms maintain weights that are not converging to 0 in the long run. Thus, after certain intervals of time, some new atoms need to be introduced. These types of algorithms are now known as sequential Monte Carlo samplers (Del Moral et al., 2006).

Demonstration. In this illustration, the mathematically precise Kalman filter is compared with a particle filter. First, generate the latent process with $x_1 \sim \mathrm{N}(0,1)$ and $x_i = x_{i-1} + z_i$, where the $\{z_i\}$ are independent standard normal for $i = 2, \ldots, 1000$. Then, simulate the observed data $y_i = x_i + \sigma\varepsilon_i$, where $\sigma = \frac{1}{2}$ and the $\{\varepsilon_i\}$ are independent standard normal for $i = 2, \ldots, 1000$.

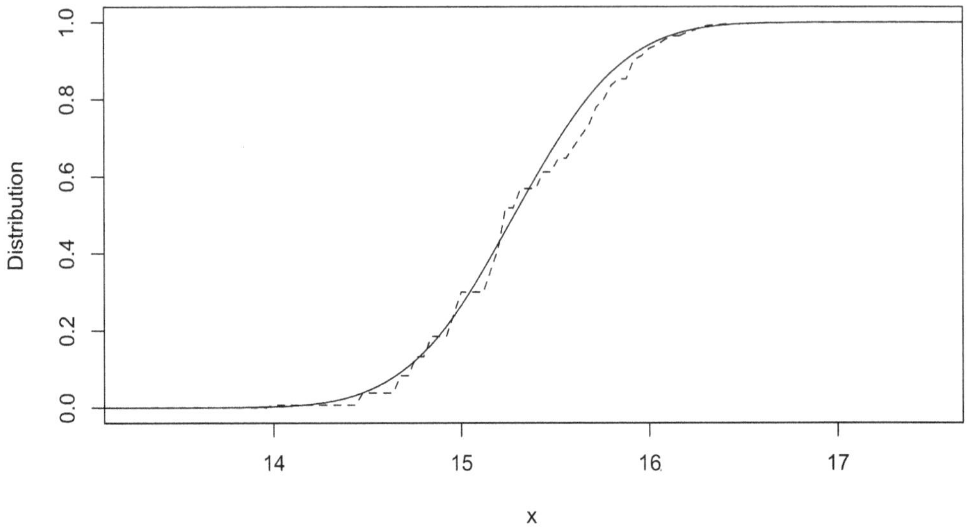

FIGURE 7.5
Predictive distributions from the Kalman filter (full line) and the particle filter (dashed line).

The Kalman filter computes μ_n and ϕ_n^2 using the updating equations in part (iii); this yields the normal distribution for predicting x_n given $\mathbf{y}_{1:n}$. Note that x_1 given y_1 is normally distributed with mean y_1 and variance $\frac{1}{2}$.

The particle filter has a set of atoms $\{\xi_j^m\}_{j=1:L}$ and weights $\{w_j^m\}_{j=1:L}$ at stage m. The set of atoms $\{\xi_j^{m+1}\}$ at stage $m+1$ are sampled from $p(\cdot \mid \xi_j^m)$ with probability w_j^m using a normal distribution with mean ξ_j^m and variance 1. The corresponding weights are given by

$$w_j^{m+1} = \frac{p(y_{m+1} \mid \xi_j^{m+1})}{\sum_{j=1}^{L} p(y_{m+1} \mid \xi_j^{m+1})}.$$

The atoms $\{\xi_j^n\}$ and weights $\{w_j^n\}$ can then be used to construct a distribution estimator of x_n given $\mathbf{y}_{1:n}$. The original samples $\{\xi_j^1\}$ come from a normal distribution with mean y_1 and variance $\frac{1}{2}$ with weights $w_j^1 = 1/L$ for $j = 1, \ldots, L$.

The distributions resulting from the Kalman and particle filters are presented in Fig. 7.5. The dashed line is the distribution using the particle filter, and the full line is the distribution using the Kalman filter. Evidently, the Kalman filter provides a more precise and smooth approximation of the normal distribution.

8

Markov Chain Monte Carlo

Despite pioneering work in the 1950s and 1970s, Bayesian analysis was infeasible due to difficulties associated with the study of posterior distributions, which were by and large mathematically intractable. However, the arrival of sampling-based approaches in the early 1990s – having missed key papers in the 1950s (Metropolis et al., 1953) and 1970s (Hastings, 1970) – rescued Bayesian analysis. Sampling approaches made the study of posterior distributions feasible and marked a massive shift in statistics – not because everyone believed that the Bayesian framework was the correct way to conduct statistical analysis, but because sampling became simpler than maximizing likelihood functions and finding corresponding confidence intervals. Through sampling, it became possible to obtain estimators and uncertainty quantification together.

Direct sampling approaches often use rejection sampling techniques; however, such techniques are infeasible for complex and high-dimensional models. An alternative approach is to set the posterior distribution as the stationary distribution of a well-constructed Markov chain. The combination of Markov chains and Monte Carlo sampling gave rise to the omnipresent class of Markov chain Monte Carlo (MCMC) algorithms.

The most common MCMC algorithm is the Gibbs sampler, where the target joint distribution is the stationary distribution. The Gibbs sampler obtains samples by iterating over sequences of conditional distributions. If the conditional distributions are not analytically tractable, one could use a Metropolis step; this algorithm proposes a candidate sample and uses a ratio of posteriors evaluated at the current sample and the proposed sample to determine whether to accept the proposal. Question 8.1.1 constructs a Gibbs sampler with a Metropolis step for a generalized linear model, where one of the parameters can not be sampled directly. Question 8.1.2 investigates how the Metropolis sampler works and how unique it is by comparing it to an alternative sampling procedure. The aim in Question 8.1.3 is to interpret the acceptance probability of a proposal using the Metropolis–Hastings algorithm, which is a generalization of the Metropolis algorithm that allows for asymmetric proposal distributions.

An active area of research for MCMC algorithms is the use of latent variables. Latent variables are often introduced to a model to simplify sampling by either avoiding a Metropolis step or making distributions tractable. Question 8.1.4 concerns the use of latent variables for a normal mixture model. Difficulties arise when there are an infinite number of components because the normalizing constant is intractable; numerous ideas have been proposed to tackle this problem while maintaining the correct target distribution. Question 8.1.5 considers an infinite mixture model, where the only unknown parameters are the weights attached to fixed components.

In addition to MCMC, other Monte Carlo techniques include importance sampling, Latin hypercube sampling, rejection sampling, and stratified sampling. Question 8.1.6 discusses how to perform Monte Carlo integration and, if multiple densities are available, how to choose which density to sample from. Then, the implications to Bayesian inference are discussed.

DOI: 10.1201/9781003493471-8

Q8.1 Questions – Markov Chain Monte Carlo

Q8.1.1 – Gibbs sampler for a Poisson regression model

Introduction. A Gibbs sampler is a MCMC algorithm that always moves. These Markov chains can be used to approximate joint and marginal distributions or compute integrals of interest, such as the expected value of random variables. This question uses a Gibbs sampler to estimate parameters in a Poisson regression model. Such models are useful for analyzing count data, rates, or ratios. This question also investigates the stationary distribution of the constructed Gibbs sampler and concludes with a discussion on model selection using the Markov chains.

Question. Consider the Poisson regression model

$$\mathrm{P}(y_i = k \mid x_i, \theta, \beta) = \frac{\left(\theta\, e^{\beta x_i}\right)^k}{k!} \exp\left(-\theta\, e^{\beta x_i}\right), \quad i = 1, \ldots, n,$$

for $k \in \{0, 1, 2, \ldots\}$. Assume that the prior for θ is $\mathrm{Ga}(a, b)$ and the prior for β is $\mathrm{N}(0, \sigma^2)$, where (σ, a, b) is specified.

(i) What is the posterior distribution for (θ, β)?

(ii) Find the conditional density functions for $\pi(\theta \mid \beta, \mathbf{y})$ and $\pi(\beta \mid \theta, \mathbf{y})$, where $\mathbf{y} = (y_1, \ldots, y_n)'$. What is the problem with sampling the latter density, and what solutions are available?

(iii) Describe a Gibbs sampler that approximately samples from the posterior. Additionally, show that the transition density for the chain has the posterior as the stationary density.

(iv) Denote the above model as M_1, and let M_2 be the same model but with $\beta = 0$ fixed. If the prior for each model is $\frac{1}{2}$ (i.e., $\mathrm{P}(M_1) = \mathrm{P}(M_2) = \frac{1}{2}$), provide a method to compute $\mathrm{P}(M_1 \mid \mathbf{y})$.

(v) Give a full interpretation of the probabilities for the models. This interpretation should include the cases when one of the models is correct and when neither is correct.

Q8.1.2 – Reversibility for the Metropolis algorithm

Introduction. MCMC methods are often used in Bayesian statistics to sample from otherwise intractable posterior distributions. The most common such methods have well-studied theoretical properties that make them appropriate for a large class of models; one such property is reversibility. This question investigates the conditions under which reversibility is satisfied for the Metropolis algorithm. Finally, an alternative algorithm is proposed and compared to the Metropolis algorithm.

Question. Suppose $\pi(x)$ is the target density for a Markov chain, and the transition density taking x to y is $p(y \mid x)$. A necessary condition for π to be the stationary density is that $\pi(y) = \int p(y \mid x)\, \pi(x)\, dx$.

(i) Show that the stationary condition is satisfied if the reversible condition

$$p(y \mid x)\, \pi(x) = p(x \mid y)\, \pi(y)$$

is satisfied for all x and y.

(ii) If $q(y \mid x)$ is an arbitrary conditional density, what condition must $\alpha(x,y)$ satisfy to ensure that

$$p(y \mid x) = \alpha(x,y)\, q(y \mid x) + (1 - r(x))1(y = x)$$

satisfies the reversible condition with respect to π? Here,

$$r(x) = \int \alpha(x,y)\, q(y \mid x)\, dy.$$

(iii) Show that the reversible condition is satisfied if

$$\alpha_M(x,y) = \min\left\{1, \frac{\pi(y)\, q(x \mid y)}{\pi(x)\, q(y \mid x)}\right\}.$$

(iv) An alternative α is given by

$$\alpha_B(x,y) = \frac{\pi(y)\, q(x \mid y)}{\pi(y)\, q(x \mid y) + \pi(x)\, q(y \mid x)}.$$

Show that $r_M(x) \geq r_B(x)$ for all x, where the subscript indicates which α to use for $r(x)$. What are the implications of this result?

(v) Consider taking some proposal y^* from $q(\cdot \mid x)$. Then, set $y = y^*$ with probability $\alpha(x, y^*)$; otherwise, set $y = x$. Show that y is a sample from $p(\cdot \mid x)$.

Q8.1.3 – Acceptance rate of a Metropolis–Hastings algorithm

Introduction. If a Markov chain is in equilibrium, the expected probability of moving is related to the distance between the proposal density function and the target density function. This question demonstrates such a connection for a chain arising from an independence Metropolis–Hastings algorithm, where the proposal distribution is independent of the current state of the chain. Notably, the probability of moving can be thought of as a measure of overlap between the proposal and target densities.

Question. The target density and the proposal density for an independence Metropolis–Hastings sampler are given by $\pi(y)$ and $q(y)$, respectively. Let x denote the current state of the chain and y denote the next state of the chain. The probability of accepting a proposed move to y using q is given by

$$\alpha = \min\left\{1, \frac{\pi(y)\,q(x)}{\pi(x)\,q(y)}\right\}.$$

(i) Show that the probability of accepting the proposed move is

$$r(x) = \pi(x)^{-1}\int \min\{q(y)\pi(x), q(x)\,\pi(y)\}\,dy$$

when the chain is in equilibrium.

(ii) Show that $|a-b| = a+b-2\min\{a,b\}$ for any a and b.

(iii) Show that the probability of moving from x is

$$r = 1 - \tfrac{1}{2}\int\int |q(y)\pi(x) - q(x)\,\pi(y)|\,dy\,dx$$

if x is distributed according to the target density.

(iv) Identify the distance

$$d(\pi, q) = \int\int |q(y)\pi(x) - q(x)\,\pi(y)|\,dy\,dx.$$

(v) Explain why r can be viewed as a measure of overlap between π and q, and provide an interpretation for it.

Q8.1.4* – MCMC for a Gaussian finite mixture model

Introduction. Mixture models are often used to describe populations in which different sub-populations are present. In such cases, it is intuitive to model the data as a combination of several components, where the components are specified to model the different sub-populations. This question considers a mixture model where each component has a normal form with different means and a common variance. From a Bayesian perspective, this Gaussian mixture model is best estimated using sampling-based methods, which often involve the use of latent indicator variables.

Question. Consider the mixture model

$$p(x \mid \boldsymbol{\theta}) = \sum_{j=1}^{M} w_j \mathrm{N}(x \mid \mu_j, \sigma^2),$$

where $\boldsymbol{\theta} = (\mu_1, \ldots, \mu_M, \sigma^2)'$ and M is finite and known. Assume a Dirichlet prior for $\mathbf{w} = (w_1, \ldots, w_M)'$ with common parameter $\alpha > 0$:

$$\pi(\mathbf{w}) \propto \prod_{j=1}^{M} w_j^{\alpha-1}\, 1(w_1 + \cdots + w_m = 1).$$

For each μ_j, consider a Gaussian prior with mean 0 and variance τ^2. Finally, assume a gamma prior for $\lambda = 1/\sigma^2$ with both the shape and scale parameters being $\epsilon > 0$.

Describe a Markov chain Monte Carlo algorithm using suitable latent variables for estimating the model parameters. The aim is to estimate the density from which the samples $\mathbf{x} = (x_1, \ldots, x_n)'$ arise. The Bayesian estimator is the predictive density function; define this function and show how the output of the chain can be used to estimate it.

Q8.1.5* – Estimating weights in an infinite mixture model

Introduction. Mixture models are commonly used in statistics and machine learning to capture complex data structures using a combination of simpler component distributions. The main idea is that the components represent sub-populations of the data. If the number of sub-populations is potentially infinite or unknown, infinite mixture models may be used, which allow for an unbounded number of components. An essential part of fitting (infinite) mixture models is the estimation of the component weights, which indicate the contribution of each component to the overall distribution of the data. This question presents a convenient Bayesian method to estimate decreasing component weights through the introduction of latent component indicator variables; this is a common model fitting technique in Bayesian nonparametrics.

Question. Consider a model for independent observations $\mathbf{y} = (y_1, \ldots, y_n)'$ of the form

$$f(y_i) = \sum_{j=1}^{\infty} w_j \, \phi_j(y_i),$$

where the weights $\{w_j\}$ sum to one and are decreasing and the $\{\phi_j\}$ are a set of known and mutually different density functions.

(i) Show that $\sum_{j=1}^{\infty} j\,(w_j - w_{j+1}) = 1$.

(ii) Show that there exists a random variable X defined on $\{1, 2, \ldots\}$ such that

$$w_j = \sum_{l \geq j} \mathrm{P}(X = l)/l.$$

(iii) By incorporating component indicator variables $\mathbf{d} = (d_1, \ldots, d_n)'$ and latent variables $\mathbf{x} = (x_1, \ldots, x_n)'$, show that the latent model

$$f(\mathbf{y}, \mathbf{d}, \mathbf{x}) = \prod_{i=1}^{n} p(x_i \mid \theta) \, \frac{1}{x_i} \, 1(d_i \leq x_i) \, \phi_{d_i}(y_i)$$

returns the correct marginal model for $\mathbf{y}$, where $p(x_i \mid \theta) = \mathrm{P}(X = x_i)$ is the model for x_i with parameter θ and $1(d_i \leq x_i)$ defines the support for d_i and x_i.

(iv) Assume the prior $\theta \sim \mathrm{Unif}(0, 1)$ and let

$$p(x_i \mid \theta) = (1 - \theta)\, \theta^{x_i - 1}, \quad i = 1, \ldots, n.$$

Is it possible to write the $\{w_j\}$ directly in terms of a computable function of θ?

(v) Describe a Gibbs sampler for estimating the posterior distribution of the weights.

Q8.1.6 – Monte Carlo integration

Introduction. Monte Carlo integration is a stochastic approach to approximating integrals. It is widely used in Bayesian statistics to approximate posterior distributions and expectations that do not have closed-form solutions. There are often several approaches to approximating an integral with Monte Carlo integration. This question compares the variance of two Monte Carlo estimators that vary in how the samples were obtained. The results can be generalized for many situations in which Monte Carlo integration is used.

Question. The aim of this exercise is to estimate the integral

$$I = \int g(x)\, h(x)\, dx$$

using Monte Carlo methods. Consider two Monte Carlo estimators

$$\widehat{I}_g = N^{-1} \sum_{i=1}^{N} h(x_i) \quad \text{and} \quad \widehat{I}_h = N^{-1} \sum_{i=1}^{N} g(y_i),$$

where the $\{x_i\}$ and $\{y_i\}$ are independent and identically distributed from g and h, respectively.

(i) Show that the variance of $\widehat{I}_g$ is smaller than that of $\widehat{I}_h$ if

$$\int h^2(x)\, g(x)\, dx < \int g^2(x)\, h(x)\, dx.$$

(ii) The chi-squared distance between density functions f and g is given by

$$d(f,g) = \int \{f^2(x)/g(x)\}\, dx - 1.$$

Show that $d(f,g) \geq 0$.

(iii) If I exists, explain why the density function $f(x) \propto g(x)\, h(x)$ also exists.

(iv) Show that the variance of $\widehat{I}_g$ is smaller than the variance of $\widehat{I}_h$ if $d(f,g) < d(f,h)$.

(v) Assume that $g \sim \mathrm{N}(0, \sigma_1^2)$, $h \sim \mathrm{N}(0, \sigma_2^2)$, and $\sigma_1^2 < \sigma_2^2$. Which density, h or g, should one sample from in order to estimate I?

S8.1 Solutions – Markov Chain Monte Carlo

S8.1.1 – Gibbs sampler for a Poisson regression model

(i) Using Bayes' theorem, the posterior distribution is

$$\begin{aligned}
\pi(\theta, \beta \mid \mathbf{y}, \mathbf{x}) &\propto \pi(\mathbf{y} \mid \theta, \beta)\pi(\theta)\pi(\beta)) \\
&\propto \prod_{i=1}^{n} \left\{ \frac{(\theta e^{\beta x_i})^k}{k!} \exp\left(-\theta e^{\beta x_i}\right) \right\} \theta^{a-1} e^{-\theta b} \exp\left(-\frac{1}{2\sigma^2}\beta^2\right) \\
&\propto \prod_{i=1}^{n} \left\{ \theta^k \exp(\beta k x_i) \right\} \exp\left(-\theta \sum_{i=1}^{n} e^{\beta x_i}\right) \theta^{a-1} e^{-\theta b} \exp\left(-\frac{1}{2\sigma^2}\beta^2\right) \\
&\propto \theta^{nk+a-1} \exp\left(\beta k \sum_{i=1}^{n} x_i - \theta \sum_{i=1}^{n} e^{\beta x_i} - \frac{1}{2\sigma^2}\beta^2 - \theta b\right),
\end{aligned}$$

where $\mathbf{y} = (y_1, \ldots, y_n)'$ and $\mathbf{x} = (x_1, \ldots, x_n)'$.

(ii) The full-conditional distribution for θ is

$$\pi(\theta \mid \beta, \mathbf{x}, \mathbf{y}) \propto \underbrace{\theta^{nk+a-1} \exp\left\{-\theta\left(\sum_{i=1}^{n} e^{\beta x_i} + b\right)\right\}}_{\text{kernel of Ga}\left(nk+a,\ \sum_{i=1}^{n} e^{\beta x_i}+b\right)},$$

which is a gamma density. The full-conditional distribution for β does not resemble a known density:

$$\pi(\beta \mid \theta, \mathbf{x}, \mathbf{y}) \propto \exp\left(\beta k \sum_{i=1}^{n} x_i - \theta \sum_{i=1}^{n} e^{\beta x_i} - \frac{1}{2\sigma^2}\beta^2\right).$$

Therefore, it will be difficult to sample directly. Instead, a Metropolis step may be used to indirectly sample from the full-conditional distribution.

(iii) From part (ii), samples for θ can be obtained directly from its full-conditional distribution. For β, consider the random-walk proposal $\beta^* \sim \mathrm{N}(\beta^{(k-1)}, \sigma_\beta^2)$, where $\beta^{(k)}$ is the state of the chain at iteration k and σ_β^2 is the tuning parameter for the random walk.

It is possible to show that the joint posterior $\pi(\theta, \beta \mid \mathbf{x}, \mathbf{y})$ is the stationary distribution:

$$\pi(\theta^*, \beta^* \mid \mathbf{x}, \mathbf{y}) = \int\int p(\theta^*, \beta^* \mid \theta, \beta)\pi(\theta, \beta \mid \mathbf{x}, \mathbf{y}) d\theta d\beta.$$

Note that $p(\theta^*, \beta^* \mid \theta, \beta) = \pi(\theta^* \mid \beta^*, \mathbf{x}, \mathbf{y}) p(\beta^* \mid \theta, \beta)$. Thus,

$$\begin{aligned}
\int\int p(\theta^*, \beta^* \mid \theta, \beta)\pi(\theta, \beta \mid \mathbf{x}, \mathbf{y}) d\theta d\beta \\
&= \int\int \pi(\theta^* \mid \beta^*, \mathbf{x}, \mathbf{y}) p(\beta^* \mid \theta, \beta)\pi(\theta, \beta \mid \mathbf{x}, \mathbf{y}) d\theta d\beta \\
&= \int\int \pi(\theta^* \mid \beta^*, \mathbf{x}, \mathbf{y}) p(\beta \mid \theta, \beta^*)\pi(\beta^* \mid \theta, \mathbf{x}, \mathbf{y})\pi(\theta \mid \mathbf{x}, \mathbf{y})\, d\theta\, d\beta \\
&= \int \pi(\theta^* \mid \beta^*, \mathbf{x}, \mathbf{y})\pi(\beta^*, \theta \mid \mathbf{x}, \mathbf{y}) d\theta \\
&= \pi(\theta^*, \beta^* \mid \mathbf{x}, \mathbf{y}),
\end{aligned}$$

where the second equality follows by the reversible condition:

$$p(\beta^* \mid \theta, \beta)\pi(\beta \mid \theta, \mathbf{x}, \mathbf{y}) = p(\beta \mid \theta, \beta^*)\pi(\beta^* \mid \theta, \mathbf{x}, \mathbf{y}).$$

Therefore, the joint posterior distribution is the stationary distribution.

(iv) The model M_2 is

$$\begin{aligned}
\mathrm{P}(y_i = k \mid x_i, \theta) &= \frac{\theta^k}{k!} \exp(-\theta), \\
\pi(\theta \mid \mathbf{x}, \mathbf{y}) &\equiv \mathrm{Ga}(\theta \mid nk + a, b + n).
\end{aligned}$$

To compute the posterior model probabilities, the two marginal likelihoods are needed: $p(M_k \mid \mathbf{y}) \propto p(\mathbf{y} \mid M_k)$ for $k = 1, 2$ because the prior probabilities for each model are $\frac{1}{2}$. The marginal $p(\mathbf{y} \mid M_1)$ is available explicitly because the model is conjugate. However, computing $p(\mathbf{y} \mid M_2)$ is difficult because the integral over β is not available in closed form. There are several ways to compute a marginal likelihood when direct integration is not available. Arguably, the most accurate approximation is

$$p(\mathbf{y} \mid M_2) = \frac{\pi(\beta^*, \theta^*) l(\beta^*, \theta^*)}{\pi(\beta^*, \theta^* \mid \mathbf{x}, \mathbf{y})} \tag{8.1}$$

for an arbitrary choice (θ^*, β^*) (i.e., those with high posterior values). The numerator comprises the prior and likelihood which are computable directly, and the denominator can be estimated using the output from the Markov chain for M_2. An alternative approach to compute the posterior model probabilities is to use a Metropolis sampler that jumps between the models; this relies on the reversible dynamics between the models to ensure that the target stationary density function is correctly sampled.

(v) Suppose the data comes from $f^*(y) = f(\mathbf{y} \mid \theta^*)$, and one is deciding between two models: $f_1(\mathbf{y} \mid \theta_1)$ and $f_2(\mathbf{y} \mid \theta_2)$. Generally, one should select the model that gets closest to f^*. The prior expresses the belief about which model can get closer to f^*. This is practically relevant because the parameters from each model will attempt to minimize

$$\int f^*(\mathbf{y}) \log\{f^*(\mathbf{y})/f_k(\mathbf{y} \mid \theta_k)\}\, d\mathbf{y},$$

which is the Kullback–Leibler divergence. If the models are nested, the probability assigned to the reduced model M_1 is the degree of belief that the extra parameter in M_2 does nothing to allow that model to get closer to f^*.

Further Reading

Historical Background. This question uses a Metropolis algorithm within a Gibbs sampler to estimate the parameters of a Poisson regression model. One of the full-conditional distributions was intractable and replaced by a Metropolis step. The reversible condition ensured that the posterior remained the stationary density of the chain. With the advent of reversible jump Metropolis algorithms, samplers were able to adopt target distributions that were historically infeasible to work with (Green, 1995). Notably, reversible jump Metropolis steps allowed MCMC algorithms to jump between models in a particular way that secured reversibility.

Points of Interest.

1. To understand reversible samplers that jump between models, consider using a standard Metropolis step to construct a target distribution. For example, suppose there are two models, M_1 and M_2, with parameters θ_1 and θ_2, respectively. The target posterior distribution is
$$p(k, \theta_k \mid \mathbf{y}), \quad k = \{1, 2\}.$$
This can be supplemented with the latent variable θ_{-k}, which is the parameter for the alternative model:
$$p(k, \theta_k \mid \mathbf{y})\, q(\theta_{-k} \mid \theta_k, \mathbf{y}).$$
Now, a Metropolis step can handle the moves between models because, for each k, there is a full set of parameters. For example, a proposed deterministic move from $k = 1$ to $k = 2$ is accepted with probability
$$\min\left\{1, \frac{p(2, \theta_2 \mid \mathbf{y})\, q(\theta_1 \mid \theta_2, \mathbf{y})}{p(1, \theta_1 \mid \mathbf{y})\, q(\theta_2 \mid \theta_1, \mathbf{y})}\right\};$$
note that this requires sampling the parameters for M_2 from $q(\theta_2 \mid \theta_1, \mathbf{y})$ first. A similar move exists from $k = 2$ to $k = 1$. Such a strategy can extend to model selection for any number of models:
$$p(k, \theta_k \mid \mathbf{y})\, q(\boldsymbol{\theta}_{-k} \mid \theta_k, \mathbf{y}),$$
where $\boldsymbol{\theta}_{-k}$ denotes the parameters for all other models. This representation simplifies the reversible jump Metropolis algorithm (Godsill, 2001). However, it is often difficult to ensure the reversibility in this algorithm, and Jacobians may be required for certain transformations (Hastie and Green, 2012).

2. An alternative approach is to compute the marginal likelihoods. Then, the only immediately unknown term is the posterior evaluated at (θ^*, β^*). If a Markov chain is run for parameters (θ, β) and the two conditional density functions are available, then the posterior at (θ^*, β^*) can be estimated using Monte Carlo approximations (Chib, 1995). Chib and Jeliazkov (2001) outlined the extension for the case when one of the conditional distributions needs to be sampled using a Metropolis step.

Demonstration. In this Poisson regression model, the most challenging part of sampling from the posterior is the conditional density for β. Consider the density function
$$\pi(\beta) \propto \mathrm{N}(\beta \mid \mu, \sigma^2)\, \exp\left\{-\sum_{i=1}^{n} e^{\beta x_i}\right\}.$$

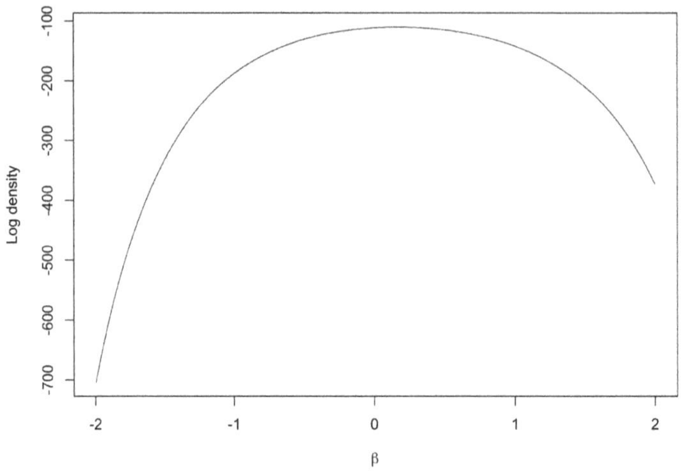

FIGURE 8.1
Illustration of the log-concave property of the density function for β.

A rejection sampler is typically difficult to implement. However, if the density function is log-concave, there is a fast adaptive rejection sampler that uses piecewise linear functions to approximate $\log \pi(\beta)$. To illustrate, consider independent $x_i \sim \mathrm{N}(0,1)$ for $i = 1,\ldots,100$, and let $\mu = \sum_{i=1}^n x_i$ and $\sigma^2 = 10$. The concavity of the log-density is seen in Fig. 8.1. Now, consider linear pieces approximating the function by placing lines just above the curve. The adaptive rejection sampler described in Gilks and Wild (1992) begins with a few lines and iteratively adds more until a sample is accepted. Note that a piecewise linear log-density is a mixture of exponential density functions, which is easy to sample.

S8.1.2 – Reversibility for the Metropolis algorithm

(i) If the reversible condition is satisfied, then $p(y \mid x)\pi(x) = p(x \mid y)\pi(y)$ for all x and y. Therefore,

$$\int p(y \mid x)\pi(x)dx = \int p(x \mid y)\pi(y)dx = \pi(y) \int p(x \mid y)dx = \pi(y),$$

which shows that the stationary condition is satisfied.

(ii) The reversible condition with respect to π is $p(y \mid x)\pi(x) = p(x \mid y)\pi(y)$. If $y = x$, the reversible condition is trivially satisfied. Now, consider the case when $y \neq x$. To satisfy the reversible condition,

$$\alpha(x,y)\, q(y \mid x)\pi(x)$$

must be symmetric in (x,y).

(iii) If

$$\alpha_M(x,y) = \min\left\{1, \frac{\pi(y)q(x \mid y)}{\pi(x)q(y \mid x)}\right\},$$

then

$$\alpha_M(x,y)\, q(y \mid x)\pi(x) = \min\{\pi(x)\, q(y \mid x), \pi(y)\, q(x \mid y)\};$$

this is symmetric in (x,y).

(iv) Note that

$$\min\{1, a/b\} \geq a/(a+b)$$

for all $a, b \geq 0$ because $a/b \geq a/(a+b)$ and $1 \geq a/(a+b)$. Therefore, $\alpha_M(x,y) \geq \alpha_B(x,y)$ for all x and y. Hence, $r_M(x) \geq r_B(x)$ for all x, implying that moves proposed by the Metropolis step will get accepted more often than moves proposed in the alternative approach with $\alpha_B(x,y)$.

(v) The aim is to sample from

$$p(y \mid x) = \alpha(x,y)\, q(y \mid x) + (1 - r(x))1(y = x).$$

To do this, one must know which component to sample from. Thus, an event is required with probability $r(x)$. Take y^* from $q(\cdot \mid x)$ and check if an independent uniform random variable is smaller than $\alpha(x, y^*)$; this event occurs with probability $r(x)$ and requires sampling from the density proportional to $\alpha(x,y)\, q(y \mid x)$. Such a sample is precisely y^*. If the required event does not occur, which arises with probability $1 - r(x)$, then take y to be x.

Further Reading

Historical Background. Reversible Markov chains are popular in Bayesian statistics because they are guaranteed to converge to the target distribution. The most well-known reversible sampler is the Metropolis algorithm. However, the Metropolis chain may get stuck or exhibit low acceptance rates for moves, leading to high autocorrelation. If the proposals are too close to the current state, then moves might get accepted more frequently, but at the cost of high autocorrelation.

Points of Interest.

1. Typically, Metropolis steps are implemented inside a Gibbs sampling framework. Consider the joint density $f(x,y)$. Using a Gibbs sampler, one moves from (x,y) to (x',y') with the transition kernel

$$k(x',y' \mid x,y) = f(x' \mid y')\, f(y' \mid x).$$

Note that $f(x,y)$ is the stationary density. If it is not possible to sample from $f(y' \mid x)$ directly, then one may sample from $p(y' \mid x,y)$, which satisfies

$$p(y' \mid x,y)\, f(y \mid x) = p(y \mid x,y')\, f(y' \mid x).$$

Then, it can be seen that

$$f(x',y') = \int k(x',y' \mid x,y)\, f(x,y)\, dx\, dy,$$

where

$$k(x', y' \mid x, y) = f(x' \mid y')\, p(y' \mid x, y).$$

For more details, see Tierney (1994), Gelman et al. (1995), Robert and Casella (2004), and Kalos and Whitlock (2009).

2. A useful application of reversible Markov chains is for Bayesian posterior sampling in which at least one of the variables to be sampled controls the dimension of the parameter space. For example, consider sampling from $p(k, \theta_1, \ldots, \theta_k)$, where the dimension of $\boldsymbol{\theta}$ depends on the value of k. Thus, a move from k to $k+1$ requires a θ_{k+1} that does not exist with the current state k. There are many applications where this joint density gets embedded into a larger model; as an example, consider a mixture model with an unknown number of components (Richardson and Green, 1997). Reversible jump MCMC algorithms are popular approaches for circumventing this problem (Green, 1995). Alternatively, a Gibbs sampler can be used to construct a joint probability model for $\boldsymbol{\theta}_{1:\infty}$. Consider the latent model
$$p(k, \boldsymbol{\theta}_{1:k})\, p(\boldsymbol{\theta}_{k+1:\infty} \mid \boldsymbol{\theta}_{1:k}).$$
Now, a standard Metropolis step can be used to propose moves between different values of k because there will be no corresponding change in dimension. Note that cancellations occur if
$$p(\boldsymbol{\theta}_{k+1:\infty} \mid \boldsymbol{\theta}_{1:k}) = \prod_{j>k} p(\theta_j \mid \theta_{j-1});$$
see Godsill (2001) for more details.

3. The Metropolis algorithm can be seen as an extension to rejection sampling. If $p(x) \propto \alpha(x)\, q(x)$ – where it is not possible to sample from p directly, but it is from q – and $0 < \alpha \leq 1$, then a rejection sampler takes x' from q (an independent uniform random variable u from $(0,1)$) and accepts x' from p if $u < \alpha(x')$. If rejected, then a new attempt is made. The Metropolis algorithm takes α to be a specific choice which, if the sample is rejected, allows for the current state of the chain to be the new state based on the set-up of the stationarity condition through the choice of α.

Demonstration. To compare the two reversible samplers using α_M and α_B, consider running two chains with the same target density, a standard normal, and the same proposal density, $q(\cdot \mid x) = \mathrm{N}(\cdot \mid x, h^2)$ with $h = \frac{1}{2}$. One chain moves with probability α_M and the other with probability α_B. The autocorrelation functions of the two samplers are shown in Fig. 8.2. The upper panel (α_M) has slightly smaller tails compared with the lower panel (α_B). Thus, the Metropolis sampler has less autocorrelation than the alternative approach for larger lags.

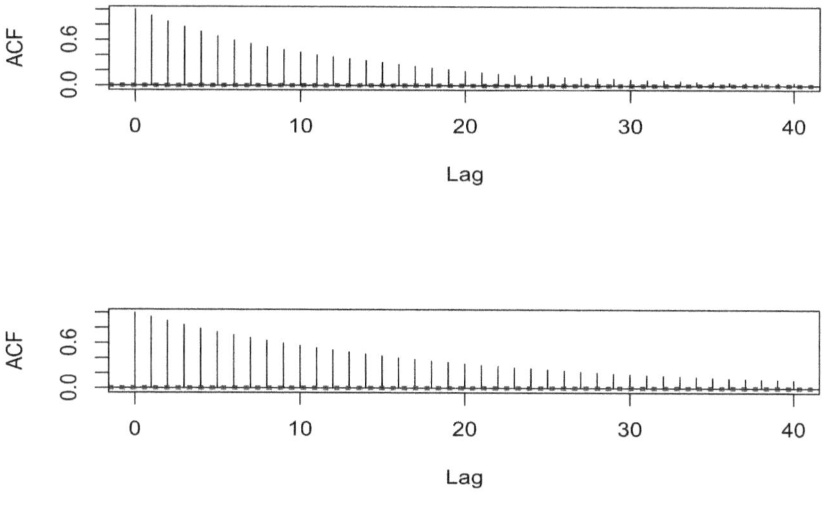

FIGURE 8.2
Autocorrelation functions for α_M (upper) and α_B (lower).

S8.1.3 – Acceptance rate of a Metropolis–Hastings algorithm

(i) At equilibrium, x is a sample from π, so the probability of moving from x is

$$\int \min\left\{1, \frac{\pi(y)\,q(x)}{\pi(x)\,q(y)}\right\} q(y)\,dy = \int \min\left\{q(y), \frac{\pi(y)q(x)}{\pi(x)}\right\} dy$$
$$= \pi(x)^{-1} \int \min\{q(y)\pi(x), q(x)\pi(y)\}dy.$$

(ii) This is a very useful equality when dealing with absolute values. If $a > b$, then $|a - b| = a - b = a + b - 2b$ and $b = \min\{a, b\}$. A similar case arises when $a < b$.

(iii) Now, there is interest in the probability of moving from any state to any state while the chain is in equilibrium. Therefore, consider

$$r = \int\int \min\left\{1, \frac{\pi(y)q(x)}{\pi(x)q(y)}\right\} \pi(x)q(y)dy\,dx$$
$$= \int\int \min\{\pi(x)\,q(y), \pi(y)\,q(x)\}\,dx\,dy.$$

From part (ii), r can be written as

$$r = \tfrac{1}{2}\int\int \pi(x)q(y) + \pi(y)q(x) - |\pi(x)q(y) - \pi(y)q(x)|dx\,dy$$
$$= \tfrac{1}{2}\int\int \pi(x)q(y)dx\,dy + \tfrac{1}{2}\int\int \pi(y)q(x)dx\,dy - \tfrac{1}{2}\int\int |\pi(x)q(y) - \pi(y)q(x)|dx\,dy,$$

where the integrals in the first two terms are equal to 1.

(iv) This is the L_1 distance between the two density functions $p_1(x,y) = q(y)\,\pi(x)$ and $p_2(x,y) = q(x)\,\pi(y)$. Therefore, it satisfies the conditions for being a distance. Note that $p_1 \equiv p_2$ only when $q \equiv \pi$.

(v) Note that r measures how close q and π are to each other because $1 - r = \frac{1}{2}d$, where d measures their distance. If $\pi \equiv q$, then $r = 1$ and the densities overlap in full. If $r = 0$, then π and q have no common points for which they are both positive.

Further Reading

Historical Background. The study of acceptance rates for the Metropolis–Hastings algorithm is based on the works of Metropolis et al. (1953) and Hastings (1970). More recent work has focused on the development of efficient Metropolis algorithms. For example, a series of papers by Andrew Gelman, Walter Gilks, and Gareth Roberts concern the efficiency of Metropolis updates (Gilks and Roberts, 1996; Gelman et al., 1996; Roberts et al., 1997). Although such developments are relatively recent, measures of overlap date back at least to Pearson (1895).

Points of Interest.

1. If two populations are represented by π and q, then r measures the probability that a sample from q can replace a sample from π, and vice versa. Thus, r represents the proportion of either population that could represent a member of the other population. To elaborate, suppose that there are two populations, and the hypothesis is that these populations are the same. The usual assessment of this hypothesis would use a p-value. However, a measure of overlap would provide the probability that a sample from one population could be accepted as a sample from the other. If the measure of overlap is 0, then no member of either population can be mistaken as a member of the other population; there is no overlap in their support. On the other hand, if the measure of overlap is 1, then the populations are the same.

2. Rom and Hwang (1996) used overlaps as a means of assessing the similarity of medical treatments. As a result, the comparative trial became a common method for assessing the efficacy between two or more treatments. In particular, Rom and Hwang (1996) considered the standard overlap coefficient

$$\mathrm{OVL}(p,q) = \int \min\{p(x), q(x)\}\, dx$$

between densities p and q; this coefficient is also known as the proportion of similar responses (PSR) and is related to the L_1 distance between p and q. Today, the overlap coefficient is a popular measure in medical data analysis (Lei and Olson, 2010; Giacoletti and Heyse, 2015). It is also used in ecology to measure niche overlap (Mason et al., 2011) and in economics to measure the overlap in income distributions (Weitzman, 1970) and the polarization between groups (Anderson, 2010).

This exercise suggests that the overlap measure

$$\int\int \min\{p(x)\, q(y), p(y)\, q(x)\}\, dx\, dy$$

can have a clearer interpretation.

S8.1.4* – MCMC for a Gaussian finite mixture model

A likelihood for the mixture model

$$p(x_i \mid \boldsymbol{\theta}) = \sum_{j=1}^{M} w_j \mathrm{N}(x_i \mid \mu_j, \sigma^2)$$

contains a product of sums and, consequently, would be difficult to work with. Fortunately, the mixture model can be expressed as the augmented model

$$\begin{aligned} p\left(x_i \mid z_i = j, \boldsymbol{\theta}\right) &\overset{ind}{=} \mathrm{N}\left(x_i \mid \mu_j, \sigma^2\right) \\ \mathrm{P}(z_i = j \mid \mathbf{w}) &= w_j, \end{aligned}$$

where the $\mathbf{z} = (z_1, \ldots, z_n)'$ are latent indicator variables. Now, denote $\boldsymbol{\theta} = (\boldsymbol{\mu}, w, \sigma^2)'$, where $\mathbf{w}$ and $\boldsymbol{\mu}$ are M-dimensional vectors. The complete data likelihood can be written as

$$\begin{aligned} p(\mathbf{x}, \mathbf{z} \mid \boldsymbol{\theta}) &= \prod_{i=1}^{n} p\left(x_i, z_i \mid \boldsymbol{\theta}\right) \\ &= \prod_{i=1}^{n} \left\{p\left(x_i \mid z_i, \boldsymbol{\theta}\right) p\left(z_i \mid \boldsymbol{\theta}\right)\right\} \\ &= \prod_{i=1}^{n} w_{z_i} \mathrm{N}(x_i \mid \mu_{z_i}, \sigma^2). \end{aligned}$$

This augmented model yields a product likelihood, which is easier to handle than a product of sums. Now, the joint posterior distribution is

$$\begin{aligned} \pi(\boldsymbol{\mu}, \mathbf{w}, \mathbf{z}, \sigma^2 \mid \mathbf{x}) &\propto p(\mathbf{x}, \mathbf{z} \mid \boldsymbol{\mu}, \mathbf{w}, \sigma^2) \pi(\mathbf{w}) \pi(\boldsymbol{\mu}) \pi(\sigma^2) \\ &= \left\{\prod_{i=1}^{n} w_{z_i} \mathrm{N}(x_i \mid \mu_{z_i}, \sigma^2)\right\} \pi(\mathbf{w}) \pi(\boldsymbol{\mu}) \pi(\sigma^2). \end{aligned}$$

A straightforward Gibbs sampler can be implemented by isolating the corresponding full-conditional density functions:

1. $\mathrm{P}(z_i = j \mid \cdots) \propto p(x_i \mid \mu_j, \sigma^2, z_i = j) p(z_i = j) = \mathrm{N}(x_i \mid \mu_j, \sigma^2) w_j$. Hence, it is seen that $p(z_i \mid \cdots) = \mathrm{Mult}(1, \boldsymbol{\pi}_i)$ with $\boldsymbol{\pi}_i = (\pi_{i,1}, \ldots, \pi_{i,M})'$ and

$$\pi_{i,j} = \frac{w_j \mathrm{N}(x_i \mid \mu_j, \sigma^2)}{\sum_{j=1}^{M} w_j \mathrm{N}(x_i \mid \mu_j, \sigma^2)}.$$

2. $\pi(\mathbf{w} \mid \cdots) \propto \prod_{j=1}^{M} w_j^{\alpha-1} \cdot w_j^{n_j}$; thus, $\pi(\mathbf{w} \mid \cdot) = \mathrm{Dir}(\mathbf{w} \mid \alpha + n_1, \ldots, \alpha + n_M)$, where $n_j = \#\{z_i = j\}$ is the number of observations belonging to the jth component.

3. $\pi(\mu_j \mid \cdots) \propto p(\mu_j) \prod_{z_i = j} p(x_i \mid \mu_j, \sigma^2) = \mathrm{N}(\mu_j \mid 0, \tau^2) \prod_{z_i = j} \mathrm{N}(x_i \mid \mu_j, \sigma^2)$. By conjugacy, $\pi(\mu_j \mid \cdots) = \mathrm{N}(\mu_j \mid \mu_{nj}, \tau_{nj}^2)$ with

$$\tau_{nj}^2 = \left(\frac{1}{\tau^2} + \frac{1}{\sigma^2}\right)^{-1} \quad \text{and} \quad \mu_{nj} = \tau_{nj}^2 \left(\frac{\mu_0}{\tau^2} + \frac{n_j \overline{y}_j}{\sigma^2}\right),$$

where $\overline{y}_j$ is the mean across the (y_i) for which $z_i = j$.

4. Finally,

$$\pi(\sigma^2 \mid \cdots) \propto p(\sigma^2) \prod_{j=1}^{M} \prod_{z_i=j} p\left(x_i \mid \mu_j, \sigma^2\right)$$
$$= \left(\frac{1}{\sigma^2}\right)^{\epsilon-1} \exp\left(-\frac{1}{\sigma^2}\epsilon\right) \left(\frac{1}{\sigma^2}\right)^{N/2} \exp\left[-\frac{1}{2\sigma^2}\left\{\sum_{j=1}^{M} \sum_{z_i=j} (x_i - \mu_j)^2\right\}\right],$$

which implies that the full-conditional distribution for $\lambda = 1/\sigma^2$ is

$$\text{Ga}\left(\lambda \mid \epsilon + \tfrac{1}{2}N, \epsilon + \tfrac{1}{2} \sum_{j=1}^{M} \sum_{z_i=j} (x_i - \mu_j)^2\right).$$

A Gibbs sampler iteratively samples from the full-conditional distributions described in 1-4. Once the chain has been sampled, estimates of the posterior parameters can be obtained. The predictive density is defined as

$$p(x_{n+1} \mid \mathbf{x}_{1:n}) = \int p(x_{n+1}, \boldsymbol{\theta} \mid \mathbf{x}_{1:n}) d\boldsymbol{\theta} = \int p(x_{n+1} \mid \boldsymbol{\theta}) p(\boldsymbol{\theta} \mid \mathbf{x}_{1:n})\, d\boldsymbol{\theta}.$$

This can be estimated via a Monte Carlo approximation that uses samples from the Markov chain:

$$\widehat{p}(x_{n+1} \mid \mathbf{x}_{1:n}) = \frac{1}{K} \sum_{k=1}^{K} \sum_{j=1}^{M} w_j^{(k)}\, \text{N}(x_{n+1} \mid \mu_j^{(k)}, \sigma^{2(k)}),$$

where the (k) superscript denotes the kth of K iterations of the Markov chain.

Further Reading

Historical Background. Finite mixture models appear in many different forms because they are used to accomplish many different tasks (Everitt, 2014). For example, some use mixture models to provide a large class of density functions for capturing a possibly unusual dataset. Alternatively, a number of researchers use mixture models to identify groups or clusters of the data; this presumes that a cluster is best represented by a single component, but it may be that one cluster requires several components to adequately capture a skewed or heavy-tailed density.

Points of Interest.

1. This question explores the most popular mixture model: the Gaussian mixture model. Posterior inference is facilitated by an introduction of latent indicator variables and obtained via a Gibbs sampler. The latent indicator variables are crucial and maintain the correct model. See Tanner and Wong (1987) for sampling strategies based on data augmentation; their approach is known to provide a recurrent chain when dealing with mixture models (Marin et al., 2005).

2. The algorithm provided in the solution is not the only available approach for estimating the model parameters. For example, the expectation–maximization (EM) algorithm can be used to maximize the likelihood function. Additionally, parameter estimation for mixture models can be executed with moment matching, spectral methods, or graphical techniques (Tarter and Lock, 1993).

3. A more general mixture model allows for $M = \infty$ and encompasses the well-known mixture of Dirichlet process models (Escobar and West, 1995). However, the parameters for this model have notable issues regarding identifiability (Ferguson, 1983). In particular, the indicator variables have an infinite support and the normalizing constant is intractable. Thus, clever procedures are needed to maintain a valid chain with the correct stationary distribution.

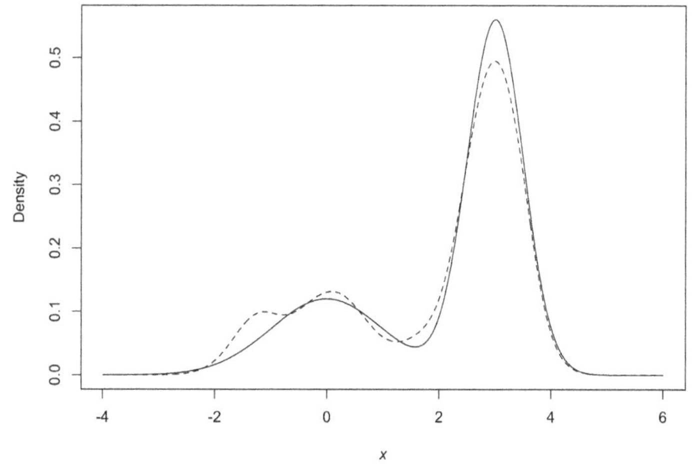

FIGURE 8.3
The predictive density function (dashed line) and the true density (full line).

Demonstration. A sample of size $n = 100$ is taken from the model

$$p(x) = 0.3\,\mathrm{N}(x \mid 0, 1) + 0.7\,\mathrm{N}\left(x \mid 3, (1/2)^2\right).$$

Though only two components are used for data generation, consider fitting a mixture model with $M = 10$ components to the data. Assume a Dirichlet prior for the weights with common parameter 1 and a gamma prior for $\lambda = 1/\sigma^2$ with parameters $(\frac{1}{2}, \frac{1}{2})$. Finally, let the prior for the normal location parameters be independent and normal with mean 0 and variance 10^2. The Gibbs sampler from this exercise is run for 1000 iterations. At each iteration, the predictive density function with the current parameters is saved. In Fig. 8.3, the average of the predictive density functions over all iterations (dashed line) is plotted alongside the true density (full line).

Note that one component has a standard deviation of 1 and the other of $\frac{1}{2}$, yet a common σ is used for all components. As a result, the posterior accumulated at the smallest of the two (Fig. 8.4). This occurred because several normal components combined to produce the component with standard deviation 1; this is possible. What is not possible is producing a component with standard deviation $\frac{1}{2}$ if the common standard deviation is estimated to be 1.

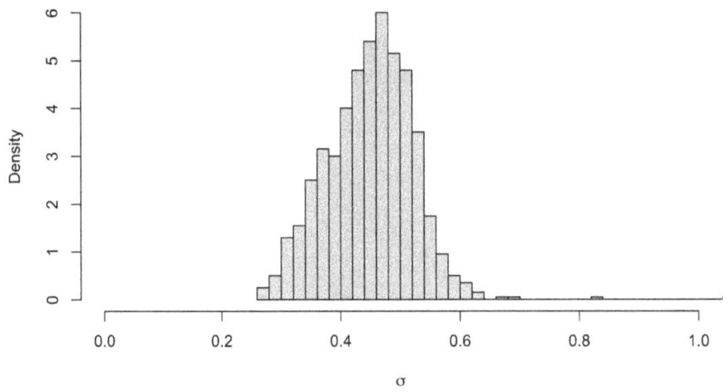

FIGURE 8.4
Posterior distribution for σ, which accumulated at the smallest standard deviation of the two components.

S8.1.5* – Estimating weights in an infinite mixture model

(i) Use the fact that the weights $\{w_j\}$ sum to 1:

$$\begin{aligned}\sum_{j=1}^{\infty} j(w_j - w_{j+1}) &= \sum_{j=1}^{\infty} jw_j - \sum_{j=1}^{\infty} jw_{j+1} \\ &= \sum_{j=1}^{\infty} jw_j - \sum_{j=1}^{\infty} (j-1)w_j \\ &= \sum_{j=1}^{\infty} \{jw_j - (j-1)w_j\} \\ &= \sum_{j=1}^{\infty} w_j = 1.\end{aligned}$$

(ii) Substitute the given expression for w_j into the result from part (i):

$$\sum_{j=1}^{\infty} j(w_j - w_{j+1}) = \sum_{j=1}^{\infty} j \underbrace{\left[\sum_{l \geq j} \mathrm{P}(X = l)/l - \sum_{l \geq j+1} \mathrm{P}(X = l)/l\right]}_{a}.$$

The only term left in a is the term for $l = j$, so $a = \mathrm{P}(X = j)/j$. Thus,

$$\sum_{j=1}^{\infty} j(w_j - w_{j+1}) = \sum_{j=1}^{\infty} \mathrm{P}(X = j) = 1,$$

using the result from part (i). Therefore, the cumulative probability of x taking on a value in $\{1, 2, \ldots\}$ is 1. Because x is implicitly defined on $\{1, 2, \ldots\}$, there exists a random variable x on $\{1, 2, \ldots\}$.

(iii) The marginal model $f(y_i)$ can be obtained by summing the latent model $f(y_i, x_i, d_i)$ over all possible values for x_i and d_i. Note that the component indicator variables d_i must be in $\{1, 2, \ldots\}$ and x_i must have the same support. Without loss of generality, consider $i = 1$:

$$\begin{aligned} f(y_1) &= \sum_{d_1=1}^{\infty} \sum_{x_1=1}^{\infty} f(y_1, d_1, x_1) \\ &= \sum_{d_1=1}^{\infty} \sum_{x_1=1}^{\infty} p(x_1 \mid \theta) \frac{1}{x_1} 1(d_1 \le x_1) \phi_{d_1}(y_1) \\ &= \sum_{d_1=1}^{\infty} \left(\sum_{x_1 \ge d_1} \frac{1}{x_1} \mathrm{P}(X = x_1) \right) \phi_{d_1}(y_1) \\ &= \sum_{d_1=1}^{\infty} w_{d_1} \phi_{d_1}(y_1), \end{aligned}$$

which is the correct marginal model for $f(y_1)$.

(iv) Now,

$$w_j = \sum_{l \ge j} \mathrm{P}(X = l)/l = \sum_{l \ge j} \frac{1}{l} (1 - \theta)\theta^{l-1}$$

is a function of θ. Generally, an infinite sum is not ideal for computation. Fortunately, this sum can be split into an infinite component from 1 to ∞ and a finite component, where the former is an infinite series with a closed-form expression and the latter is more computationally feasible than an infinite summation. For $j > 1$, write

$$w_j = \sum_{l=1}^{\infty} \frac{1}{l} (1 - \theta)\theta^{l-1} - \sum_{l=1}^{j-1} \frac{1}{l} (1 - \theta)\theta^{l-1}.$$

The first term can be manipulated to resemble an infinite series with a closed-form expression: $\sum_{k=1}^{\infty} z^k/k = -\log(1 - z)$. In particular,

$$\begin{aligned} \sum_{l=1}^{\infty} \frac{1}{l} (1 - \theta)\theta^{l-1} &= \frac{1 - \theta}{\theta} \sum_{l=1}^{\infty} \frac{\theta^l}{l} \\ &= -\frac{1 - \theta}{\theta} \log(1 - \theta). \end{aligned}$$

Therefore, w_j can be written as a function of θ with a finite summation:

$$w_j(\theta) = -\frac{1 - \theta}{\theta} \log(1 - \theta) - \frac{1 - \theta}{\theta} \sum_{l=1}^{j-1} \theta^l / l.$$

(v) First, note that the weights $\{w_{d_i}\}$ are a deterministic function of $\{x_i\}$ and $\{d_i\}$, which need to be sampled. Given the joint latent model $f(y_i, d_i, x_i)$, the full-conditional distributions for x_i and d_i are

$$\begin{aligned} p(x_i \mid \cdots) &\propto p(x_i \mid \theta) \frac{1}{x_i} 1(d_i \le x_i) \\ &\propto \theta^{x_i - 1} \frac{1}{x_i} 1(d_i \le x_i), \\ p(d_i \mid \cdots) &\propto 1(d_i \le x_i) \phi_{d_i}(y_i), \end{aligned}$$

where $p(x_i \mid \cdots)$ denotes the distribution of x_i conditioned on everything else in the model. The normalizing constant for $p(x_i \mid \cdots)$, denoted as c, can be obtained from

$$\frac{1}{c} = \sum_{x_i=d_i}^{\infty} \theta^{x_i-1}/x_i = -\frac{\log(1-\theta)}{\theta} - \frac{1}{\theta}\sum_{x_i=1}^{d_i-1} \theta^{x_i}/x_i,$$

where the intermediate steps for this computation are similar to the steps in part (iv). Thus, the full-conditional distribution for x_i is

$$p(x_i \mid \cdots) = \frac{c\theta^{x_i-1}}{x_i} 1(d_i \leq x_i),$$

which can be sampled from directly. Note that $p(x_i \mid \cdot)$ depends on θ, so θ must be updated in the Gibbs sampler. The full-conditional distribution for θ is

$$\begin{aligned} p(\theta \mid \cdots) &\propto \left(\prod_{i=1}^{n} p(x_i \mid \theta)\right) \pi(\theta) \\ &\propto \prod_{i=1}^{n} (1-\theta)\theta^{x_i-1} \\ &\propto (1-\theta)^n \theta^{\sum_{i=1}^{n} x_i - n}, \end{aligned}$$

which is the kernel of a Beta$(\sum_{i=1}^{n} x_i - n + 1, n+1)$ distribution.

The full-conditional distribution for each d_i is proportional to the terms containing d_i in the known density functions ϕ_{d_i} truncated above by the x_i. Therefore, the d_i can be sampled directly because the possible values for each d_i are finite. Finally, each w_{d_i} can be deterministically computed given x_i and d_i.

Further Reading

Historical Background. The first version of the infinite mixture model appeared in Lo (1984). However, Bayesian inference for the model was infeasible until sampling algorithms started to appear several years later. The breakthrough work of Escobar (1988) effectively used a Gibbs sampler to sample from the posterior parameters in an infinite mixture model. Since the early 1990s, many more algorithms have been proposed to sample from these models.

Points of Interest.

1. Mixture models are often used to model heterogeneity when the data comprise distinct sub-populations or groups because each component distribution can be thought of as representing one of these groups. These models are commonly used for handling multi-modal data or for clustering, where each component corresponds to a cluster in which data points are assigned. Infinite mixture models are generally adopted when the number of groups, modes, or clusters is unknown, and they frequently find application in biology, finance, image analysis, and nature language processing. Although the infinite mixture model may accommodate an infinite number of components, typically only a finite number of components have substantial weights.

2. The marginal model $f(y_i)$ intuitively defines the mixture model mechanism: $f(y_i)$ is a mixture of distinct components $\{\phi_j\}$ with component weights $\{w_j\}$. However, the likelihood $f(\mathbf{y})$ is rarely written as

$$f(\mathbf{y}) = \prod_{i=1}^{n} \sum_{j=1}^{\infty} w_j \phi_j(y_i)$$

because fitting the product of a sum is computationally prohibitive and may result in analytically intractable estimators for the component weights. Fortunately, a joint latent likelihood can be used in place of the marginal data model. The introduction of the latent variables naturally follows because there is often a presumed component membership d_i associated with each data point y_i. Additionally, it is common to define w_{d_i} as a function of latent variables (i.e., x_i and d_i) that are easier to update. A classic example of the introduction of latent variables to improve computational feasibility is found in probit regression (Albert and Chib, 1993).

3. Because infinite mixture models contain an infinite number of component weights, it is natural to wonder how one might model all of them. The most prevalent estimation scheme for an infinite number of weights makes use of the Dirichlet process, a stochastic process that generalizes the Dirichlet distribution to an infinite-dimensional parameter space. In Bayesian nonparametrics, the Dirichlet process is often used as the prior distribution on the component weights (Hjort et al., 2010). The weights are then estimated through the stick-breaking process, where one can imagine breaking a stick of unit length into an infinite number of pieces; here, the broken pieces represent the weight of each mixture component. Despite the Dirichlet process mixture model (DPMM) being the most common choice within Bayesian nonparametrics, there are a number of possible ways to handle the weights. For example, this question defines the weights as a deterministic function of latent random variables $\{x_i\}$ and $\{d_i\}$.

4. In a standard infinite mixture model, the only assumption on the component weights is that they sum to 1. Without placing more structure on the weights, the model may be unidentifiable. In this question, the weights are identifiable because they are assumed to be decreasing. Hatjispyros et al. (2023) discussed the importance of weight ordering for model identifiability and convergence. In particular, they considered the inferential implications of a geometric process mixture model, which is a simpler alternative to DPMMs that assumes the weights are decreasing. Such an assumption improves identifiability and does not decrease the flexibility of infinite mixture models. Without this assumption, the estimation of the infinite component indicator variables is problematic due to the lack of an analytically tractable normalizing constant. By assuming that the weights decrease, the analytical tractability of the $\{x_i\}$ and $\{d_i\}$ are exchanged, so the component indicator variables may be easily sampled, as shown in part (iv). Note that the difficulty in estimating the $\{x_i\}$ imposed by the decreasing weight assumption is circumvented because the $\{x_i\}$ are not used to estimate the component weights $\{w_j\}$ in part (v).

5. The approach in this exercise can be used to sample from many probability mass functions for which the normalizing constant is unknown. For example, consider

$$p(d) = \pi(d)\, \tau^d, \quad d \in \{1, 2, \ldots\},$$

where $0 < \tau < 1$ and the sum

$$\sum_{d=1}^{\infty} \pi(d) \tau^d$$

is not available. The lack of a normalizing constant renders direct sampling from $p(d)$ awkward. As seen in this exercise, one possible remedy is to introduce the latent variable k such that the joint mass function with d is

$$p(d,k) = (1-\tau)\,\pi(d)\,\tau^k\,1(d \le k).$$

It is easy to check that the marginal mass function for d is $p(d)$. It can also be seen that the two conditional mass functions

$$p(k \mid d) \propto 1(k \ge d)\,\tau^k \quad \text{and} \quad p(d \mid k) \propto \pi(d)\,1(d \le k)$$

are easy to sample from directly.

6. There is an entire field of statistical inference called Bayesian nonparametrics that concerns itself with infinite-dimensional parameter spaces. Introductions to Bayesian nonparametrics include Hjort et al. (2010) and Müller et al. (2015).

S8.1.6 – Monte Carlo integration

(i) The variance of $\widehat{I}_g$ is

$$\begin{aligned}\operatorname{Var}(\widehat{I}_g) &= \frac{1}{N^2}\sum_{i=1}^{N}\operatorname{Var}(h(X_i))\\ &= \frac{1}{N}\left\{\int h^2(x)g(x)dx - \left(\int h(x)g(x)dx\right)^2\right\}.\end{aligned}$$

Similarly,

$$\operatorname{Var}(\widehat{I}_h) = \frac{1}{N}\left\{\int g^2(x)h(x)dx - \left(\int g(x)h(x)dx\right)^2\right\}.$$

Therefore, $\operatorname{Var}(\widehat{I}_g) < \operatorname{Var}(\widehat{I}_h)$ if $\int h^2(x)\,g(x)\,dx < \int g^2(x)\,h(x)\,dx$.

(ii) Because $g(x)$ and $f(x)$ are density functions,

$$\int \frac{\{f(x)-g(x)\}^2}{g(x)}dx \ge 0.$$

Therefore,

$$\begin{aligned}\int \frac{\{f(x)-g(x)\}^2}{g(x)}dx &= \int \left\{\frac{f^2(x)}{g(x)} - 2f(x) + g(x)\right\}dx\\ &= \int \frac{f^2(x)}{g(x)}dx - 1 \ge 0.\end{aligned}$$

(iii) Note that $I = \int g(x)h(x)dx$ is the normalizing constant for the density function $f(x)$, i.e.,

$$f(x) = \frac{g(x)h(x)}{\int g(x)h(x)dx}.$$

Therefore, if $I < \infty$ exists, then $f(x)$ also exists.

(iv) The chi-squared distance between f and g is

$$d(f,g) = \int \frac{f^2(x)}{g(x)} dx - 1 = \frac{1}{I^2} \int \frac{g^2(x)h^2(x)}{g(x)} dx - 1 = \frac{1}{I^2} \int g(x)h^2(x)dx - 1.$$

Similarly,

$$d(f,h) = \frac{1}{I^2} \int g^2(x)h(x)dx - 1.$$

If $d(f,g) < d(f,h)$, then

$$\int g(x)h^2(x)dx < \int g^2(x)h(x)dx,$$

and so by part (i), the variance of $\widehat{I}_g$ must be smaller than that of $\widehat{I}_h$.

(v) Now,

$$f(x) \propto \exp\left(-\frac{x^2}{2\sigma_1^2}\right) \cdot \exp\left(-\frac{x^2}{2\sigma_2^2}\right) = \exp\left\{-\frac{x^2}{2}\left(\frac{1}{\sigma_1^2} + \frac{1}{\sigma_2^2}\right)\right\},$$

which is the kernel of a normal distribution with mean 0 and variance $\sigma^2 = \sigma_1^2\sigma_2^2/(\sigma_1^2 + \sigma_2^2)$. Because $\sigma^2 < \sigma_1^2 < \sigma_2^2$, it must be that $d(f,g) < d(f,h)$. Therefore, the best estimator for I is $\widehat{I}_g$ because the variance is smaller. Thus, one should sample from g.

Further Reading

Historical Background. The first recorded Monte Carlo integration was used to solve Buffon's needle problem in 1777 (Schuster, 1974). The underlying idea was to estimate fixed quantities using random mechanisms; historically, this was done using the law of large numbers. Further reading on Monte Carlo methods can be found in Hammersley and Handscomb (2013) and Rubinstein and Kroese (2016).

Many modern Monte Carlo methods often make use of Markov chains; one such method is aptly named Markov chain Monte Carlo, which is fundamental enough to receive an entire chapter in this text. The work of Stanisław Ulam and John von Neumann in the 1940s is often credited as the first Markov chain Monte Carlo method (Robert and Casella, 2004).

Points of Interest.

1. It is worth thinking about how to optimally conduct Monte Carlo integration when several approaches are available. For example, consider estimating the marginal probability of the data in a Bayesian linear regression model:

$$m(\mathbf{y}) = \int \mathrm{N}(\mathbf{y} \mid \mathbf{X}\boldsymbol{\beta}, \sigma^2\mathbf{I})\pi(\boldsymbol{\beta})d\boldsymbol{\beta},$$

where π is the prior density function on the p-dimensional regression coefficients $\boldsymbol{\beta}$. Assume that σ^2 is known. Typically, Monte Carlo integration is achieved by sampling from the prior $\pi(\boldsymbol{\beta})$ and then weighting the samples using the likelihood function. That is, sample $\boldsymbol{\beta}_j \sim \pi(\boldsymbol{\beta})$ independently for $j = 1, \ldots, M$, and estimate the integral with

$$\widehat{m}(\mathbf{y}) = M^{-1} \sum_{j=1}^{M} \mathrm{N}(\mathbf{y} \mid \mathbf{X}\boldsymbol{\beta}_j, \sigma^2 \mathbf{I}).$$

Alternatively, one could sample the $\{\boldsymbol{\beta}_j\}$ independently from the density proportional to the likelihood:

$$\pi^*(\boldsymbol{\beta}) \propto \exp\left\{-\tfrac{1}{2}(\boldsymbol{\beta} - \widehat{\boldsymbol{\beta}})'(\mathbf{X}'\mathbf{X})^{-1}(\boldsymbol{\beta} - \widehat{\boldsymbol{\beta}})/\sigma^2\right\}.$$

Now, the estimator for the marginal likelihood is

$$\widehat{m}(\mathbf{y}) = c(\sigma, |\mathbf{X}'\mathbf{X}|, p) \exp\left\{-\tfrac{1}{2}\mathbf{y}'(\mathbf{I} - \mathbf{H})\mathbf{y}/\sigma^2\right\} M^{-1} \sum_{j=1}^{M} \pi(\boldsymbol{\beta}_j),$$

where c is a constant and $\mathbf{H}$ is the usual hat matrix. The variance for this marginal likelihood may have better properties than the one obtained by sampling from the prior because the variability from the prior may be larger if the prior is assigned a large variance; note that priors are often assigned large variances to be "vague" or "non-informative."

2. A simpler, univariate version of the idea in the previous point is to estimate

$$m(y) = \int \mathrm{N}(y \mid \theta, \sigma^2)\, \pi(\theta)\, d\theta,$$

where σ is known and of order $1/\sqrt{n}$. Then,

$$\widehat{m}(y) = \pi(y) + \tfrac{1}{2}\sigma^2\, \pi''(y) + o(1/n).$$

This estimator is often more accurate than estimators arising from Laplace approximations (Tierney and Kadane, 1986). The following demonstration uses this model to illustrate the point.

Demonstration. The aim is to estimate

$$m(y) = \int \mathrm{N}(y \mid \theta, \sigma^2)\, \pi(\theta)\, d\theta,$$

where $\pi(\theta)$ is the standard Cauchy density and $\sigma = 0.1$. Fig. 8.5 illustrates two Monte Carlo estimators of size $M = 100$. The dashed line is the Monte Carlo estimator based on sampling from the Cauchy density (the prior) and weighting according to the normal density, and the full line is the Monte Carlo estimator based on sampling from the normal density (the likelihood) and weighting according to the Cauchy density. The latter provides a better approximation, which can be verified via a large Monte Carlo sample. Thus, one should not always sample from the prior when conducting Monte Carlo integration.

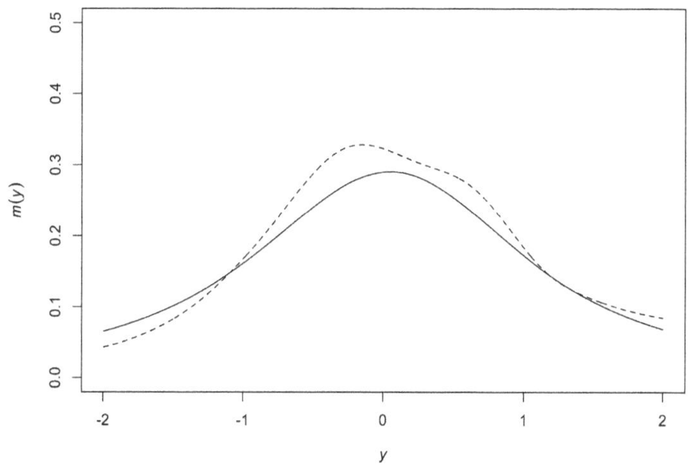

FIGURE 8.5
A comparison of two Monte Carlo estimators for the marginal density $m(y)$. The Monte Carlo estimator sampling from the Cauchy (prior) is the dashed line, and the Monte Carlo estimator sampling from the normal (likelihood) is the full line. The full line is a closer representation of the true $m(y)$, which can be verified via a large Monte Carlo sample.

Notation

Notation	Description
$\mathbb{R}$	Real numbers
$\mathbb{N}$	Natural numbers
n	Sample size
$n!$	Factorial
$X_1, \ldots, X_n$ or $\{X_i\}_{i=1}^n$	Random variables
$\overline{X}$	Population mean
σ^2	Population variance
$x_1, \ldots, x_n$ or $\{x_i\}_{i=1}^n$	Observed values of random variables
$\mathbf{x} = (x_1, \ldots, x_K)'$	Column vector
$\overline{x}$	Sample mean
$f(\cdot)$	Probability density/mass function
$F(\cdot)$	Cumulative distribution function
$X \sim f$	X has density f
θ	Parameter
θ^*	True value of a parameter
$\mathrm{E}(X)$	Expectation of random variable X
$\mathrm{Var}(X)$	Variance of random variable X
$\mathrm{Cov}(X, Y)$	Covariance between X and Y
$l(\cdot)$	Likelihood function
$L(\cdot)$	Log-likelihood function
$\mathcal{L}(\cdot)$	Likelihood ratio
$m(\cdot)$	Marginal likelihood
$\phi(\cdot)$	Laplace transform
$I(\cdot)$	Fisher information
$1(\cdot)$	Indicator function
H_0, H_1	Null, alternative hypotheses
$\Phi(\cdot)$	CDF for $\mathrm{N}(0,1)$
z_α	z-score at level α
$x_{(k)}$	k-th ordered statistic
$X_n = o(a_n)$	$X_n/a_n \to 0$ as $n \to 0$
$X_n = O(a_n)$	$\|X_n/a_n\|$ is bounded for large n
$I_k(\kappa)$	Modified Bessel function of the first kind of order k

Abbreviations	Description
AIC	Akaike information criterion
AR(p)	Autoregressive model of order p
a.s.	Almost surely
CDF	Cumulative distribution function
CLT	Central limit theorem
DAR	Discrete autoregressive model
EM algorithm	Expectation–maximization algorithm
KL divergence	Kullback–Leibler divergence
LSE	Least squares estimator
MCMC	Markov chain Monte Carlo
MSE	Mean squared error
MLE	Maximum likelihood estimation
PDF	Probability density function
PMF	Probability mass function
UMP	Uniformly most powerful

Symbols	Name	Support	PDF/PMF	Mean	Variance
$\mathrm{Bern}(p)$	Bernoulli	$\{0,1\}$	$f(x) = p^x(1-p)^{(1-x)}$	p	$p(1-p)$
$\mathrm{Bin}(n,p)$	Binomial	$\{0,1,\ldots,n\}$	$f(x) = \binom{n}{x}p^x(1-p)^{(n-x)}$	np	$np(1-p)$
$\mathrm{Ca}(x_0,\gamma)$	Cauchy	$\mathbb{R}$	$f(x) = \left[\pi\gamma\left\{1+\left(\frac{x-x_0}{\gamma}\right)^2\right\}\right]^{-1}$	undefined	undefined
χ^2_p	Chi-squared	$[0,\infty)$	$f(x) = \Gamma^{-1}(p/2)2^{-p/2}x^{p/2-1}e^{-x/2}$	p	2p
$\mathrm{Dir}(\alpha_1,\ldots,\alpha_K)$	Dirichlet	$K-1$ simplex	$f(x) = \frac{\Gamma(\alpha_0)}{\prod_{i=1}^K\Gamma(\alpha_i)}\prod_{i=1}^K x_i^{\alpha_i-1},\ \alpha_0 = \sum_{i=1}^K\alpha_i$	$\mathrm{E}(X_i) = \frac{\alpha_i}{\alpha_0}$	$\mathrm{Var}(X_i) = \frac{\alpha_i/\alpha_0(1-\alpha_i/\alpha_0)}{\alpha_0+1}$
$\mathrm{Exp}(\theta)$	Exponential	$[0,\infty)$	$f(x) = \frac{1}{\theta}e^{-x/\theta}$	θ	θ^2
$\mathrm{Fisk}(\theta)$	Fisk ($\alpha = 1$)	$[0,\infty)$	$f(x) = \frac{\theta x^{\theta-1}}{(1+x^\theta)^2}$	$\frac{\pi/\theta}{\sin(\pi/\theta)}(\theta > 1)$	(see online)
$\mathrm{Ga}(a,b)$	Gamma	$(0,\infty)$	$f(x) = \frac{b^a}{\Gamma(a)}x^{a-1}e^{-bx}$	a/b	a/b^2
$\mathrm{N}(\mu,\sigma^2)$	Gaussian	$\mathbb{R}$	$f(x) = \frac{1}{\sigma\sqrt{2\pi}}\exp\left\{-\frac{1}{2\sigma^2}(x-\mu)^2\right\}$	μ	σ^2
$\mathrm{Laplace}(\mu,b)$	Laplace	$\mathbb{R}$	$f(x) = \frac{1}{2b}\exp(-\lvert x-\mu\rvert/b)$	μ	$2b^2$
$\mathrm{Pois}(\lambda)$	Poisson	$\mathbb{N}$	$f(x) = \lambda^x e^{-\lambda}/x!$	λ	λ
t_ν	Student–t	$\mathbb{R}$	$f(x) = \frac{\Gamma(\frac{\nu+1}{2})}{\sqrt{\pi\nu}\Gamma(\frac{\nu}{2})}(1+\frac{x^2}{\nu})^{-\frac{\nu+1}{2}}$	$0\ (\nu > 1)$	$\frac{\nu}{\nu-2}\ (\nu > 2)$
$\mathrm{vM}(\mu,\kappa)$	von Mises	$[a-\pi, a+\pi),\ a\in\mathbb{R}$	$f(x) = \frac{\exp\{\kappa\cos(x-\mu)\}}{2\pi I_0(\kappa)}$	μ	$1 - I_1(\kappa)/I_0(\kappa)$

Bibliography

Aalen, O. (1978). Nonparametric inference for a family of counting processes. *Annals of Statistics*, 6:701–726.

Akaike, H. (1974). A new look at the statistical model identification. *IEEE Transactions on Automatic Control*, 19:716–723.

Al-Osh, M. A. and Aly, E. A. A. (1992). First order autoregressive time series with negative binomial and geometric marginals. *Communications in Statistics, Series A*, 21:2483–2492.

Albert, J. H. and Chib, S. (1993). Bayesian analysis of binary and polychotomous response data. *Journal of the American Statistical Association*, 88(422):669–679.

Aldous, D. J. (1985). Exchangeability and related topics. In *École d'Été de Probabilités de Saint-Flour XIII — 1983*, volume 1117 of *Lecture Notes in Mathematics*, pages 1–198. Springer, Berlin.

Aldrich, J. (1997). R. A. Fisher and the making of maximum likelihood (1912-1922). *Statistical Science*, 12(3):162–176.

Amari, S. (1972). Learning patterns and pattern sequences by self-organizing nets of threshold elements. *IEEE Transactions on Computers*, 100(11):1197–1206.

Anderson, G. (2010). Polarization of the poor: Multivariate relative poverty measurement sans frontiers. *Review of Income and Wealth*, 56(1):84–101.

Apostol, T. M. (1991). *Calculus*, volume 1. John Wiley & Sons.

Arnak, D. and Olivier, C. (2012). Wilks phenomenon and penalized likelihood ratio tests for nonparametric curve registration. In *Proceedings of the 15th International Conference on Artificial Intelligence and Statistics*, pages 264–272.

Arnold, B. C., Balakrishnan, N., and Nagaraja, H. N. (2008). *A First Course in Order Statistics*. Society for Industrial and Applied Mathematics.

Asmussen, S. (2003). *Applied Probability and Queues*. Applications of Mathematics: Stochastic Modelling and Applied Probability. Springer.

Athreya, K. B. and Lahiri, S. N. (2006). *Measure Theory and Probability Theory*, volume 19. Springer, New York.

Athreya, K. B. and Ney, P. E. (2004). *Branching Processes*. Courier Corporation.

Barron, A., Schervish, M. J., and Wasserman, L. (1999). The consistency of posterior distributions in nonparametric problems. *The Annals of Statistics*, 27(2):536–561.

Barron, A. R. (1988). The exponential convergence of posterior probabilities with implications for Bayes estimators of density functions. Technical Report 7, University of Illinois, Urbana-Champaign.

Bartlett, M. S. (1957). A comment on D. V. Lindley's statistical paradox. *Biometrika*, 44:533–534.

Battiti, R. (1992). First- and second-order methods for learning: Between steepest descent and Newton's method. *Neural computation*, 4(2):141–166.

Bayarri, M. J. et al. (2012). Criteria for Bayesian model choice with application to variable selection. *Annals of Statistics*, 40:1550–1577.

Beran, R. (1977). Minimum Hellinger distance estimators for parametric models. *Annals of Statistics*, 5:445–463.

Berger, J. O. (2013). *Statistical Decision Theory and Bayesian Analysis*. Springer Science & Business Media.

Berti, P., Pratelli, L., and Rigo, P. (2004). Limit theorems for a class of identically distributed random variables. *Annals of Probability*, 32:2029–2052.

Bhattacharyya, A. (1943). On a measure of divergence between two statistical populations defined by their probability distribution. *Bulletin of the Calcutta Mathematical Society*, 35:99–110.

Bickel, P. J. and Li, B. (2006). Regularization in statistics. *Test*, 15(2):271–344.

Billingsley, P. (1968). *Convergence of Probability Measures*. Wiley Series in Probability and Statistics. Wiley, New York, second edition.

Bock, R. D. and Aitkin, M. (1981). Marginal maximum likelihood estimation of item parameters: Application of an EM algorithm. *Psychometrika*, 46(4):443–459.

Broniatowski, M. (2021). Minimum divergence estimators, maximum likelihood and the generalized bootstrap. *Entropy*, 23:185.

Bruck, J. (1990). On the convergence properties of the Hopfield model. *Proceedings of the IEEE*, 78:1579–1585.

Casella, G. and Berger, R. L. (2021). *Statistical Inference*. Cengage Learning.

Chandra, T. K. (2012). *The Borel–Cantelli Lemma*. Springer.

Chapman, D. G. and Robbins, H. (1951). Minimum variance estimation without regularity assumptions. *The Annals of Mathematical Statistics*, 22(4):581–586.

Chatterjee, A. and Lahiri, S. N. (2011). Strong consistency of lasso estimators. *Sankhya A*, 73:55–78.

Chernoff, H. (1954). On the distribution of the likelihood ratio. *The Annals of Mathematical Statistics*, 1:573–578.

Chib, S. (1995). Marginal likelihood from the Gibbs output. *Journal of the American Statistical Association*, 90(432):1313–1321.

Chib, S. and Jeliazkov, I. (2001). Marginal likelihood from the Metropolis–Hastings output. *Journal of the American Statistical Association*, 96:270–281.

Choy, S. T. B. and Smith, A. F. M. (1997). On robust analysis of a normal location parameter. *Journal of the Royal Statistical Society: Series B (Statistical Methodology)*, 59:463–474.

Christensen, R. (2002). *Plane Answers to Complex Questions*. Springer.

Codling, E., Plank, M., and Benhamou, S. (2008). Random walks in biology. *Journal of the Royal Society Interface*, 5:813–34.

Costa, S. I. R., Santos, S. A., and Strapasson, J. E. (2015). Fisher information distance: A geometrical reading. *Discrete Applied Mathematics*, 197:59–69.

Cox, D. R. (1972). Regression models and life–tables. *Journal of the Royal Statistical Society, Series B*, 34:187–202.

Cox, D. R. and Isham, V. I. (1980). *Point Processes*. Chapman & Hall, London.

Cramer, H. (1946). *Mathematical Methods of Statistics*. Princeton University Press.

Cressie, N. and Wikle, C. K. (2015). *Statistics for Spatio-Temporal Data*. John Wiley & Sons.

Curry, H. B. (1994). The method of steepest descent for non-linear minimization problems. *Quarterly of Applied Mathematics*, 2:258–261.

Darmois, G. (1935). Sur les lois de probabilité à estimation exhaustive. *Comptes rendus de l'Académie des Sciences*, 260(1265):85.

David, H. A. and Nagaraja, H. N. (2004). *Order Statistics*. John Wiley & Sons.

de Moivre, A. (1738). *The Doctrine of Chances: or, a Method for Calculating the Probabilities of Events in Play*. Woodfall.

Del Moral, P., Doucet, A., and Jasra, A. (2006). Sequential Monte Carlo samplers. *Journal of the Royal Statistical Society: Series B (Statistical Methodology)*, 68:411–436.

Dempster, A. P., Laird, N. M., and Rubin, D. B. (1977). Maximum likelihood from incomplete data via the EM algorithm. *Journal of the Royal Statistical Society, Series B*, 39:1–22.

Diaconis, P. and Stroock, D. (1991). Geometric bounds for eigenvalues of Markov chains. *Annals of Applied Probability*, 1(1):36–61.

Diaconis, P. and Ylvisaker, D. (1979). Conjugate priors for exponential families. *Annals of Statistics*, 7:269–281.

Diffey, S. M. et al. (2017). A new REML (parameter expanded) EM algorithm for linear mixed models. *Australian & New Zealand Journal of Statistics*, 59(4):433–448.

Doob, J. L. (1940a). Application of the theory of martingales. In *Le calcul des probabilités et ses applications*, CNRS International Colloquia, pages 23–27. Centre National de la Recherche Scientifique, Paris.

Doob, J. L. (1940b). Regularity properties of certain families of chance variables. *Transactions of the American Mathematical Society*, 47(3):455–486.

Doob, J. L. (1953). *Stochastic Processes*. Wiley, New York.

Dormann, C. F. et al. (2013). Collinearity: A review of methods to deal with it and a simulation study evaluating their performance. *Ecography*, 36(1):27–46.

Dudley, R. M. (2014). *Uniform Central Limit Theorems*, volume 142. Cambridge University Press.

Dudley, R. M., Giné, E., and Zinn, J. (1991). Uniform and universal Glivenko–Cantelli classes. *Journal of Theoretical Probability*, 4(3):485–510.

Durrett, R. (1999). *Essentials of Stochastic Processes.* Springer.

Efron, B. (2012). Bayesian inference and the parametric bootstrap. *The Annals of Applied Statistics*, 6(4):1971.

Efron, B. (2022). *Exponential Families in Theory and Practice.* Cambridge University Press.

Escobar, M. D. (1988). *Estimating the Means of Several Normal Populations by Nonparametric Estimation of the Distribution of the Means.* PhD thesis, Yale University.

Escobar, M. D. and West, M. (1995). Bayesian density estimation and inference using mixtures. *Journal of the American Statistical Association*, 90:577–588.

Etemadi, N. (1981). An elementary proof of the strong law of large numbers. *Zeitschrift für Wahrscheinlichkeitstheorie und verwandte Gebiete*, 55:119–122.

Ethier, S. N. and Griffiths, R. C. (1993). The transition function of a Fleming–Viot process. *Annals of Probability*, 21:1571–1590.

Everitt, B. S. (2014). Finite mixture distributions. In *Wiley StatsRef: Statistics Reference Online*. Wiley.

Fan, J., Zhang, C.-H., and Zhang, J. (2001). Generalized likelihood ratio statistics and Wilks phenomenon. *Annals of Statistics*, 29:153–193.

Feller, W. (1966). *An Introduction to Probability Theory and Its Applications*, volume 1. Wiley.

Ferguson, T. S. (1983). Bayesian density estimation by mixtures of normal distributions. In *Recent Advances in Statistics*, pages 287–302. Academic Press, New York.

Finch, H. (2022). Regularized methods for generalized linear models. In *Applied Regularization Methods for the Social Sciences*. Chapman and Hall/CRC.

Fischer, H. (2011). *A History of the Central Limit Theorem: From Classical to Modern Probability Theory*. Springer.

Fisher, R. A. (1922). On the mathematical foundations of theoretical statistics. *Philosophical Transactions of the Royal Society of London: Series A*, 222(594-604):309–368.

Fleming, W. H. and Viot, M. (1979). Some measure valued Markov processes in population genetics theory. *Indiana University Mathematics Journal*, 28:817–843.

Fong, E., Holmes, C., and Walker, S. G. (2024). Martingale posterior distributions. *Journal of the Royal Statistical Society Series B: Statistical Methodology*, 85(5):1357–1391.

Fudenberg, D., Romanyuk, G., and Strack, P. (2017). Active learning with a misspecified prior. *Theoretical Economics*, 12:1155–1189.

Gallager, R. G. (1997). Discrete stochastic processes. *Journal of the Operational Research Society*, 48(1):103.

Galton, F. (1894). *Natural Inheritance*. Macmillan.

Gelman, A. et al. (1995). *Bayesian Data Analysis*. Chapman and Hall/CRC.

Gelman, A. and Meng, X. L. (1998). Simulating normalizing constants: From importance sampling to bridge sampling to path sampling. *Statistical Science*, 13:163–185.

Gelman, A., Roberts, G. O., and Gilks, W. R. (1996). Efficient Metropolis jumping rules. In *Bayesian Statistics 5: Proceedings of the Fifth Valencia International Meeting*. Oxford University Press. Online edition, Oxford Academic, 31 Oct. 2023.

Geman, S. and Hwang, C. R. (1982). Nonparametric maximum likelihood estimation by the method of sieves. *The Annals of Statistics*, 10:401–414.

Ghosal, S., Ghosh, J. K., and Ramamoorthi, R. V. (1999). Posterior consistency of Dirichlet mixtures in density estimation. *The Annals of Statistics*, 27(1):143–158.

Ghosal, S. and van der Vaart, A. (2017). *Fundamentals of Bayesian Nonparametric Inference*. Cambridge Series in Statistical and Probabilistic Mathematics. Cambridge University Press.

Giacoletti, K. E. and Heyse, J. (2015). Using proportion of similar response to evaluate correlates of protection for vaccine efficacy. *Statistical Methods in Medical Research*, 24(2):273–286.

Gilks, W. and Roberts, G. (1996). Strategies for Improving MCMC. In *Markov Chain Monte Carlo in Practice*, pages 89–114. Chapman & Hall.

Gilks, W. R. and Wild, P. (1992). Adaptive rejection sampling for Gibbs sampling. *Journal of the Royal Statistical Society: Series C (Applied Statistics)*, 41(2):337–348.

Godsill, S. J. (2001). On the relationship between Markov chain Monte Carlo methods for model uncertainty. *Journal of Computational and Graphical Statistics*, 10(2):230–248.

Golub, G. H., Heath, M., and Wahba, G. (1979). Generalized cross-validation as a method for choosing a good ridge estimator. *Technometrics*, 21:215–223.

Goodman, S. (2008). A dirty dozen: Twelve p-value misconceptions. In *Seminars in Hematology*, volume 45, pages 135–140. WB Saunders.

Gordon, N., Salmond, D. J., and Smith, A. F. M. (1993). Novel approach to nonlinear/non-Gaussian Bayesian state estimation. *IEEE Proceedings F – Radar and Signal Processing*, 140:107–113.

Green, P. J. (1995). Reversible jump Markov chain Monte Carlo computation and Bayesian model determination. *Biometrika*, 82(4):711–732.

Greff, K., van Steenkiste, S., and Schmidhuber, J. (2017). Neural expectation maximization. In *31st Conference on Neurips*.

Grenander, U. (1981). *Abstract Inference*. Wiley.

Grimmett, G. R. and Stirzaker, D. R. (1982). *Probability and Random Processes*. Clarendon Press, Oxford.

Grunwald, G. K., Hyndman, R. J., and Tedesco, L. M. (1996). A unified view of linear AR(1) models. Research report, Department of Statistics, University of Melbourne.

Hamilton, J. D. (2020). *Time Series Analysis*. Princeton University Press.

Hammersley, J. M. (1950). On estimating restricted parameters. *Journal of the Royal Statistical Society: Series B (Methodological)*, 12:192–229.

Hammersley, J. M. and Handscomb, D. C. (2013). *Monte Carlo Methods*. Springer.

Harris, T. E. (1964). *The Theory of Branching Processes*. Rand Corporation.

Hastie, D. I. and Green, P. J. (2012). Model choice using reversible jump Markov chain Monte Carlo. *Statistica Neerlandica*, 66.

Hastings, W. K. (1970). Monte Carlo sampling methods using Markov chains and their applications. *Biometrika*, 57(1):97–109.

Hatjispyros, S. J., Merkatas, C., and Walker, S. G. (2023). Mixture models with decreasing weights. *Computational Statistics & Data Analysis*, 179:107651.

He, Y. and Liu, C. (2012). The dynamic "expectation conditional maximization either" algorithm. *Journal of the Royal Statistical Society: Series B (Statistical Methodology)*, 74:313–336.

Held, L. and Ott, M. (2018). On p–values and Bayes factors. *Annual Review of Statistics and Its Application*, 5(1):593–419.

Hjort, N. L. et al. (2010). *Bayesian Nonparametrics*. Cambridge University Press.

Hoerl, A. E. and Kennard, R. W. (1970). Ridge regression: Applications to nonorthogonal problems. *Technometrics*, 12:69–82.

Hopfield, J. J. (1982). Neural networks and physical systems with emergent collective computational abilities. *Proceedings of the National Academy of Sciences*, 79:2554–2558.

Hyvarinen, A. (2005). Estimation of non–normalized statistical models by score matching. *Journal of Machine Learning Research*, 6:695–709.

Ibrahim, J. G., Chen, M. H., and Sinha, D. (2001). *Bayesian Survival Analysis*, volume 2. Springer, New York.

Jacod, J. and Protter, P. (2000). *Probability Essentials*. Springer.

Jeffreys, H. (1935). Some tests of significance, treated by the theory of probability. *Proceedings of the Cambridge Philosophy Society*, 31:203–222.

Joe, H. (1996). Time series models with univariate margins in the convolution-closed infinitely divisible class. *Journal of Applied Probability*, 33:664–677.

Joshi, V. M. (1976). On the attainment of the Cramér–Rao lower bound. *The Annals of Statistics*, 4(5):998–1002.

Joyce, J. M. (2011). Kullback–Leibler divergence. In *International Encyclopedia of Statistical Science*, pages 720–722. Springer, Berlin.

Jäntschi, L. (2020). Detecting extreme values with order statistics in samples from continuous distributions. *Mathematics*, 8(2):216.

Kalman, R. E. (1960). A new approach to linear filtering and prediction problems. *Transactions of the ASME: Journal of Basic Engineering*, 82:35–45.

Kalman, R. E. (1961). New methods and results in linear filtering and prediction theory. *ASME Journal of Basic Engineering*, 83.

Kalos, M. H. and Whitlock, P. A. (2009). *Monte Carlo Methods*. John Wiley & Sons.

Kaplan, E. L. and Meier, P. (1958). Nonparametric estimation from incomplete observations. *Journal of the American Statistical Association*, 53:457–481.

Karlin, S. and Rubin, H. (1956). The theory of decision procedures for distributions with monotone likelihood ratio. *Annals of Mathematical Statistics*, 27:272–299.

Kass, R. E. and Raftery, A. E. (1995). Bayes factors. *Journal of the American Statistical Association*, 90:773–795.

Kempe, J. (2003). Quantum random walks: An introductory overview. *Contemporary Physics*, 44(4):307–327.

Klein, J. P. et al. (2014). *Handbook of Survival Analysis*. CRC Press.

Koopman, B. O. (1936). On distributions admitting a sufficient statistic. *Transactions of the American Mathematical Society*, 39(3):399–409.

Lafontaine, F. and White, K. J. (1986). Obtaining any Wald statistic you want. *Economics Letters*, 21(1):35–40.

Lalley, S. P. (2009). Convergence rates of Markov chains.

Laplace, P. S. (1812). *Théorie analytique des probabilités*. Courcier.

Last, G. and Penrose, M. (2017). *Lectures on the Poisson Process*. IMS Textbooks, Ann Arbor.

Lawler, G. F. and Limic, V. (2010). *Random walk: A modern introduction*, volume 123. Cambridge University Press.

Lawrance, A. J. (1992). Uniformly distributed first-order autoregressive time series models and multiplicative congruential random number generators. *Journal of Applied Probability*, 29:896–903.

Le Cam, L. (1986). The central limit theorem around 1935. *Statistical Science*, Institute of Mathematical Statistics, pages 78–91.

Lehmann, E. L. (1998). *Nonparametrics: Statistical Methods Based on Ranks*. Prentice Hall.

Lehmann, E. L. and Romano, J. P. (2005). *Testing Statistical Hypotheses*. Springer-Verlag, New York, 3rd edition.

Lei, L. and Olson, K. (2010). Evaluating statistical methods to establish clinical similarity of two biologics. *Journal of Biopharmaceutical Statistics*, 20(1):62–74.

Lewis, P. A. W., McKenzie, E., and Hugus, D. K. (1989). Gamma processes. *Communications in Statistics: Stochastic Models*, 5:1–30.

Liang, F. et al. (2008). Mixtures of g priors for Bayesian variable selection. *Journal of the American Statistical Association*, 103:410–423.

Lindley, D. V. (1957). A statistical paradox. *Biometrika*, 44:187–192.

Little, W. A. (1974). The existence of persistent states in the brain. *Mathematical Biosciences*, 19(1-2):101–120.

Liu, C. and Rubin, D. B. (1994). The ECME algorithm: A simple extension of EM and ECM with faster monotone convergence. *Biometrika*, 81:633.

Ljung, G. M. and Box, G. E. (1978). On a measure of lack of fit in time series models. *Biometrika*, 65(2):297–303.

Lo, A. Y. (1984). On a class of Bayesian nonparametric estimates I: Density estimates. *Annals of Statistics*, 12:351–357.

Lobato, I. N., Nankervis, J. C., and Savin, N. E. (2002). Testing for zero autocorrelation in the presence of statistical dependence. *Econometric Theory*, 18(3):730–743.

MacEachern, S. N. (1993). A characterization of some conjugate prior distributions for exponential families. *Scandinavian Journal of Statistics*, 20:77–82.

Machin, D., Cheung, Y. B., and Parmar, M. (2006). *Survival Analysis: A Practical Approach*. John Wiley & Sons.

Mai, J. F. and Scherer, M. (2017). *Simulating Copulas: Stochastic Models, Sampling Algorithms, and Applications*. Imperial College Press, London.

Mann, H. B. and Wald, A. (1943). On stochastic limit and order relationships. *The Annals of Mathematical Statistics*, 14:217–226.

Mardia, K. V. and Jupp, P. E. (1999). *Directional Statistics*. John Wiley & Sons.

Marin, J.-M., Mengersen, K., and Robert, C. P. (2005). Bayesian modelling and inference on mixtures of distributions. In *Handbook of Statistics*, volume 25, pages 459–507. Elsevier.

Marrelec, G. and Giron, A. (2024). Estimating the concentration parameter of a von Mises distribution. *Communications in Statistics: Simulation and Computation*, 53:117–129.

Mason, N. et al. (2011). Niche overlap reveals the effects of competition in experimental grassland communities. *Journal of Ecology*, 99:788–796.

McKenzie, E. (1985). Some simple models for discrete variate time series. *Water Resources Bulletin*, 21:645–650.

McKenzie, E. (1988). Some ARMA models for dependent sequences of Poisson counts. *Advances in Applied Probability*, 20:822–835.

Mclachlan, G. J. and Krishnan, T. (2007). *The EM Algorithm and Extensions*. Wiley & Sons.

Meng, X. L. and Rubin, D. B. (1993). Maximum likelihood estimation via the ECM algorithm: A general framework. *Biometrika*, 80:267–278.

Metropolis, N. et al. (1953). Equation of state calculations by fast computing machines. *Journal of Chemical Physics*, 21(6):1087–1092.

Mitchell, A. F. S. (1994). A note on posterior moments for a normal mean with double exponential prior. *Journal of the Royal Statistical Society: Series B (Statistical Methodology)*, 56:605–610.

Moraes, C. P. A., Fantinato, D. G., and Neves, A. (2021). Epanechnikov kernel for PDF estimation applied to equalization and blind source separation. *Signal Processing*, 189:108251.

Mukhopadhyay, N. (2014). Sufficient statistics. In *International Encyclopedia of Statistical Science*, pages 1569–1572. Springer.

Müller, P. et al. (2015). *Bayesian Nonparametric Data Analysis*. Springer.

Murray, I., Ghahramani, Z., and MacKay, D. J. C. (2006). MCMC for doubly-intractable distributions. In *Proceedings of the 22nd Annual Conference on Uncertainty in Artificial Intelligence.*

Nakano, K. (1972). Associatron: A model of associative memory. *IEEE Transactions on Systems, Man, and Cybernetics*, (3):380–388.

Nelsen, R. B. (2007). *An Introduction to Copulas.* Springer Science.

Newton, M. A. and Raftery, A. E. (1994). Approximate Bayesian inference with the weighted likelihood bootstrap. *Journal of the Royal Statistical Society: Series B (Statistical Methodology)*, 56(1):3–26.

Neyman, J. and Pearson, E. S. (1933). On the problem of the most efficient tests of statistical hypotheses. *Philosophical Transactions of the Royal Society of London, Series A*, 231(694-706):289–337.

Norris, J. R. (1997). *Markov Chains.* Cambridge Series in Statistical and Probabilistic Mathematics. Cambridge University Press.

Parmigiani, G. and Inoue, L. Y. T. (2009). *Decision Theory: Principles and Approaches.* John Wiley & Sons.

Parzen, E. (1962). On estimation of a probability density function and mode. *The Annals of Mathematical Statistics*, 33(3):1065–1076.

Pearson, K. (1895). X. Contributions to the mathematical theory of evolution, II: Skew variation in homogeneous material. *Philosophical Transactions of the Royal Society of London, Series A.*, 186:343–414.

Pearson, K. (1896). VII. Mathematical contributions to the theory of evolution, III: Regression, heredity, and panmixia. *Philosophical Transactions of the Royal Society of London, Series A.*, 187:253–318.

Pearson, K. (1905). The problem of the random walk. *Nature*, 72(1867):342–342.

Perrichi, L. R. and Smith, A. F. M. (1992). Exact and approximate posterior moments for a normal location parameter. *Journal of the Royal Statistical Society: Series B (Statistical Methodology)*, 54:793–804.

Pitman, E. J. G. (1936). Sufficient statistics and intrinsic accuracy. In *Mathematical Proceedings of the Cambridge Philosophical Society*, volume 32, pages 567–579. Cambridge University Press.

Pitt, M. K., Chatfield, C., and Walker, S. G. (2002). Constructing first order stationary autoregressive models via latent processes. *Scandinavian Journal of Statistics*, 29:657–663.

Pollard, D. (1984). *Convergence of Stochastic Processes.* Springer, New York.

Polson, N. G. (1996). *Convergence of Markov Chain Monte Carlo Algorithms.* Oxford University Press.

Prokhorov, Y. V. (1956). Convergence of random processes and limit theorems in probability theory. *Theory of Probability & Its Applications*, 1(2):157–214.

Raftery, A. E. et al. (1995). Hypothesis testing and model. In *Markov Chain Monte Carlo in Practice*, pages 165–187. Chapman and Hall/CRC.

Rao, C. R. (1945). Information and the accuracy attainable in the estimation of statistical parameters. *Bulletin of the Calcutta Mathematical Society*, 37:81–89.

Rao, C. R. (2005). Score test: Historical review and recent developments. *Advances in Ranking and Selection, Multiple Comparisons, and Reliability: Methodology and Applications*, pages 3–20.

Rényi, A. (1961). On measures of information and entropy. In *Proceedings of the Fourth Berkeley Symposium on Mathematics, Statistics and Probability 1960*, pages 547–561. University of California Press.

Richardson, S. and Green, P. J. (1997). On Bayesian analysis of mixtures with an unknown number of components (with discussion). *Journal of the Royal Statistical Society: Series B (Statistical Methodology)*, 59:731–792.

Robert, C. P. (2007). *The Bayesian Choice: From Decision-Theoretic Foundations to Computational Implementation*, volume 2. Springer, New York.

Robert, C. P. and Casella, G. (2004). *Monte Carlo Statistical Methods*. Springer.

Roberts, G. O., Gelman, A., and Gilks, W. R. (1997). Weak convergence and optimal scaling of random walk Metropolis algorithms. *The Annals of Applied Probability*, 7(1):110–120.

Rogers, L. C. G. and Williams, D. (2000). *Diffusions, Markov Processes and Martingales*. Cambridge Mathematical Library. Cambridge University Press.

Rom, D. M. and Hwang, E. (1996). Testing for individual and population equivalence based on the proportion of similar responses. *Statistics in Medicine*, 15(14):1489–1505.

Rosenbaum, P. R. (2005). An exact distribution-free test comparing two multivariate distributions based on adjacency. *Journal of the Royal Statistical Society: Series B (Statistical Methodology)*, 67(4):515–530.

Rubin, D. B. (1981). The Bayesian bootstrap. *The Annals of Statistics*, pages 130–134.

Rubinstein, R. Y. and Kroese, D. P. (2016). *Simulation and the Monte Carlo Method*. Wiley, 3rd edition.

Sato, K. I. (1999). *Lévy Processes and Infinitely Divisible Distributions*. Cambridge Studies in Advanced Mathematics. Cambridge University Press.

Schuster, E. F. (1974). Buffon's needle experiment. *The American Mathematical Monthly*, 81(1):26–29.

Schwartz, L. (1965). On Bayes procedures. *Zeitschrift für Wahrscheinlichkeitstheorie und Verwandte Gebiete*, 4:10–26.

Shiryaev, A. N. (1999). *Essentials of Stochastic Finance: Facts, Models, Theory*. Advanced Series on Statistical Science & Applied Probability. World Scientific.

Shumway, R. H. and Stoffer, D. S. (2000). *Time Series Analysis and Its Applications*, volume 3. Springer, New York.

Sklar, A. (1959). Fonctions de répartition à n dimensions et leurs marges. *Publications de l'Institut de Statistique de l'Université de Paris*, 8:229–231.

Stanton, J. M. (2001). Galton, Pearson, and the peas: A brief history of linear regression for statistics instructors. *Journal of Statistics Education*, 9(3):1–13.

Steutel, F. and van Harn, K. (2004). *Infinite Divisibility of Probability Distributions on the Real Line*. Number 259 in Pure and Applied Mathematics. CRC Press.

Stigler, S. M. (2007). The epic story of maximum likelihood. *Statistical Science*, Institute of Mathematical Statistics, pages 598–620.

Stroock, D. W. (2010). *Probability Theory: An Analytic View*. Cambridge University Press.

Sundberg, R. (2019). *Statistical Modelling by Exponential Families*, volume 12. Cambridge University Press.

Tanner, M. A. and Wong, W. H. (1987). The calculation of posterior distributions by data augmentation. *Journal of the American Statistical Association*, 82(398):528–540.

Tarter, M. E. and Lock, M. (1993). *Model-free curve estimation*. Chapman & Hall, New York.

Tavaré, S. (1984). Line of descent and genealogical processes and their applications in population genetic models. *Theoretical Population Biology*, 26:119–164.

Tibshirani, R. O. (1996). Regression shrinkage and selection via the lasso. *Journal of the Royal Statistical Society: Series B (Statistical Methodology)*, 58:267–288.

Tierney, L. (1994). Markov chains for exploring posterior distributions. *The Annals of Statistics*, pages 1701–1728.

Tierney, L. and Kadane, J. B. (1986). Accurate approximations for posterior moments and marginal densities. *Journal of the American Statistical Association*, 81(393):82–86.

van de Geer, S. (1993). Hellinger consistency of certain nonparametric maximum likelihood estimators. *Annals of Statistics*, 21:14–44.

van der Vaart, A. W. (2000). *Asymptotic Statistics*. Cambridge University Press.

Vuong, Q. H. (1989). Likelihood ratio tests for model selection and non-nested hypotheses. *Econometrica: Journal of the Econometric Society*, 57(2):307–333.

Walker, A. M. (1969). On the asymptotic behaviour of posterior distributions. *Journal of the Royal Statistical Society Series B: Statistical Methodology*, 31(1):80–88.

Walker, G. (1931). On periodicity in series of related terms. *Proceedings of the Royal Society of London, Series A*, 131:518–532.

Walker, S. G. (2004). New approaches to Bayesian consistency. *Annals of Statistics*, 32:2028–2043.

Walker, S. G., Hatjispyros, S. J., and Nicoleris, T. (2007). A Fleming–Viot process and Bayesian nonparametrics. *Annals of Applied Probability*, 17:67–80.

Wasserman, L. (2004). *All of Statistics*. Springer.

Wasserstein, R. L. and Lazar, N. A. (2016). The ASA statement on p-values: Context, process, and purpose. *The American Statistician*, 70(2):129–133.

Wei, G. C. G. and Tanner, M. A. (1990). A Monte Carlo implementation of the EM algorithm and the poor man's data augmentation algorithms. *Journal of the American Statistical Association*, 85:699–704.

Weiss, G. H. (1983). Random walks and their applications: Widely used as mathematical models, random walks play an important role in several areas of physics, chemistry, and biology. *American Scientist*, 71(1):65–71.

Weitzman, M. S. (1970). *Measures of Overlap of Income Distributions of White and Negro Families in the United States*, volume 22. US Bureau of the Census.

White, K. J. (1992). The Durbin–Watson test for autocorrelation in nonlinear models. *The Review of Economics and Statistics*, MIT Press, pages 370–373.

Wilks, S. S. (1938). The large-sample distribution of the likelihood ratio for testing composite hypotheses. *Annals of Mathematical Statistics*, 9:60–62.

Williams, D. (1991). *Probability with Martingales*. Cambridge University Press.

Williams, J. (1973). *Laplace Transforms: Problem Solvers*. George Allen & Unwin.

Young, K. D. S. and Pettit, L. I. (1996). On priors and Bayes factors. *Journal of Econometrics*, 75:113–119.

Ypma, T. J. (1995). Historical development of the Newton–Raphson method. *SIAM Review*, 37(4):531–551.

Yu, Q. (2021). The MLE of the uniform distribution with right censored data. *Lifetime Data Analysis*, 27:662–678.

Yule, G. U. (1927). On a method of investigating periodicities in disturbed series, with special reference to Wolfer's sunspot numbers. *Philosophical Transactions of the Royal Society of London, Series A*, 226:267–298.

Zellner, A. (1986). On assessing prior distributions and Bayesian regression analysis with g prior distributions. In *Bayesian Inference and Decision Techniques: Essays in Honor of Bruno de Finetti*, volume 6 of *Studies in Bayesian Econometrics and Statistics*, pages 233–243. Elsevier, New York.

Zhang, X. D. (2022). *Modern Signal Processing*. Walter de Gruyter GmbH & Co KG.

Zou, H. and Hastie, T. (2005). Regularization and variable selection via the elastic net. *Journal of the Royal Statistical Society: Series B (Statistical Methodology)*, 67:301–320.

Index